P9-CNI-807

physics without math

A DESCRIPTIVE INTRODUCTION

physics

PRENTICE-HALL, INC., *Englewood Cliffs, New Jersey 07632*

without math

A DESCRIPTIVE INTRODUCTION

GILBERT SHAPIRO

University of California, Berkeley

Library of Congress Cataloging in Publication Data

Shapiro, Gilbert
Physics without math.

Bibliography: p.
Includes index.
1. Physics. I. Title.
QC23.S516 530 78-27288
ISBN 0-13-674317-X

for pk

physics without math
A DESCRIPTIVE INTRODUCTION
Gilbert Shapiro

Editorial/production and interior design by M. L. McAbee
Title Page and Openings by Judith Winthrop
Cover design by Allyson Everngam
Manufacturing buyer: Trudy Pisciotti

Printed in the United States of America

10 9 8 7 6 5

Front cover: Top, left to right: Copernicus,
Newton, and Galileo (all courtesy Bettman Archive,
Inc.) Bottom, left to right: Einstein (courtesy of
American Institute of Physics), Maxwell, and
Archimedes (both Bettman Archive, Inc.)

Prentice-Hall International, Inc., *London*
Prentice-Hall of Australia Pty. Limited, *Sydney*
Prentice-Hall of Canada, Ltd., *Toronto*
Prentice-Hall of India Private Limited, *New Delhi*
Prentice-Hall of Japan, Inc., *Tokyo*
Prentice-Hall of Southeast Asia Pte. Ltd., *Singapore*
Whitehall Books Limited, *Wellington, New Zealand*

contents

7 *Heat energy*

8 *Chemical energy:* ATOMS AND PHOTOSYNTHESIS

9 *Heat engines:* THE SECOND LAW

section
three

10

Beyond the solar system

preface

Just as one can appreciate the beauty of a Beethoven quartet without being able to read a note of music, it is possible to learn about the scope and power and, yes, beauty of a scientific explanation of nature without solving equations.

For many years the University of California at Berkeley has offered the course Physics 10 "Descriptive Introduction to Physics." This is a one-quarter course designed to acquaint non-science students with some of the ideas of modern physics. It is taken as an elective by students with widely varying backgrounds, whose major interests lie in such diverse fields as business administration, physical education, or landscape architecture. Few of these students are required to take any physics course at all. They choose it, from among many options, to help fulfill a loose "cultural breadth" requirement in the liberal-arts curriculum. Similar standards exist at most American colleges and universities.

Of the thousands of students who have taken this course at Berkeley, there is a good chance that one of them will one day be a representative in the U.S. Congress. There are some who will become judges and other government officials, elected or appointed. There are certainly many who will be business executives and public administrators. All of these people will be called upon to make many decisions, and some of the decisions may involve scientific considerations. This course is probably their last, perhaps their only, formal exposure to scientific ideas.

The person who designs such a course faces a two-fold challenge. He must make it attractive enough that the students will choose to take the course. He must teach those who do elect the course something lasting and relevant. He must convey to these students some ideas about the nature of modern science that will remain with them for the rest of their lives.

Many students at American schools arrive with a built-in bias against science. They understand that mathematics is involved. Their last encounter with math is likely to have been a high-school algebra class that they disliked and whose subject matter is fading rapidly from their consciousness. These are the students at whom this course and this text are aimed.

There is much in science that can be communicated without invoking the full mathematical apparatus. We can hope to inform the student how scientists operate, what they can do and what they cannot, using verbal description and qualitative reasoning only.

This is not a trivial course. The student is asked to think hard about the subject, and the line of reasoning often leads to unexpected and profound conclusions. When necessary, we do not shy away from the methods of logical deduction, of drawing graphs, or of doing numerical calculations. These are usually confined to "boxes" outside of the main stream of the text, however. These can be skipped by the student, or instructor, who is interested in pressing ahead with the basic ideas. The questions at the end of each chapter can normally be answered with a few sentences of thoughtful reasoning. There are several coherent threads that run through this book. We examine the process by which new scientific ideas gain acceptance. We learn that our "common sense" can be deceptive when we begin to deal with regimes far from the realm of ordinary experience. We discover that even great and intelligent scientists can be mistaken some of the time.

If there is one lesson that I hope the students will retain, it is the idea of the limits that scientific laws sometimes place on our abilities. We show, in our study of relativity, that there is a limit to how fast we can travel. The laws of thermodynamics show that the energy our bodies and our societies require cannot be fashioned out of nothing. Even when energy is present, it cannot be used with perfect efficiency. The Uncertainty Principle of quantum mechanics sets limits on our very knowledge of the state of the world around us. If students in this course will remember nothing more than the message of this paragraph, I would consider that we have succeeded in our teaching.

A course that attempts to survey all of physics in one quarter necessarily has to leave something out. Short shrift is given to Newton's Principles of Mechanics, although the role they play in the development of physics is made clear. Such concepts as acceleration and momentum, the core of conventional elementary physics courses, are barely mentioned. Some "practical" applications, like geometric optics or electrical circuits, are omitted entirely. The philosophy adopted in this text is that we must not get bogged down in the mire of problem-solving and "plug-in" equations, so that we lose sight of the whole edifice of science.

The course embodied in this text is both easier and more difficult than the conventional elementary physics course as it is taught in American colleges and high schools. The absence of detailed calculations makes it more accessible to the student with a weak mathematical background. But the course demands of the student a degree of intellectual effort no less rigorous than that required for the study of, let us say, the plays of Shakespeare. The process of becoming educated is not a simple task.

At an average of one chapter per week, this entire course can be taught in one semester. In the ten-week quarter allotted at Berkeley, I have never managed to cover the entire text. I found that Chapter VIII, on Chemical Energy, can be omitted without breaking the stream, and that any of the last four chapters can be taught out of sequence without relying on the others. Chapter VI, on Wave Motion and Sound, can probably be omitted if desired, but I have not done so in my own teaching of the course.

Metric units are used exclusively and without apology in this text. Students at this level are as unfamiliar with pounds (not to mention slugs and poundals!) as they are with kilograms. With the minimum of quantitative calculations employed here, it makes little sense to worry about conversion factors from English units. We proceed, rather, as if "yards" and "inches" simply do not exist (as they would not, for a child learning measurements in, say, France or Sweden). The question, "What is a meter?" is best answered, "It is the length of this meter stick."

In our non-mathematical spirit, we eschew scientific notation. When it is necessary to display a particularly large, or particularly small number, we exhibit it with all the zeroes. This is done rarely; such numbers are equally meaningless to the students no matter how denoted. In accordance with American usage, one billion will mean here 1 000 000 000, and one trillion is 1 000 000 000 000.

There are many excellent books extant in popular science, in biography, and the history of science, and about the problems of science and society. I have customarily asked the students to read two such books in the quarter and to submit written reports on them. This practice has enriched the students' experience and has enabled those who are interested in some particular aspect of the course to learn more about that topic. The reading list used at Berkeley is appended in the section, "List of Suggested Readings."

GILBERT SHAPIRO

section

WHERE SCIENTIFIC THEORIES COME FROM

1

the nature of an explanation

Introduction

We are always making up explanations. We can't help it. It is the way our minds work. The things we see and hear and feel are complex and varied. To deal more easily with our experiences, we use our intelligence to sort them out.

Often it is enough just to give a name to something. A farmer hears a loud sound from the sky and says, "It is thunder." A doctor examines a sick patient and concludes, "He has influenza." The name itself does not always tell us any more about the subject. But being able to give it a name is reassuring. It means that we have had a similar experience before, and can recognize it. It shows that other people have also seen or heard or felt the same sort of things.

We can use our experience to help us tell what is likely to happen next. The farmer says, "It will rain soon." The doctor thinks, "This patient will get better in a few days." We have been able to organize our sense impressions so that we recognize common patterns. We have learned to use information to deal with what we find in our everyday lives.

But humans are not usually satisfied with just recognizing patterns. We try to put our knowledge together into wider groupings. We want to know what is the *reason* something happens. We ask ourselves what is the relation between an experience of one kind and some other experience.

The peasant may say, "Heaven is angry, and is throwing thunderbolts at us. Heaven will then feel sorry, and send us rain." On other occasions the same Heaven may send sunshine, or wind, or snow.

The physician may conclude, "A virus has infected your body." Other viruses may cause other diseases.

The kind of explanation that satisfies the farmer might not appeal to the doctor. A lot depends on what we have already learned. Or on what we think we have learned. Once we are used to a certain way of thinking, it is hard to change our mental habits. This is as true for a learned scientist as it is for a superstitious peasant.

There is good reason why not to give up old, established ways of thought. If we had to "unlearn" every day what we had just learned with such difficulty the day before, we would soon become quite confused. Just because someone comes up with a new idea, there is no reason to drop everything and accept the new doctrine right away. So it is quite natural that we are slow to change our minds.

The history of human progress is, nevertheless, full of cases where people were persuaded to accept new ideas. This is particularly true of *science*, the accumulated body of observed facts, explanations, and uses of things that happen in nature. New ideas are always coming out in science. Many of them fall by the wayside. Once in a while a theory is so convincing that scientists are ready to accept it. This doesn't happen often. When it does, the changes in the ways of scientific thinking can be dramatic.

What we call science is not true because your teacher, or some famous scientist, says it is so. It is not true just because it is written in this textbook. Parts of what we now accept as scientific knowledge may turn out not to be true at all. The only validity science has is based on its ability to explain facts of nature in a simple and useful way.

In this chapter we shall explore what we mean by a scientific explanation of the events of nature. We shall try to see what features a new scientific idea must have to help it become accepted. Through examples that we shall give in this and later chapters, we hope to show what science is, how it has grown, and what may be expected of it.

A simple experiment We begin this course with a simple demonstration. Most of you have seen something like this before. If you haven't, you might be a bit surprised at what happens.

Try to make an *objective* description of what takes place. That is, describe the events without putting your own interpretation on them. We just want the facts.

You might not realize that you already have a lot of built-in notions. For example, you will probably want to talk about something called "air." The idea of "air" is not really objective. Air is not something we see or hear or feel. We have been taught about "air" since childhood, to help explain other facts that we *can* sense directly. The fact that "everybody knows" about air doesn't matter. At this stage we must be careful to keep our direct experience separate from our explanations of it.

Demonstration: Hot-Air Balloon

A bag is prepared, made of a light, plastic material. The bag has a rather wide mouth at the open end, perhaps supported by a hoop around the mouth. The open end of the bag is held over something hot, like a hot plate or a Bunsen burner.

The bag stands up straight and swells out as much as it can. The teacher then lets go of it. The bag rises to the ceiling of the room. It remains there for a moment, and then slowly falls to the floor.

What happened? Why did it happen?

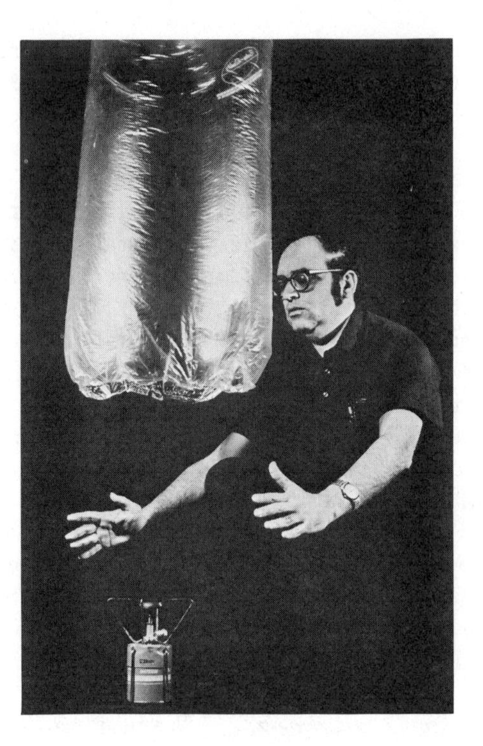

Figure 1-1

Types of explanation

There are many different ways to explain events in nature. The kind of explanation we accept depends on what we already know and what we have come to expect. An explanation that one person likes may not satisfy somebody else at all. Perhaps we can show, by the following examples, what sort of explanations have been acceptable to some people at various times in history, even to various people alive today, or to the same person at different stages of life. Perhaps we can show why some of these explanations are more acceptable to a scientist than others.

Very well, then. The question is, "*Why does the bag rise to the ceiling*?"

The Animistic Answer. Animism: The belief that natural objects (such as rocks, trees, wind) are alive and have souls.

"*Because it wanted to get away from the fire.*" This answer would satisfy most small children, and many primitive tribesmen. To them, all things are alive, and have a will of their own. The fact that the bag is able to rise is no more surprising than that the child is able to walk to the door.

The animistic world is a chaotic world. All natural objects are behaving according to their natural whims. You can no more predict what the bag will do than you can tell what the child himself, changeable creature that he is, will want to do five minutes from now.

Survival in this world requires you to be able to know what is likely to happen next, and if possible, control it to some extent. The child, and the primitive man, soon progress to the next level of explanation.

The Magical Explanation. Magic: The art of producing effects or controlling events by charms, spells, and rituals.

"*The lecturer is a wizard, who knows a secret trick.*" There is some incantation, or special formula, or sleight-of-hand, that the teacher knows, that enables him to astound the audience. The image of the scientist-as-parlor-magician, capable of doing all sorts of tricks, is probably widespread even today. A century or more ago, it was common for scientists to make money by giving lecture demonstrations with lots of sparks flying and chemicals foaming.

The primitive man had to deal with situations he could not control. He came to feel that if he could only learn the right secrets, all things would be possible to him. Experience eventually led him to admit that there were some things beyond his control. He continued to believe that there existed superior beings who could do things that a human could not.

The Supernatural Explanation. Supernatural: Caused by supposed forces beyond the normal, known forces of nature.

"*A miracle has occurred.*" In other words, there exists some Being or Power that is able to intervene in natural affairs, and to cause events to occur that would not ordinarily come to pass. This Being, like some arbitrary medieval king, has the ability to violate the Laws of Nature, even if the Laws were originally decreed by the Being itself. A more orderly, if more passive, version of this type of explanation is:

The Theistic Explanation. Theism: Belief in a God who is the creator and ruler of the universe.

"*It is the will of God.*" Whether or not there has been a miracle, all things happen according to a scheme ordained by a Supreme Being. If there are Natural Laws obeyed by all nature, it is He who has established them. And if events take place in violation of these laws, it is He alone who is responsible. In the end, all actions that take place do so because He willed them, and directly or indirectly made them happen.

Many sincere people, including scientists, hold these or similar beliefs. It is not the purpose of this course to criticize them, or to discuss them at any length.

From a scientist's point of view, however, this type of explanation *by itself* explains too little because it explains too much. If everything is due to Divine will, and if that will is beyond human knowing, then we know nothing at all.

If, on the other hand, a deity exists or existed that originally established the natural system, and the laws of nature, however established, are now universally obeyed, then it is the goal of the scientist to figure out the nature of the laws. It is an unspoken tenet of science that such laws do exist, and that they are basically simple.

The Teleological Explanation. Teleological: Directed toward a definite end, or having an ultimate purpose.

"*It is the nature of warm air to rise.*" This quotation is almost direct from Aristotle. It is the first of our explanations that deals directly with the details of the experiment itself. Indeed it was only when we warmed the bag that it began to rise. After it moved away from the heat, and had a chance to cool down again, the bag began to fall. By various experiments we can verify that indeed the factor of heat is associated with the rise of the bag.

At this point we begin to speak about a strange substance, which we can never see, and not always feel or hear or smell, a substance called *air*. It is air that fills the bag, causing it to stand erect and swell. It is the air in the bag that absorbs heat and gets warm and carries the bag containing it to the ceiling. Why are we inclined to accept the existence of such an invisible substance? It is not because we can sense it directly, but because it helps to explain so many other things that we can sense.

Not every such concept survives the test of time. The ancient Greeks and their followers also thought there was another material called *fire*. The modern view of the nature of fire is, as we shall see in Chapter 8, quite different.

According to the ancient view, a warm vapor is a mixture of air and fire. The warmer the vapor, the more fire it was supposed to have in it. Many observations led to a generalization about the behavior of such vapors. Warm vapors tend to rise; cooler vapors tend to fall.

The explanation reported by Aristotle of *why* the warm vapor rises has to do with the purpose of things. In this scheme each element seeks its natural place. The natural place for fire is upward. When the vapor cools, some of the fire supposedly leaves it. The remaining air is then free to seek its own natural place, which is lower down.

In a teleological scheme, we find further explanations by seeking more general purposes. Thus, we may say that warm air rises so that the earth may be cooled, so that plants may grow, so that we may have food to eat, and so forth. Explanations of this sort offer a fruitful ground for philosophical discussions. They do not, however, reflect the direction that has been followed by modern physical science.

An Empirical Explanation. Empirical: Based on practical experience, ("fitting the data"), without trying to understand why it should be so.

Archimedes' principle: "*A body immersed in a fluid is buoyed up by a force equal to the weight of the displaced fluid.*"

Gay-Lussac's (or Charles') law: "*A gas that is heated* (*or cooled*) *at constant pressure will expand* (*or contract*) *in direct proportion to the increase* (*or decrease*) *in temperature.*"

These two principles differ from the previous explanations in two important respects. They are quantitative; they tell *how much* of an effect to expect. And they can be applied to more experiments than just the one we are talking about.

Applying these two principles to the hot air bag is straightforward. When the air is heated, according to Gay-Lussac, it expands. Thus, it displaces a quantity of cooler air, outside the bag, that is heavier than the warm air inside. Then, according to Archimedes' principle, the immersed body (namely, the bag of warm air) is buoyed up by a force that is greater than its own weight. Therefore, the bag rises to the ceiling. When the air inside it cools, it becomes as heavy as the surrounding air. So the extra weight of the bag itself is enough to make it descend.

Archimedes' principle and the law of Gay-Lussac are examples of *empirical* laws. They have the form: "Whenever A happens, B follows." An empirical law describes, as accurately as it can, *what* takes place. It usually does not give much insight into *why* it should be true. This type of explanation is characteristic of relatively new fields of science.

Figure 1-2 Archimedes, 287–212 B.C. (THE BETTMAN ARCHIVE INC.)

ARCHIMEDES AND HIS PRINCIPLE

Archimedes lived in the then-Greek city of Syracuse, about 200 BC. The story goes that he discovered his famous principle when his king, Hiero, asked him to find a way to decide whether a new crown he had bought was truly made of gold. Such a valuable item clearly had to be treated nondestructively. No bending, melting, clipping off samples, or the like, was to be permitted. Legend has it that the idea came to Archimedes as he lay immersed in the public bath. So excited was he by his discovery that he ran through the streets of Syracuse shouting, "Eureka!" ("I have found it."), clad in the usual manner of men in the Greco-Roman public baths, which is not at all.

Suppose the crown contained one kilogram (about 2.2 pounds). Lead is a lighter metal than gold, and so a kilogram of lead takes up more space than a kilogram of gold.

A kilogram of gold displaces about 50 grams of water (nearly 2 ounces), whereas a kilogram of lead displaces about 80 grams of water. So the test Archimedes devised was to weigh the crown first in the air, and then under water.

A gold crown should "lose" 5% of its weight (50 grams out of 1000) under water, whereas a lead crown would lose 8%. Even in Archimedes time the scales were good enough to tell that much difference. Archimedes proved that Hiero's crown was a "ringer," and won undying fame. The fate of the goldsmith was less fortunate.

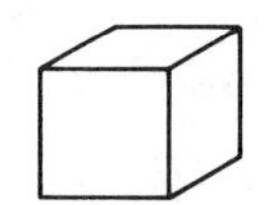

1 kilogram of gold

1 kilogram of lead

Figure 1-3

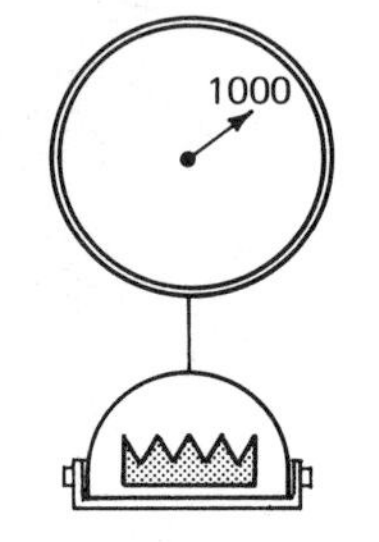

Crown weighs 1000 grams in air . . .

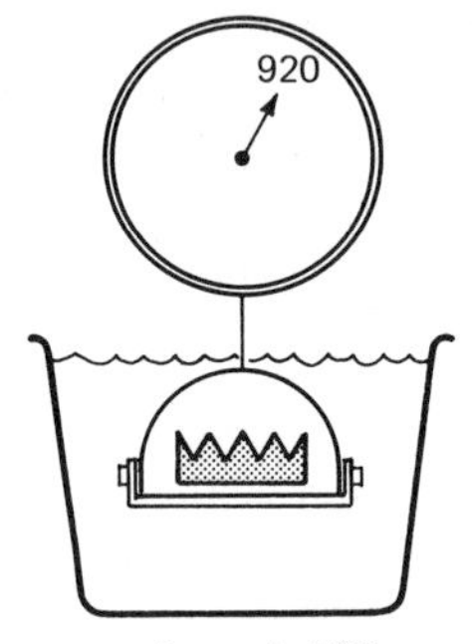

. . . but only 920 grams under water.

Figure 1-4 Was it gold or lead?

A deeper explanation of Archimedes' principle

In the case of Archimedes' principle, we can give an explanation of the principle itself that goes a bit deeper than the empirical level. We are about to show that the law is a simple consequence of the nature of a *fluid.* In logic this sort of proof is called a *reduction.* In it we show that whenever certain simple conditions hold, the more complicated-sounding principle will always be true. This is satisfying to the scientist. We would like to keep our assumptions about nature as simple as we can.

Proof: *that Archimedes' principle is a simple consequence of the nature of a fluid.*

We define a *fluid* as a material that is free to flow. It can and will change its shape or size to adjust to unbalanced forces. Liquids and gases are fluids.

Since a fluid will move around in response to unbalanced forces, the position in which it finally settles will be one in which the forces are balanced everywhere within the fluid. Take a region of any shape inside the fluid. Suppose we imagine this region surrounded by a lightweight plastic bag.

There are two types of forces acting on the fluid inside the imaginary plastic bag: (1) The weight of the fluid inside the bag is a force pulling downward. (2) The pressure of the fluid outside the bag is pushing in many directions. The net effect of the second type must be upward. In fact it must just balance the weight of the fluid inside the bag. Otherwise there would be an imbalance of forces, and the fluid would readjust.

Suppose we replace the fluid inside the bag with some other object that has the same shape as the bag. The fluid outside the object will have the same distribution as before. The outside fluid will exert the same pattern of pressures on the object as they did on the imaginary bag. The net upward force will be the same as before, namely, equal to the weight of the displaced fluid. But this is just the buoyant force specified in Archimedes' principle.

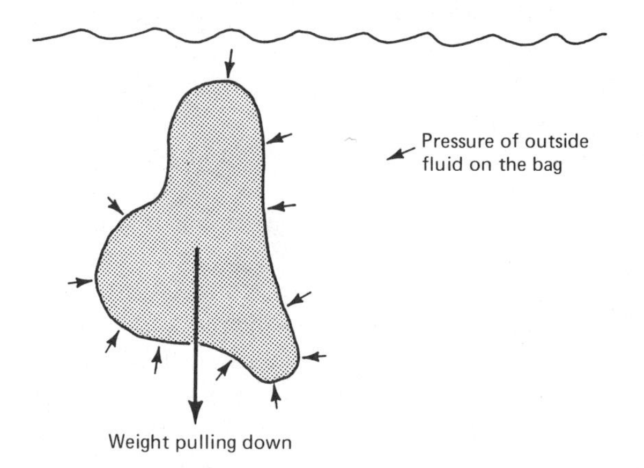

Figure 1-5

Hence, we have derived Archimedes' principle from some rather simple assumptions that involve the nature of a fluid and the notion of balanced forces. We have been rather loose in our use of terms like *weight*, *force*, and *pressure*, leaving them to your intuitive sense, rather than supplying rigid definitions. This was on purpose. We are trying to show how one law can be derived from simpler assumptions. Logical exactness will come later.

Having made this reduction, we can now rephrase the empirical explanation of why the bag rises.

1. Gay-Lussac's law, as before.
2. *Air is a fluid.*

The kinetic theory of gases

We hesitate to label this as the ultimate *scientific* explanation. It is the most advanced explanation we shall offer of why the bag rises.

> "*A gas consists of a large number of very small molecules in rapid motion, moving independently of each other, colliding frequently with each other and with the container walls, and obeying the laws of mechanics in all their motions.*"

At first sight this does not seem to be any explanation at all. But we can quickly show that Archimedes' principle (at least applied to gases) and Gay-Lussac's law are both immediate consequences of the kinetic theory of gases. A large assembly of independently moving molecules will clearly behave like a fluid (think of a bag of marbles, or better, a bag of very fine sand). Archimedes' principle, as we have shown, applies to this as to any fluid.

If we identify higher temperature with faster motion of the molecules, we can also derive Gay-Lussac's law from the kinetic theory. If we heat the gas, the molecules will move faster. They will hit the container walls more often and harder. Thus, the walls are forced to expand. So the same number of molecules fill up a larger volume. This is just what Gay-Lussac's law says.

The kinetic theory of gases has two parts: a *model* and a set of *principles*. The model, in this case "a gas as a large collection of molecules in motion," is what we think the physical situation is for this problem. The set of principles, in this case "the principles of mechanics," are supposed to be a simple set of rules by which we can tell what will happen next in any situation, in particular the one described by the model.

The model for any theory may be as complicated as we need. A complete model for Geology, the theory of the structure of the Earth, would require us to know where every rock, every drop of water, on the Earth is. This is clearly not possible. Our information about the Earth is incomplete. Even if we had it, our computers couldn't handle all the data. So we must resort to simplified models, that will not tell the whole story, but, we hope, will describe the main features. But then, if the theory fails to predict an earthquake, we would blame this failure on the insufficiency of the model. We would still feel that we understand the principles involved.

Most scientists hope and believe that a simple and universal set of principles can be found and applied to *all* theories and models of physical nature. The "principles of mechanics," that is, Newton's principles, came close to fulfilling this hope. Using them, one could unify such diverse fields as mechanics, heat, sound, astronomy, fluids, and others. In the 20th century other principles have had to be introduced to deal with new regions of experience (the very small, the very distant, the very fast). These principles all reduce to Newton's laws in the region of common experience. But as yet there is no single unifying concept.

We need not be satisfied with the kinetic theory of gases as the final explanation of why the bag rose to the ceiling. We might ask, for example: "Why do molecules exist? What are they made of?" While scientists may offer answers to these questions, the answers themselves would probably suggest still further questions. As anyone who has listened to a child's questions knows, there is no end to asking "Why?"

From a philosophic point of view, there is probably no answer to the question, "Why does the bag rise?" or to any other question about nature, that will satisfy everybody. Perhaps we have indicated here, by illustration, the sort of answer that might be satisfactory to a modern scientist.

The criteria for acceptance

The question now arises, "What does it take to convince somebody that a given explanation is correct?"

The answer depends, of course, on which type of explanation we are dealing with. Someone who is inclined to accept a theistic explanation might be convinced by being shown that the explanation agrees with certain Holy Books. Another person might accept an idea if it is endorsed by somebody else who is considered learned and wise. These arguments would probably not persuade a scientific audience.

There can be more than one scientific explanation of something. Sometimes one explanation can include another, the way the kinetic theory of gases encompassed the law of Gay-Lussac. Sometimes a new theory, even if it is scientific in nature, can be incorrect. Even if a theory is right, it may be slow to gain acceptance. What does it take, then, to convince the scientific community that a given theory of nature is likely to be correct?

A survey of cases in which new theories of nature came to be accepted may shed some light on the process. Some examples will be given in later chapters. We can identify some features that help to make a theory persuasive.

Suppose that at some stage of history, there is a generally accepted theory of nature that gives an explanation of events, and that seems to work well in many cases. Let us call this the *Previously Accepted Theory*. We presume that this theory has some shortcomings that are recognized by at least some scientists. Under these conditions a *New Theory* may be proposed, partly to fix the shortcomings. The following factors will be effective in helping to persuade the scientific community to accept the New Theory:

Reducibility. *The new theory must be at least as good as the old.* There is a large body of information that is more or less well-accounted for by existing theories. A new theory, to be acceptable, must do equally well in accounting for this material.

The usual way to satisfy this criterion is to show that in the pertinent situations the new theory *reduces* to the old. Thus, once we show that Archimedes' principle follows from the properties of a fluid, we don't have to prove that each experiment that had been used to verify Archimedes' principle will also follow from the new theory. On the other hand, if the new theory makes some prediction that contradicts a well-verified consequence of the old theory—if it said, for example, that a block of wood should sink in water—there would be strong grounds for rejecting the theory.

When a large body of information has accumulated in some field, any attempt to improve on it must also account for all the successes of the existing theories. This means that anyone who wishes to add to the knowledge in such a field must first learn all about it. This may explain why a theory advanced by an untrained individual usually has such difficulty getting a hearing.

Any prominent scientist or teacher often gets letters from strangers proposing in all sincerity to overturn all the laws of physics. A typical proposal might insist that the earth is hollow. We could point to the seismic waves that seem to come directly through the Earth at such high speed that they must be moving through a solid. But how can we explain this to someone who does not understand the nature of wave propagation? The "outsider" may not even be aware of the body of knowledge that already exists, and cannot comprehend in what respect his ideas fail to take account of problems that are dealt with by the existing scientific doctrine.

To summarize the first criterion: A new theory must lead to the same results as previously accepted theories in all cases where the latter have proved successful.

Novelty. *The new theory must be better than the old.* Giving the same results as a previous theory is not enough. If the old and the new theories give the same predictions in all cases, they are said to be equivalent. The choice between them is a matter of taste. If the new idea is really to supersede the old, it must add something new to the picture.

Thus, the kinetic theory of gases not only reproduces all the successes of Gay-Lussac and Archimedes, but also helps to explain phenomena about which the earlier theories had nothing to say. On the basis of kinetic theory, we can calculate the speed of sound and the resistance of air to projectiles passing through it.

Most noteworthy is the case of a new theory that is able to make correct predictions where the old theory was known to be wrong. This is the classical situation in which a new theory comes to replace a clearly inadequate older one. Thus, the "classical" theories of physics, as they were understood in the year 1900, said that an electrically charged electron in orbit about an atom must radiate away all its energy and spiral down into the center. The experiments showed that atoms *did* consist of electrons in orbit about a small, heavy nucleus, and that the atoms were nevertheless stable. It took the 20th century theory of *quantum mechanics* to account for the stability of the atom.

The word "prediction" is sometimes used in a peculiar sense in this context. One speaks of a theory "predicting" the results of experiments that have already been done. It would make better semantics to speak of a theory as *accounting* for results that are already known.

A new theory must account for facts that are either not encompassed by previously accepted theories or are predicted incorrectly by the latter.

Testability. *It should be possible to tell whether a theory is right.* Physics is an experimental science. This means that a theory of physics must stand or fall on its ability to predict the results of experiments. Usually the convincing argument in favor of a new theory is that it make a real prediction of the result of an experiment that has not yet been performed. Maxwell's theory of electricity and magnetism, proposed in 1873, predicted that electromagnetic signals could be sent across empty space at a speed equal to the speed of light. The discovery of *radio waves* by Heinrich Hertz, and their application to "wireless telegraphy" by G. Marconi, constituted the final verification of this great theory.

There are some theories that are so vague or complicated that they resist experimental testing. The Delphic oracle of ancient Greece claimed to be able to predict the future. But the wording of its predictions was always so contrived that, if the apparent prediction failed to come true, it could be argued that it had been misinterpreted. A theory that can always be readjusted to account for whatever experimental results are found is clearly not in a satisfactory state.

There are some theories that are so phrased that they cannot be tested at all. For example, the statement that "there exists a parallel universe which has absolutely no contact with our own," may amuse science-fiction writers, but it is not subject to experimental test. Such a theory is not a theory of physics; it is said to be *unphysical.*

It must be possible to derive consequences of a new theory which differ from those of competing theories, and which can be tested by experiment.

Elegance. *A theory of nature should be beautiful.* There is another element—we may call it "elegance" or "beauty"—that helps to gain acceptance for a new theory. Scientists believe that the basic laws of nature are simple in form. The most powerful physical theories of the past have been simple. This quality is not easy to define, though easy to recognize. Often it takes the form of a particularly brief mathematical expression, such as "$F = ma$" or "$E = mc^2$."

A new theory should incline toward a particularly brief, powerful, and universal statement of its principles.

We want to call attention to the sort of criteria that are *not* to be applied. It is not necessary for a scientific theory to be in accord with the writings of V.I. Lenin, for example, or the sayings of Chairman Mao, or for that matter with the Biblical account of Creation. A theory need not be ratified by a vote of the Academy of Sciences, or approved by the editor of the Physical Review. It cannot be argued that personal feelings do not ever enter into scientific discussions. But it is expected that, in the long run, scientific ideas will be accepted or rejected on the basis of the above, generally objective criteria.

In following chapters we will show how specific theories have come to gain scientific acceptance.

Glossary

air—invisible fluid which supposedly fills up the space around us. The "air" model helps explain such facts as why balloons rise.

animism—the belief that natural objects (such as rocks, trees, wind) are alive and have souls.

Archimedes' principle—a body immersed in a fluid is buoyed up by a force equal to the weight of the displaced fluid.

elegance—the criterion that a new theory should incline toward a particularly brief, powerful, and universal statement of its principles.

empirical—based on practical experience ("fitting the data"), without trying to understand why it should be so.

fire—supposed by the ancient Greeks to be one of the "elements" out of which matter is made up. The modern view of what fire is is discussed in Chapter 8.

fluid—a material that is free to flow. A fluid can and will change its shape or size to adjust to unbalanced forces. Liquids and gases are fluids.

force—an effect, such as a push or pull, that can make something begin to move. A more precise definition of force will be given in Chapter 5.

gas—a light-weight fluid that can change its size and shape easily. Air and steam are examples of gases.

Gay-Lussac's law—a gas that is heated (or cooled) at constant pressure will expand (or contract) in direct proportion to the increase (or decrease) in temperature.

kinetic theory of gases—a gas consists of a large number of very small molecules, moving independently of each other, colliding frequently with each other and with the container walls, and obeying the laws of mechanics in all their motions.

liquid—a fluid, usually much heavier than most gases, that can change its shape easily, but not its volume. Water and oil are examples of liquids.

magic—the art of producing effects or controlling events by charms, spells, and rituals.

model—the part of a theory in which we describe what we think the physical situation is.

novelty—the criterion that a new theory must account for facts that are either not encompassed by previously accepted theories, or are predicted incorrectly by the latter.

objective—dealing with real facts that all observers would agree on, regardless of our personal interpretations or feelings.

pressure—the forces, per unit area, that are exerted by (and on) a fluid on (and by) its walls and objects immersed in it. A more precise definition of pressure will be given in Chapter 5.

principles—a simple set of rules, part of any theory, by which we can tell what will happen next in any situation.

quantum mechanics—a 20th century theory that accounts for the stability of the atom. We will say more about quantum mechanics in Chapter 11.

radio waves—electromagnetic signals that can be sent across empty space at a speed equal to the speed of light. They were predicted by Maxwell in 1873, found later by Heinrich Hertz, and used by Marconi.

reducibility—the criterion that a new theory must lead to the same results as previously accepted theories in all cases where the latter have proved successful.

reduction—a logical proof that a complicated-sounding principle will always be true whenever certain simpler conditions hold. For example, Archimedes principle can be *reduced* to being a consequence of the properties of a fluid.

supernatural—caused by supposed forces beyond the normal, known forces of nature.

teleological—directed toward a definite end, or having an ultimate purpose.

testability—the criterion that it must be possible to derive consequences of a new theory that differ from those of competing theories and can be tested by experiment.

theism—belief in a God who is the creator and ruler of the universe.
vapor—see "gas"
weight—the downward-pulling force that the Earth exerts on a given object. Weight will be discussed more fully in Chapter 4.

Questions

1. Which of the explanations of the bag rising do you find acceptable? Why do you reject the others?
2. Why is the idea of *air* not objective? Would you have guessed that such a thing existed, if you had not been taught about it?
3. Give three different kinds of explanation of why it rains.
4. What kind of explanations are the following?
   (a) "It rains so that plants can grow."
   (b) "It always rains after there is thunder."
5. Why is cool air heavier than warm air?
6. What uses can you think of for Archimedes' principle? Why should a swimmer keep her head down in the water?
7. It has been suggested that, if certain laws of physics had been slightly different, a planet like the Earth could never have formed, and intelligent creatures like ourselves would not be here to talk about it. What class of explanation of nature is this?
8. Can you give examples of empirical rules in other fields of study?
9. Is molasses a fluid? Is air?
10. On the basis of the kinetic theory of gases, how would you expect the pressure to change if (a) you raised the temperature of a gas confined in a rigid container; (b) you compressed the gas to half its original volume, while keeping the temperature fixed? In kinetic theory we identify the pressure of a gas with the rate and the impact of molecules colliding with the walls of the container.
11. Identify the *model* and the *principles* in some theories you know, not necessarily in physical science. (Suggestions: The Theory of Evolution, The Law of Supply and Demand, Grimm's Law of Linguistics)
12. The following simplified model of rush-hour traffic is supposed to make reasonable predictions of the number of cars on the highway: Let us say there are 80,000 drivers working in the central city. Beginning at 4:30 p.m. each of them flips a coin, and then repeats the coin flip every 15 minutes until arriving home. The first time the coin comes up heads, the driver takes the car onto the highway. The next time heads is flipped, the driver gets off at the nearest exit.
    How much traffic does the model predict at any given time, say, 4:50 p.m.? Because a model gives accurate predictions, does it necessarily

give a true picture of the state of affairs? What features of rush-hour traffic might this model fail to predict accurately?

13. Can you think of any other criteria, not necessarily scientific, that might be applied to a new theory of nature?

14. It has been argued that human events can be predicted from the positions of the heavenly bodies ("astrology"). What arguments can you marshal for and against such a theory? Would an astrologer be willing to give up this theory on the basis of a wrong prediction? What is the mechanism for adding new ideas to the field? Where did the original wisdom come from?

15. Do the following questions have physical answers? What happened before the world began? Has anyone ever disappeared without a trace? Do miracles really happen?

2

the motion of the planets

AN EXAMPLE OF A SCIENTIFIC THEORY

Introduction

In this chapter we will show how one specific theory came to be accepted. We don't expect you to learn all the details. What we want you to understand is how it came into being.

We will be talking about the motions of the sun, the moon, the stars, and the planets. People have been looking at them for thousands of years. Long ago many accepted the idea that maybe the events in the heavens had something to do with what took place on Earth. If this were so, and if some men were able to predict the motions of the heavenly bodies, that would mean that those men could foretell the future. So wise men through the ages made up schemes that helped them know in advance the positions of the stars.

The movement of the planets is tricky, full of speed-ups and slow-downs, zigzags and back-tracking. The models that were used to explain these motions became more and more complicated, full of circles within circles, needing long tables of numbers. In the 16th century, Copernicus showed how much simpler the calculations were when he assumed that the sun, not the earth, was the center of the world.

Others, notably Galileo, Kepler, and Newton, developed Copernicus' idea. The crowning triumph of the age came with the publication, in 1686, of Newton's *Principia*. In it he showed that the motions of all the heavenly bodies could be predicted from a very simple scheme. The rules of this scheme could be stated in three or four sentences.

The remarkable feature of these rules for the motion of planets is that they are the same rules that apply to the motion of objects on Earth. The laws of *mechanics*, the science that deals with machines and falling bodies, with pendulums and wheels, were shown to apply to the motion of the moon and of Mars. No wonder that thinkers of the century or two after Newton came to regard the universe as a giant clockwork. They looked on the world as a great piece of machinery that, once set in motion, would continue to move, predictably and unpreventably, according to the unchanging rules of the science of mechanics.

The motion of the stars

It is obvious to anyone who spends much time out doors at night that the stars appear to move in a rigid pattern across the sky. The position of the moon with respect to this pattern changes from night to night. The sun also shifts its position, though more slowly than the moon. We know this because we see different stars in the night sky at different times of year. We can guess that the stars we miss at any season are the ones that are near the sun's position, and are outshone by its brilliance.

If you look closely night after night you might notice that there are a few star-like objects that don't stay put against the heavenly background. These objects were called *planets,* from the Greek word for "wanderer." We know five of them today by the names of Roman gods: *Mercury, Venus, Mars, Jupiter, Saturn.* There are others known today, but these five planets were the only ones that could be seen before telescopes were invented.

The ancient idea that the positions of the heavenly bodies had something to do with what happened on Earth was not completely wrong. The spring rains usually began soon after the sun passed from the region of the sky known as *Pisces* to the region called *Aries.* The ocean tides were known to be higher at the times of new moon and of full moon than at other times of month.

Perhaps, the ancients thought, the little planets also had some subtle influence on earthly affairs. Whether or not this belief was correct, and today most scientists would say it is not, it certainly put a premium on being able to predict the motion of the sun, moon, and the planets. And so the early civilizations set out to learn all they could about star motions.

The sun and the moon The motions of the sun and the moon are easy to predict in advance. The sun repeats itself in position once a year. So, if you spend a year noting each day where the sun is among the stars, you can prepare a table that can be used year after year to tell you which stars are near the sun on any given day.

The moon has a repeating cycle of just under thirty days. Once you have it well measured, you can prepare a similar table for the position of the moon.

The phases of the moon are equally simple to predict, once we accept that the moon shines by light reflected from the sun. This fact was understood long ago. When the moon is near the sun's position in the sky, the moon's sunlit face is turned away from us. We see only a thin crescent edge of it, if we see the moon at all. This time of month is known as the *new moon* phase. When the moon is directly opposite the sun, we see the fully illuminated face. This phase is called the *full moon.* At times in between we can see only a fraction of the moon's face, always on the side of the moon nearest the sun. These phases are called the first and last *quarter moon.*

The moon does not follow the same path as the sun across the starry back-

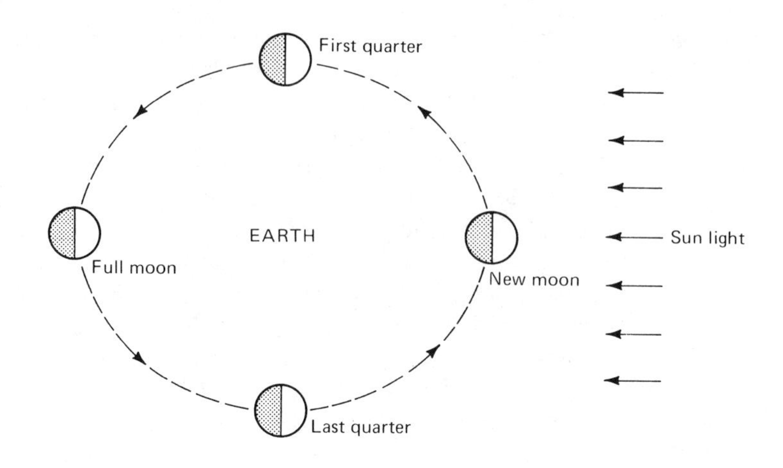

Figure 2-1 Phases of the moon

ground. If it did, the moon would pass in front of the sun at every new moon, blocking out its light, and causing the condition known as an *eclipse of the sun*. Conversely, at every full moon Earth would block the sun's light from reaching the moon, thus dimming the moon's appearance, a condition called *eclipse of the moon*. The moon's path across the stars makes an angle of 5 degrees with the path of the sun. In most months the moon passes over or under the position of the sun, avoiding an eclipse. Only very rarely does the new moon occur at the time of month when the moon's path crosses that of the sun, and that is when an eclipse can occur. The alignment must be very exact, so much so that the total eclipse is visible from only a limited part of the Earth.

A total eclipse of the sun is a dramatic and rare event. It was often supposed to foretell some disaster on earth: the death of a king, or the loss of a battle, for instance. Think then of the exalted state of an astronomer who could know in advance exactly when the next eclipse would take place.

The planets The positions of the planets are not nearly as simple as those of the sun and the moon. Each planet appears to have some overall period of repetition, usually much longer than a year. But the motion is uneven, full of halts and jerks and zigzags. If we follow the motion of the planet Mars for a few years, we will see that for most of that time the planet marches smoothly across the star pattern on nearly the same path as the sun, but at a slower pace. Every so often it will slow down in its track, stop, and begin to retrace its path back past the stars it has just passed. After many weeks of this back-tracking, the planet will stop again, turn, and resume its normal advance past the fixed stars.

Figure 2-2 Total eclipse of the sun (LICK OBSERVATORY PHOTOGRAPHS)

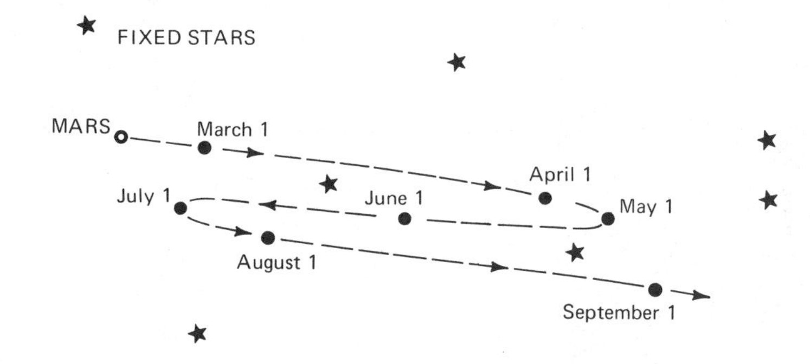

Figure 2-3 Erratic course of Mars among the fixed stars

Consider, then, the problem of an old-time court astrologer who was asked to foretell the next time that Jupiter and Venus would pass each other in the region of Capricorn. There was no simple table in which he could look it up. He had to make the calculation on his own. To do this he needed to know how to go about it.

Every worker who has to make many new calculations soon learns to use memory aids and tricks of the trade, so that he doesn't have to memorize too many things. You have, no doubt, done this yourself. In the same way the old court astrologer needed an easy way to remember how to calculate the motions of the planets.

The Ptolemaic system

Over the centuries a model had been developed for the motion of the heavenly bodies. It was used by most Western scholars from the time of the ancient Greeks until the 16th century and later. The model went something like this:

Figure 2-4 The Ptolemaic system (THE BETTMAN ARCHIVE INC.)

The heavenly bodies are supposed to be embedded in a series of perfectly clear, hollow spheres that surround the Earth. The *fixed stars*, that is, nearly all of the objects in the sky, were in a single outer layer. The sun, the moon, and the planets each had a shell. Each of these spheres was supposed to fit snugly inside the next, like the layers of an onion, with the Earth in the middle.

According to this model, there was a prime mover that pushed on the outermost sphere, the one with all the fixed stars, and turned it once about the Earth each 24 hours. The shell next to the outer one held one of the planets. It also turned, but it slipped a little. Thus, the planet in it appeared to shift position with respect to the fixed stars in the outer layer. The inner levels, each with one planet (or the sun or the moon), slipped a bit more. So each of these bodies appeared to have its separate motion.

The principle of motion for this theory was *uniform circular motion.* Since a circle was a perfect figure in geometry, and the heavenly bodies were assumed to be perfect, it was natural to suppose that they moved in perfect circles at an unchanging rate.

This model had a fair success as an aid to calculation. You could look up how fast any sphere was turning, and at what angle. Then you could figure out approximately where the sun or the moon would be on any given day.

Some people took the model a bit too seriously. They made guesses about what material the spheres were made of. They asked whether they made sounds as they rubbed against each other, the so-called "music of the spheres." But this sort of speculation had little to do with the real value of the model as an aid to calculation.

It was recognized right away that the simple model of the spheres could not explain the back-and-forth motion of the planets. The theory needed some "patching." A very clever change was proposed by the Greek mathematician Hipparchus, who lived in the 2nd century B.C. Hipparchus' idea was the basis for a book, called the *Almagest*, written by Claudius Ptolemy in the 2nd century A.D. The ideas outlined in that text have been known as the *Ptolemaic system.* We do not use the Ptolemaic system any more, but for nearly 2000 years it was the most widely accepted theory of astronomy.

In Hipparchus' model, a planet like Mars did not stay rigidly attached to the sphere assigned to it. Instead, there was supposed to be an "ideal" Mars that was stuck to the sphere of Mars. The ideal Mars would move smoothly about the heavens, but it could not be seen by humans. The real, visible planet Mars was supposed to revolve in a separate circle centered on the ideal Mars. Real Mars would then move in a looping fashion, shifting now ahead, now behind its ideal position. Its motion was as if you attached a flashlight to the spokes of a bicycle wheel, and watched the light loop and swirl as a rider came by in the dark. In this way the motion of Mars, and the other planets, could be calculated as the sum of two circular motions.

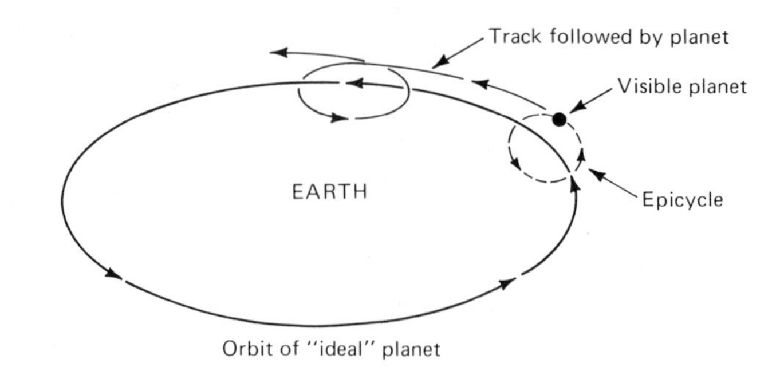

Figure 2-5 Cycle and epicycle in the Ptolemaic system

This, of course, was more complicated than the idea of simple rotating spheres. For each planet men then had to list not only the motion of the ideal sphere, but also the description of the little circles, the so-called *epicycles*, in which the planets moved around their ideal positions. They needed tables that gave the size of each epicycle, how long it took to go around, and which way the circle was tilted. Given all this, they could then figure out where to expect to see the planet on any given night. The system worked well enough to satisfy the needs of the astrologers all through the Middle Ages.

When sailing ships began to make long voyages of discovery far from home, they needed better information about the planets. They had to know exactly where they were, what time it was, and other similar questions. The hope was to guide the ships by the positions of the stars. It became clear that the old tables were not quite right, and corrections would have to be made.

At first men simply tried to put better numbers into the tables. If that didn't work, they patched the model again. They moved the centers of circles away from the ideal position, for example. They even went Ptolemy one better, and invented a second level of epicycles. In such a scheme, there was the ideal Mars as before. There also was a second invisible object, a pseudo-Mars, going about the ideal Mars in an epicycle. The visible Mars then circled the pseudo-Mars in a second epicycle. If this didn't work well enough, they added still more little circles.

Given enough patchwork, men could probably have fit the motions of the planets as well as was needed. But soon it was no longer a simple model. Long tables of numbers were needed, and not many people could understand them. Some thinkers even wondered why God had designed such a complicated system.

Copernicus

Nikolaus Kupfernick (1473–1543), a Polish-German clergyman better known by his Latinized name, *Copernicus*, thought he had a better idea. He lived at a time when the voyages of Columbus and Magellan had convinced some people that the Earth was not flat, but a globe. Why not, he began, suppose that the Earth, and not the heavenly spheres, turned once on its axis every day? This turning would explain the most obvious motion of the heavens, and leave only much slower changes to be accounted for. In particular, it would mean that all the fixed stars really were fixed, and didn't have to move at all. This first suggestion was not very radical; it had occurred to others.

Figure 2-6 Nikolaus Kupfernick (Copernicus) 1473–1543
(THE BETTMAN ARCHIVE INC.)

Copernicus' big idea was to propose that the Earth was not the center of the world, but that it and the other planets all moved in circles about the sun.

This model easily explained the back-and-forth apparent motion of the planets. Take, for example, the case of the Earth and Mars. Let us suppose that both circle the sun in the same direction, the Earth in a smaller circle than Mars, and moving faster. We can figure out the apparent position of Mars on a given day by drawing a straight line from where the Earth is to where Mars is on that day. Continue this line outward toward the fixed stars. They form a constant backdrop to this drama, at a very long distance away.

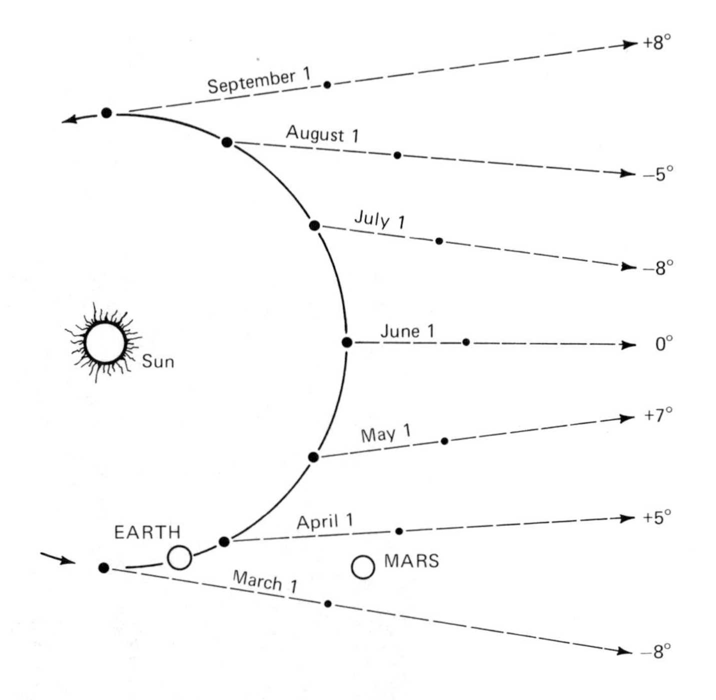

Figure 2-7 Explanation for Mars' apparent erratic motion, according to Copernicus' scheme

When the line of sight from Earth to Mars passes near a fixed star, it will look to us on Earth as if Mars is passing that star. If we draw such lines of sight on succesive days, we can trace the progress of Mars across the fixed stars.

Mars circles the sun in 687 of our days, while the Earth does so in 365 days. In a little over two of our years, the Earth gains a lap on Mars and overtakes it in its orbit. During most of these two years the two planets are fairly far apart. As both progress around the sun, the line of sight between them marches rather evenly around the sky. But as the Earth begins to overtake Mars, the projected line flexes backward. So Mars appears to retrace its path back past some of the stars it has just passed.

Thus, with a very simple assumption, the back-and-forth motion of the planets was easily explained. The heavy burden of the Ptolemaic theory, with all its patchwork, could be thrown out. Sailors and astrologers could calculate from a much simpler model. They needed only to assume that all the planets, and the Earth, revolved about the sun.

But the Copernican model was slow to gain acceptance. His work was published, but it was largely ignored for nearly a century. There are several reasons why.

- The work was presented as an aid to calculation, a "hypothesis," It was suggested that the calculations might be easier if they were done "as if" the Earth revolved about the sun. But the reader didn't have to take this idea seriously. (The book could probably not have been published without such a caution.)
- Europe at this time became embroiled in the great turmoil of Reformation and Counter-Reformation. A man's position on scientific matters could become a touchstone against which his loyalty was to be judged.
- Copernicus had some rather poor data on which to base his tables. Calculations based on them did not make a very good fit to the observed positions of the planets.
- In the end Copernicus felt it was necessary to bring back a few of the epicycles that had made the Ptolemaic system so cumbersome.

Galileo, inertia, and relativity

Galileo Galilei (1564–1642), an Italian, had read Copernicus' book, and took up the ideas in it even more wholeheartedly than Copernicus himself. This position put him in great personal danger, and the story of his conflict with the authorities is well known. His many scientific accomplishments, dealing

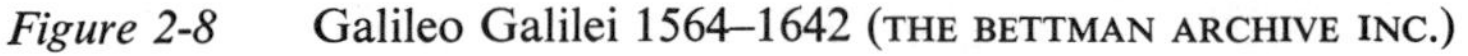

Figure 2-8 Galileo Galilei 1564–1642 (THE BETTMAN ARCHIVE INC.)

with such ideas as the telescope, the pendulum, falling bodies, etc., could fill many chapters. We shall deal here with just one contribution of Galileo: the principle of *inertia.*

One of the main *scientific* objections to the Copernican model was that, if the Earth is hurtling through space at the high speed implied by the model (30 kilometers per *second*, as we now know), why didn't everybody fall off?

The answer Galileo put forth is that the natural behavior of objects in motion is to *keep on moving* at the same speed in the same direction. All the people on Earth are moving along with it. Unless something happens to change their state, they will keep moving with the Earth.

Aristotle had said that the natural state of objects was to come to rest. Galileo's point was that it took a *force* to slow down an object. According to the principal of inertia, it doesn't take any force to keep an object going if it is already in motion. Only *changes* in the state of motion—slowing down, speeding up, turning—require an outside force.

If we roll a ball on the sand, it soon stops. If we roll it on a smooth table, it continues rolling for a while. If we slide it on a flat, icy pond, it keeps going even longer. The slowing down comes because of the roughness of the surfaces, which exert a force called *friction* on the ball. The smoother the surface, the more nearly the Galilean ideal is approached.

It was Galileo who first introduced the idea of *relativity* into physics. According to this idea, a system in steady motion should appear to people moving with it just as if the whole system was standing still.

Have you ever eaten a meal in an airplane? If the ride was not bumpy, the food sat perfectly still in the plate, even though the plane was travelling through the air at a speed of 1000 kilometers per hour. Galileo's example was similar, though keyed to the speeds available in his time. Could a prisoner below decks on a fast boat, he asked, tell whether the boat was moving or becalmed, by the way water spilled? If he faced forward, and the boat was moving, would the water splash on his clothes? Of course not, he answered, because the water in the cup was also partaking of the boat's motion.

Tycho and Kepler

We now skip a few years backward in time to Tycho Brahe (1546–1601), the Danish royal astronomer, who was assisted in Bohemia by Johannes Kepler (1571–1630). Tycho set out to prove Copernicus wrong by taking very careful sightings of the stars and planets. He worked without a telescope—telescopes had not yet been invented—but his work was very accurate. Kepler was later to discard one of his own theories because it disagreed with Tycho's measurements by 1/15 of a degree. He knew that Tycho could not have been that far off.

Tycho had his own pet theory, a compromise between Ptolemy and Copernicus. In Tycho's scheme the sun and moon revolved about the Earth, but all the other planets revolved around the sun. Tycho died before he could

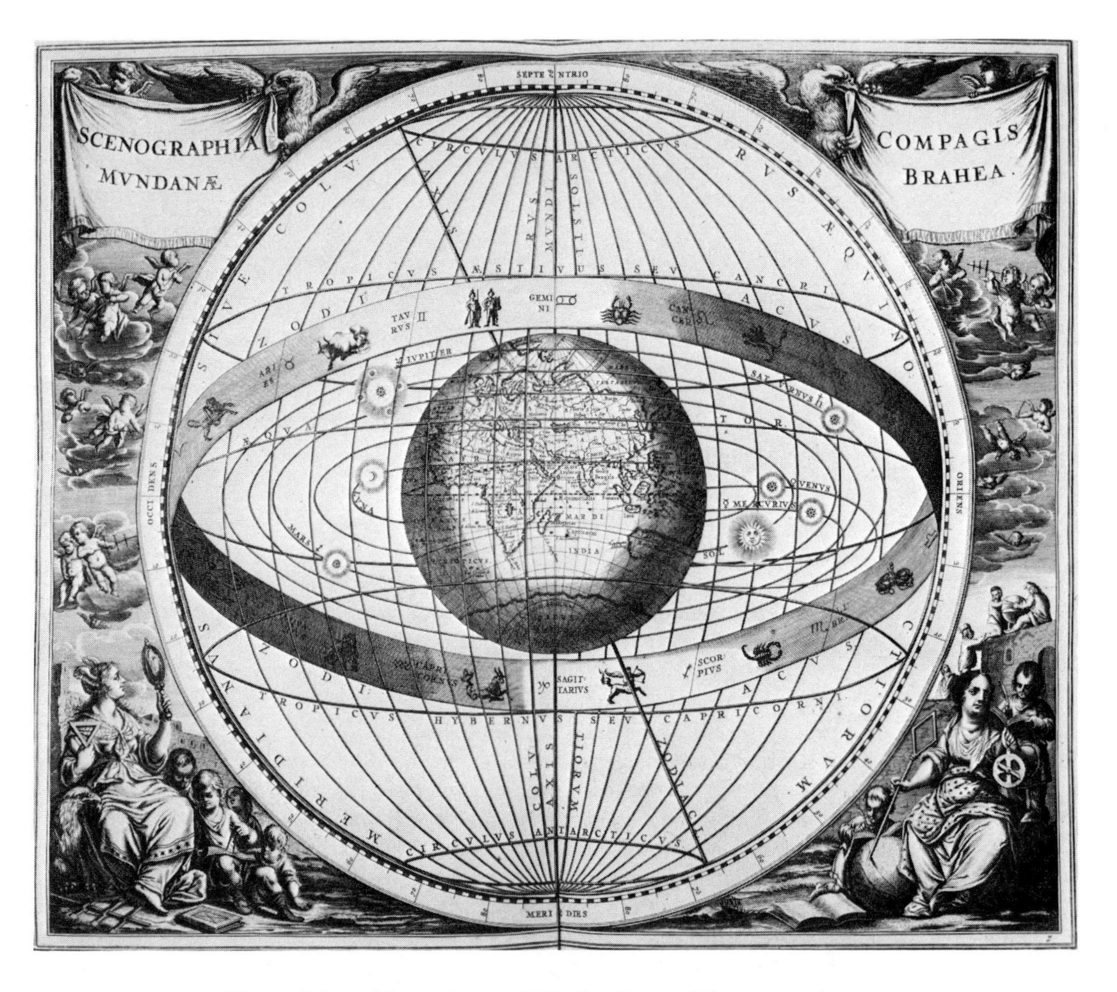

Figure 2-9 The scheme of Tycho Brahe: The sun and moon revolve about Earth, but the planets revolve about the sun. (THE BETTMAN ARCHIVE INC.)

do the calculations. Kepler was left with all of Tycho's data books. He spent years doing the math work over and over again. Unlike Tycho, Kepler had the feeling that Copernicus was right.

Kepler's first try was to fit all the observations with orbits of the planets that went in perfect circles around the sun. He hoped that, with the good data he now had, he wouldn't need to use any epicycles. To his disappointment, he couldn't make things work out that way. Then, with a flash of insight, he decided to try the next best thing to a circle. He would fit the orbits of the planets to a curve known as an *ellipse*.

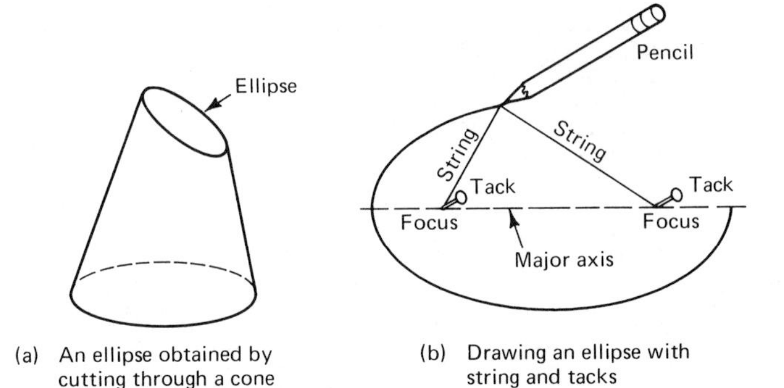

Figure 2-10 (*a*) An ellipse obtained by cutting through a cone; (*b*) Drawing an ellipse with string and tacks

THE ELLIPSE

An ellipse is an oval-shaped curve, known to the Greek mathematicians. If we make a slanted cut through a cone, the curve that emerges is an ellipse.

We can draw an ellipse with two thumbtacks and a piece of string. The points where we stick the thumbtacks into the paper will be the *foci* (singular, *focus*) of the ellipse we draw. Attach the ends of the string to the tacks. Hook the string with a pencil and pull the string taut. Slide the pencil along the string, keeping the string always taut, and trace the curve on the paper. In this way the sum of the distances from the pencil tip to each of the tacks stays the same all the way around the curve.

The length of the string used is equal to the length of the *major axis*, the longest straight line that can be drawn inside the curve. When the two foci are close together, the ellipse is very close to being a circle. When they are apart, the ellipse becomes cigar-shaped and is called *eccentric*.

An ellipse has some amusing properties. Suppose we have a hockey rink or a pool table with ellipse-shaped walls. If we start a ball at one focus and send it rolling toward the wall in any direction, it will bounce in such a way as to always reach the other focus. A sound echo will do the same thing. The rotunda of the Capitol building in Washington has an elliptical dome for a ceiling. A whisper spoken at one focus can be heard quite clearly across the room at the other focus.

The laws of planetary motion

Kepler's discoveries can be summarized in his three laws of planetary motion. These are empirical laws. They fit the data well, but Kepler had no idea why they should be true, or how they could be applied to anything but the planets. Nevertheless they are important. They describe, simply and accurately, the motion of all the planets. Scientists after Kepler did not have to compare their theories painstakingly with the data, as Kepler had. They only had to show that the new theory gave the same results as Kepler's laws.

Kepler's First Law. Each planet moves about the sun in an ellipse, with the sun at one focus.

Under this law no more cycles and epicycles are needed. The ellipse is not as "perfect" a figure as a circle, but using it is a lot simpler than the old Ptolemaic system, and more accurate.

Since the orbit of a planet is not a circle, its distance from the sun changes. Sometimes it swoops a bit closer to the sun, at other times it ranges a bit farther away. Its speed does not have to stay the same all the way around its orbit. Thus a second law is needed to tell how the planet speeds up and slows down as it goes around the sun.

Kepler's Second Law. The speed of a planet changes in such a way that the line from the sun to the planet sweeps out equal areas in equal times.

When the planet is closer to the sun it moves faster than when it is farther away.

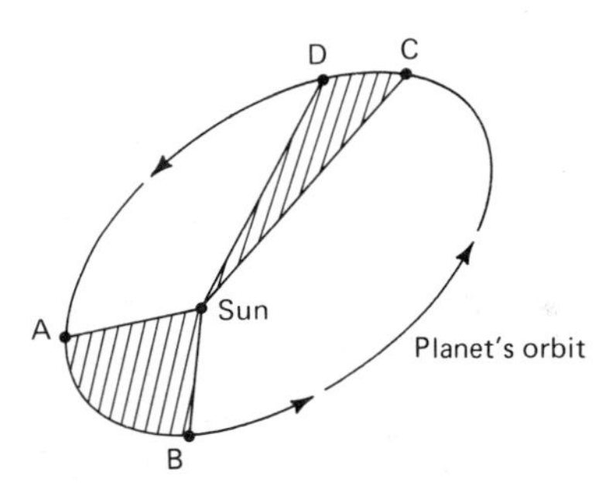

Figure 2-11 Kepler's second law: If the shaded areas are equal, it takes as long for the planet to go from *C* to *D* as it does from *A* to *B*.

If the shaded areas are equal, it takes as long for the planet to go from *C* to *D* as it does from *A* to *B*.

The second law only tells how to relate the motion of the same planet to itself at different parts of its orbit. Different planets sweep out areas at different rates. In order to relate the motion of one planet to another, a third law is needed.

Let us define the *distance* of a planet from the sun as the average distance over its orbit. More precisely, it is half the length of the major axis of its ellipse. The *period* of a planet is the time it takes to revolve once around the sun.

Kepler's Third Law. The ratio of the distance cubed divided by the period squared is the same for all planets.

To see how well the third law works, look at Table 2-1. We use as a convenient unit of distance the "astronomical unit" (A.U.), which is the average distance from the Earth to the sun, about 150 million kilometers. Our unit of time is the year, the period of revolution of the Earth around the sun.

Table 2-1 KEPLER'S RATIO FOR THE PLANETS

Planet	*Distance from sun (in A.U.'s)*	*Period of revolution (in years)*	*Kepler's ratio (in these units)*
Earth	1 (*by definition*)	1 (*by definition*)	1
Mercury	0.387	0.2408	$\frac{0.387 \times 0.387 \times 0.387}{0.2408 \times 0.2408} = 1.0008$
Mars	1.524	1.881	1.0004
Jupiter	5.203	11.862	1.001

The remarkable thing about Table 2-1 is how precisely Kepler's ratios are the same for all the planets. They are equal within one part in a thousand. Even the small differences among them are probably due to uncertainty about the exact values of the planets' distances and periods.

One feature should not be missed in this discussion. The fact that Kepler's ratio for the Earth is the same as that for Mercury, Mars, and Jupiter (and the others) shows in a numerical way that the *Earth is one of the planets*. The Ptolemaic system, or even Tycho's scheme, which set the Earth in a special place, could not explain, except as a remarkable coincidence, why the Earth's distance and period should agree so well with the Kepler ratio for all the other planets.

The moon does not fit in at all with the planets. Its distance from the earth is about 1/400 of an A.U. Its period of revolution is about 1/12 of a year. So Kepler's ratio, in these units, is about 0.000003. The fact that the moon revolves about the Earth, and not about the sun, has something to do with this big difference.

Satellites and comets

During Kepler's lifetime, Galileo discovered that there were several *satellites* revolving about the planet Jupiter. Later, moons were found around Saturn and some of the other planets. It was found that all the moons of Jupiter obeyed Kepler's third law with a ratio that was smaller than that for the planets, but larger than that of the Earth's moon. The value of Kepler's ratio depends on the central body being rotated about. The larger the central body, the greater is Kepler's ratio for its satellites.

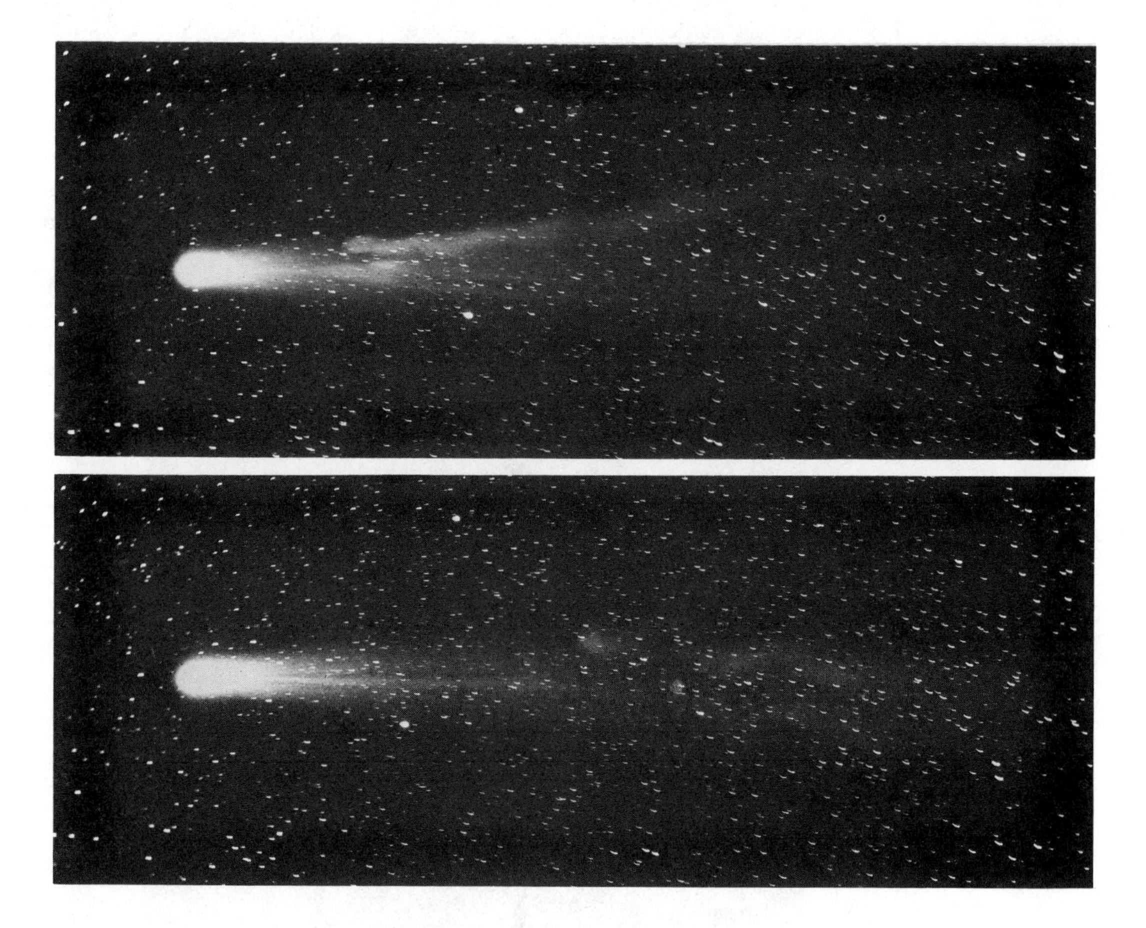

Figure 2-12 Halley's comet photographed in 1910 (LICK OBSERVATORY PHOTOGRAPH)

Comets, those spectacular objects with long fiery tails that appear once in a while in the sky, have been noted since history began. Usually a comet remains visible for only a few weeks, being brightest when it is near the sun, and dimming out as it moves farther away. In the light of Kepler's findings, it was suggested that perhaps comets, like planets, follow elliptical paths. Unlike the planets, whose orbits are not very different from circles, the comets travel in very eccentric ellipses. The royal astronomer of England, Edmund Halley, tracked a comet that appeared in 1682. He figured out how long the major axis should be, and used Kepler's laws to calculate that this comet should come back every 76 years. He checked old records that showed a comet like this one had been seen before at just about the expected times. He then predicted that it would return again in 1758, 1834, 1910, 1986, etc. Halley's comet has come back every time as predicted. Many other comets have since been found to follow repeating, elliptical orbits around the sun.

Newton

Isaac Newton (1642–1727) was born the same year Galileo died. He studied math at Cambridge, and also spent a good part of his life studying the Bible. During the plague year, 1666, the university was closed, and the 23-year-old Newton went back to his mother's farm where he had little to do but think. It was during this time that his great ideas on mechanics and gravity first came to him.

Figure 2-13 Isaac Newton 1642–1727
(THE BETTMAN ARCHIVE INC.)

The story has it that the idea came to him while he was watching an apple fall. Most likely he wondered how far up it was still true that all bodies fall at the same rate. Galileo had shown that on Earth all objects, apples, stones, raindrops, cats, when released, picked up speed downward at the same rate. Why then, Newton wondered, didn't the moon fall? At what height does gravity stop working? Does it ever stop?

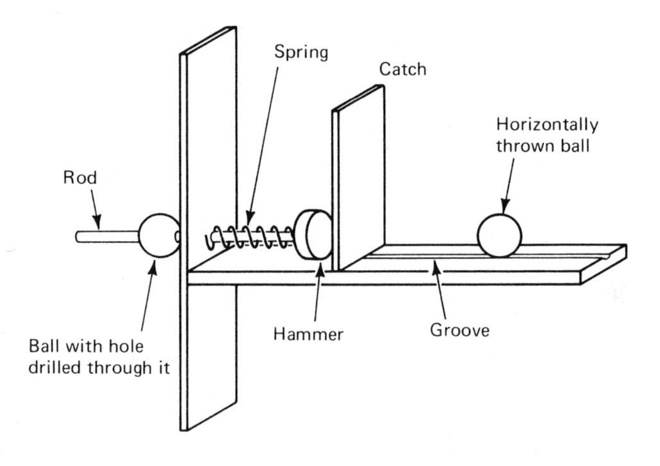

Figure 2-14 Apparatus for launching horizontally moving ball at the same time as a second ball is released at rest. When catch is removed, coiled spring drives the rod and hammer toward the ball at right. The length of the rod and the position of this ball are set up so that as the hammer reaches this ball, the far end of the rod passes out of the hole drilled through the ball at left. So the ball at left begins to fall just as the ball at right is launched.

Newton knew that if you throw a stone horizontally, the time it takes to hit the ground is the same as if it were released from rest (see the box). It takes 1 second for the stone to drop about 5 meters, starting from rest. Suppose you throw the stone, at a speed of 2 meters per second, sideways from the top of a ladder 5 meters high. It will still take 1 second to reach the ground. In the meantime it will have travelled 2 meters to the side. If you throw it at 10 meters per second, it will move 10 meters horizontally in 1 second before it hits the ground. You can use a slingshot, or a cannon, to get the stone to go faster. It will still take 1 second to fall 5 meters. The faster its horizontal speed, the farther it will go in one second.

THE INDEPENDENCE OF HORIZONTAL AND VERTICAL MOTION

A consequence of relativity The rocket-satellite that circles the Earth was something that Newton could only imagine, not make. His calculation of its orbit depended on a crucial assumption: that it would fall at the same rate no matter how fast it was moving horizontally.

This idea can be tested easily for slower projectiles.

Experiment Two metal balls are placed on a high stand. A trigger is set up in such a way that one of the balls is released to fall straight down at the same instant that the second ball is fired horizontally. Listen for the sound of the balls hitting the floor. Can you tell if one hits the floor ahead of the other? According to our assumption, the balls should hit at exactly the same time.

Repeat the experiment at a different height, or with a different speed for the horizontal ball.

The fact that the two balls should fall at the same rate follows from Galileo's principle of relativity. This principle says that there is no way for a passenger in a fast airplane (for example) to tell how fast he is moving without comparing himself to things outside the plane.

Suppose a passenger drops a fork and times how long the fork takes to hit the floor. From the point of view of a man on the ground, the fork is not falling straight down. It is moving in a long arc. By the time it hits the floor of the plane, it has moved hundreds of meters from its starting point.

If there were the slightest difference in how long the fork takes to fall, we could design a new kind of speedometer. Airplanes could carry a falling-body setup, with some photoelectric cells for fast timing. By checking exactly how long the weight takes to fall a set distance, the pilot might be able to figure out his ground speed without looking outside the cabin. But this is just the sort of thing that the relativity principle says we cannot do.

So you see how we can start with a negative-sounding idea: *You cannot tell how fast you are moving without comparing yourself to things outside.* And we use it to reach a positive result: *The time it takes for something to fall a certain distance is the same no matter how fast it is moving horizontally.*

We know that the Earth is not flat. If you take a horizontal straight line and project it for 8000 meters, the straight line will be five meters higher off the ground than when it began. This is because the earth is curving away from the line.

So if we sent off a rocket horizontally at just 8 kilometers per second, the rocket will fall 5 meters in the first second. But it won't get any closer to the earth! Its rate of fall will just match the curvature of the earth. At the end of one second the rocket will be just as far off the ground as when it started. It will still be moving horizontally. If there were no air resistance, it would still be going at the same speed.

In the next second, the rocket would repeat this sequence. It would move 8 more kilometers to the side. It would fall 5 more meters without getting

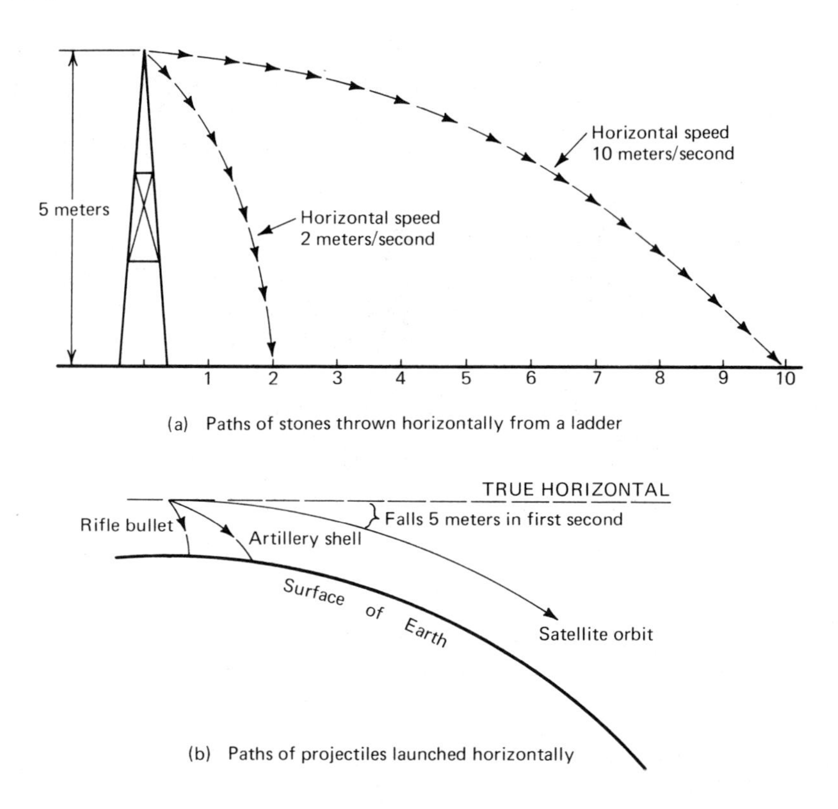

Figure 2-15 (*a*) Paths of stones thrown horizontally from a ladder; (*b*) Paths of projectiles launched horizontally

closer to the ground. In the third second it would do the same thing. The rocket could continue in this way all the way around the earth.

The Earth is 40,000 kilometers around. At 8 kilometers per second, it would take 5000 seconds for this rocket to circle the earth. This is 83 1/3 minutes, or 1.4 hours. In fact, we know today that this is just about how long it takes a low-orbit, man-made satellite (say, 200 kilometers above the ground) to go around the earth.

Newton couldn't make such a rocket in his day, but he could figure out how fast it had to go.

Let us now prepare a Kepler's third law table, to compare the moon with this low-flying satellite. For fun, we include a second satellite 35,900 kilometers above the earth's surface ("Telstar"). We will use different units of distance and time than in Table 2-1: thousands of kilometers for distance, and hours for time. So the value of Kepler's ratio will come out differently than before. The only thing that is important about this ratio is that it will be the same for all the satellites.

Table 2-2 shows that the moon is a satellite of the Earth, in the same sense that the low-flying rocket is. The same forces must be responsible for keeping both in their orbit. Gravity, the pull of the Earth, makes the rocket fall. It must be the same gravity that keeps the moon in its orbit.

Table 2-2 KEPLER'S RATIO FOR SATELLITES OF EARTH

Satellite	***Distance from center of earth (thousands of kilometers)***	***Period of revolution about earth (hours)***	***Kepler's ratio (these units)***
Moon	384.41	655.7	132.1
Satellite at surface of Earth	6.37	1.4	131.9
"Stationary" satellite	42.4	24.0	132.3

According to the principle of inertia, the moon would fly off at a tangent if there were no forces acting on it. The force of gravity is just enough to keep the moon forever falling at a rate that just matches the curvature of its orbit around the Earth.

Is the force of the Earth's gravity the same at the moon as on the Earth's surface? It turns out to be quite a bit weaker. Newton could do this calculation, too; it is the inverse of the one he could do about the low-orbit satellite. Knowing the speed of the moon in its orbit and the radius of its orbit, he could figure the rate of fall, due to earth's gravity, in the neighborhood of the moon. It turns out to be 1/3600 as great as near the surface of the earth. If you released an object from rest, out in space at 400,000 kilometers from the Earth, it would fall only 5/3600 meters in the first second.

The moon is 60 times as far from the Earth's center as we are on the Earth's surface. Noticing that 60 times 60 is 3600, Newton supposed that the force of gravity falls off as the square of the distance from the attracting body. If we go twice as far away, the force is only 1/4 as strong. This is just the rate of falloff needed to give exactly Kepler's third law. So the "inverse-square" law for gravity is shown to be true out to distances at least as far as the farthest planets and comets in our solar system.

Newton's law of universal gravitation *Every object in the universe attracts every other object with a force that increases in proportion to the amount of material in the two objects, and decreases in proportion to the square of the distance they are apart.*

THE PRINCIPLES OF MECHANICS

Newton's three laws of mechanics are very simple to state. Their consequences are so widespread that entire courses of physics are devoted to studying them. That is not the intention here. But no text in physics would be complete without at least stating what these principles are.

Newton's First Law. The same as Galileo's principle of inertia: *In the absence of any outside forces, a body at rest will remain at rest, and a body in motion will remain in motion at the same speed in the same direction.*

Newton's Second Law. When forces do act, the effect is to produce changes in the state of motion. We define *acceleration* as the rate of change of the velocity of an object. Acceleration can be a speeding up, a slowing down, or a change in direction.

The more force is applied, the faster the velocity changes. The more material in a body, the more force is needed to make the same acceleration.

The second law can be summarized by the formula

$$F = m \times a$$

The force needed to make a certain rate of change of the velocity of an object is equal to the product of the amount of material in it times the acceleration produced in that body.

Newton's Third Law. "Action and Reaction." *If one body exerts a force on a second body, then the second object exerts a force on the first that is equal in strength but opposite in direction.*

Making use of this law, together with his three principles of mechanics (see box), Newton was able to show that each of Kepler's laws could be derived from Newton's laws. But they go further than just the description of planetary motion.

In order to get further with his theory, Newton had to invent calculus. With this mathematical invention, he was able to answer many questions raised by his theory. Some examples;

Question: If, in a large body like the Earth, every stone, every drop of water, is pulling on an apple, the pulls being in different directions with different strengths, what is the net force on the apple? *Answer*: If the Earth is perfectly spherical, the force on objects outside it is the same as if all the mass of the Earth were concentrated at the center.

Question: What happens if the satellite rocket is started off a little faster than 8 kilometers per second, or if it is aimed slightly upward instead of exactly horizontally? *Answer*: It rises to a maximum height, then falls a bit lower than its launch height, returning eventually to its starting point. The exact orbit is an ellipse (!), with the center of the Earth at one focus.

Question: Can we account for Kepler's second law? *Answer*: It is a direct consequence of the principles of mechanics and the fact that the attraction of gravity is always directed at the central body. It is the same effect as when a figure-skater begins to rotate more rapidly when she draws in her arms.

The fact that a satellite speeds up when it gets closer to the central body can also be understood as a consequence of its having "fallen" from a greater distance. The speed of a falling body is always fastest at the bottom of its fall. Conversely, the speed of a projectile is least at the top of its climb. A satellite in an elliptical orbit can be thought of as rising to its maximum height ("apogee"), where it is moving at its slowest, then falling to its minimum ("perigee"), where it is going fastest, overshooting the Earth and rising again to apogee, and so forth.

Other consequences of gravitation.

The Tides. Because of the "Action and Reaction" principle, the moon also exerts a gravitational attraction on the Earth. The parts of the Earth closer to the moon feel a stronger pull than the parts farther away. So the Earth as a whole is under a stress.

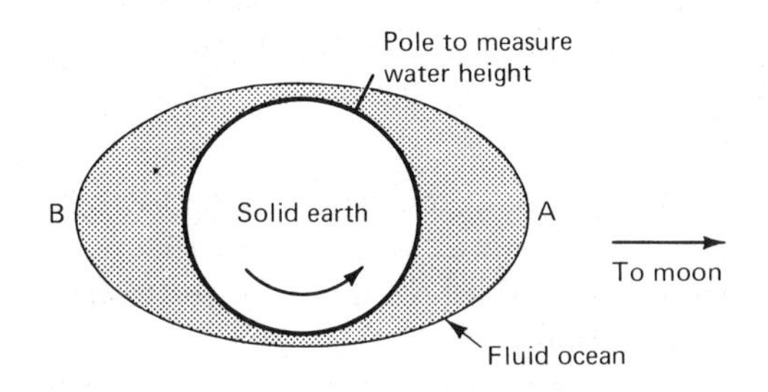

Figure 2-16 Cause of the tides. The unequal pull by the moon causes the water to "pile up" at *A*. But it also pulls the solid earth away from the water at *B*, forming another pile-up there.

The part of the Earth that is fluid, especially the ocean, has to change shape to adjust to this stress. On the side facing the moon, the water "piles up" a few meters higher than average. There is also a pile-up on the side opposite the moon. This happens because the moon's force tends to pull the solid earth away from the water. So the surface of the ocean assumes a stretched-out shape, pointing both toward and away from the moon.

If you stick a pole into the ocean bottom near the shore, you can observe the water level rise and fall. As the Earth rotates, the pole will pass through the region of the water bulges, and the tide will rise. This happens twice a day, when the pole passes the side facing the moon, and again on the opposite side. At times in between, the water level is not so high, and we have low tide.[1]

The sun also contributes to the tides. Since the sun is farther away than the moon, the difference in the sun's force from front to back of the Earth is less pronounced than in the moon's case. When the sun, moon, and earth all lie in a straight line, the stretching due to the sun reinforces the tidal effect of the moon. So at new moon and full moon, the tides are higher than usual, the so-called *spring tides*. When the sun is at right angles to the moon, during the quarter-moon phases, the tides are less pronounced than usual, the so-called *neap tides*. The moon's force tends to stretch the ocean surface toward and away from the moon. But at the neap tides, the sun is stretching it in a different way, tending to pull the ocean back toward a spherical shape.

Planets Acting on Each Other. The attractions of the planets for each other cause small changes from the Kepler orbits. These can be calculated. Sometimes such calculations show that there is an extra force acting on one of the outer planets that is not due to any known planet. So astronomers begin to look for new planets that cannot be seen by eye.

In 1781 Sir William Herschel found the planet Uranus. He knew where to look for it from the changes it made in Saturn's orbit. When Uranus had been tracked long enough to see the changes in its orbit, it was possible to find the planet Neptune. This discovery was made in 1846. In 1930 a similar study of Neptune's orbit led to the discovery of the planet Pluto. So the use of Newton's laws has repeatedly allowed astronomers to predict the positions of new planets.

Measurements on Earth. The force of gravity is actually a rather weak one. It takes very large objects, like the Earth and the sun, to exert such a total force that the effect is noticeable. In principle, every object is attracting everything else. You are attracting, and being attracted by, other people, the buildings around you, your pencil, etc. But the force is so weak that you cannot feel it at all.

In 1783 Henry Cavendish, using a very sensitive balance, was able to measure the gravitational attraction of two lead balls for each other. He thus confirmed what Newton had predicted about the universal nature of gravity. By that time there was no serious doubt that Newton's theories were basically correct.

[1] The details of the daily tides in any given place depend on the features of local geography, water depth, and wind patterns.

Glossary

Almagest—A book written in the 2nd century A.D. by Claudius Ptolemy, based on Hipparchus' ideas, and setting forth what has been known as the Ptolemaic system.

apogee—That point in the motion of a satellite (of the Earth) when it is farthest from the Earth.

Aries—The constellation of the "ram." In Greek and Roman times the sun entered the region of Aries at the beginning of spring. Nowadays the sun's position passes across Aries during the month of May.

astronomical unit—The average distance from the Earth to the sun, 150 million kilometers. A convenient unit for reckoning distances within the solar system.

Capricorn—The constellation of the "goat." The sun's position is seen in Capricorn during the winter.

Cavendish, Henry—British physicist who in 1783 showed that two lead balls attract each other. This confirmed Newton's idea of Universal gravitation, that everything attracts everything else.

comets—Small members of the solar system, spectacular in appearance, that travel in highly eccentric orbits around the sun.

Copernicus—Full name, Nikolaus Kupfernick (1473–1543). The first person to publish the idea that Earth and the planets revolve around the sun.

eccentric—The property of not being circular. The farther apart two foci of an ellipse are, for a given length of major axis, the more eccentric an ellipse is.

eclipse—A dramatic event when the moon passes directly between the sun and Earth. During a total eclipse of the sun the sky becomes dark for a few minutes in the middle of the day. An eclipse of the moon takes place when the Earth's shadow covers the moon.

ellipse—A geometric figure resembling a football or a cigar. An ellipse has two foci. The sum of the distances from the two foci to each point on the curve is the same all the way around.

epicycle—In the Ptolemaic system, the small circle in which each planet supposedly looped around its ideal position.

first quarter (moon)—The phase of the moon about one week after the new moon, when half of the moon is illuminated.

fixed stars—The great majority of stars, whose position relative to each other is the same night after night.

focus—One of the two points inside an ellipse that determines its exact shape. The sun is at one focus of the elliptic path of each planet. The other focus has no physical importance. It may even change position slowly, over the centuries, mainly because of the attraction of the other planets.

full moon—The phase of the moon when the moon is in the opposite direction from the sun, as seen from the Earth. Spring tides, and eclipses of the moon occur at full moon.

Galileo Galilei (1564–1642)—Italian physicist who took up Copernicus' idea that the earth and planets revolve around the sun. He put forth the principle of inertia and the idea of relativity to explain why we didn't fall off the moving Earth. Galileo is also well known for such inventions as the telescope and the pendulum clock.

gravity—The attraction of all massive objects for each other. (See Newton's law of universal gravitation.) In particular, the attraction by the Earth for all objects on its surface and nearby it.

Halley's comet—A comet that comes close to the sun every 76 years. Its predicted return in 1758 was a confirmation of Newton's theories.

Hipparchus—A Greek mathematician of the 2nd century B.C. whose ideas about epicycles became part of the Ptolemaic system.

inertia—The inability of matter, by itself, to change its state of motion. The principle of inertia states that, if no force is acting on it, a body at rest will remain at rest, and a body in motion will remain in motion, at the same speed in the same direction.

Jupiter—The largest of the planets. It circles the sun at a distance of 5.203 astronomical units, once every 11.862 years.

Kepler, Johannes (1571–1630)—Assistant to Tycho Brahe, who used Tycho's data to help formulate an accurate description of the motion of the planets around the sun.

Kepler's (empirical) *Laws of Planetary Motion*—(1) The orbit of each planet is an ellipse with the sun at one focus. (2) The speed of a planet changes in such a way that the line from the sun to the planet sweeps out equal areas in equal times. (3) The period of revolution and the distance from the sun of the different planets are related to each other in such a way that Kepler's ratio (which see) is the same for all of them.

Kepler's ratio—The cube of the distance divided by the square of the period of revolution, of a planet or satellite. "Distance" means one-half the major axis of the elliptical orbit. Kepler's ratio, when expressed in the same units, is the same for all the planets or satellites that revolve around a given central body. The larger the central body, the larger is Kepler's ratio for its satellites.

last quarter (moon)—The phase of the moon about one week after a full moon. Half of the face of the moon is illuminated.

major axis—The longest straight line that can be drawn within a given ellipse. It passes through both foci.

Mars—The planet that passes nearest to the Earth. It circles the sun at a distance of 1.524 astronomical units every 687 days.

mechanics—The scientific study that deals with the motion of objects when they are subjected to various forces. Newton showed that the principles of mechanics (which see) applied to the planets as well as to machines and falling bodies on earth.

Mercury—The smallest of the planets and closest to the sun, only 0.2408 astronomical units away. Its period is 88 days.

moon—A satellite of the Earth, held in orbit by the same gravitational forces that make an apple fall. The pull of the moon raises tides in the Earth's oceans. Its distance is 384,000 kilometers and its period is 27.3 days.

neap tide—Less pronounced tides that occur at the time of quarter moon, when the effects of the sun and moon in raising tides tend to cancel.

Neptune—The eighth planet from the sun. It was discovered in 1846, after a study of the effects it had on the planet Uranus. Its distance from the sun is 30 astronomical units, and its period is 164 years.

new moon—The phase of the moon when the moon is nearest the sun. Very little, if any of the face toward Earth is illuminated at this time. Spring tides occur at this time. An eclipse of the sun can occur at new moon if the moon's path is crossing the sun's apparent path at that time. In most months the moon passes over or under the sun.

Newton, Isaac (1642–1727)—English physicist well known for his unification of the scientific studies of mechanics, planetary motion, and gravitation. He also did research into the nature of light and of sound.

Newton's Law of Universal Gravitation—Every object in the universe attracts every other object with a force that increases in proportion to the amount of material in the two objects, and decreases in proportion to the square of the distance they are apart.

Newton's Principles of Mechanics—(1) The Principle of Inertia (which see); (2) $F = m \times a$; (3) Action and Reaction.

perigee—That point in the motion of a satellite (of Earth) when it is closest to the earth.

period—The time it takes a planet or satellite (or any system undergoing repetitive motion) to return to its initial position.

Pisces—The constellation of the "fish." In Greek and Roman times the sun was seen in Pisces in late winter. Nowadays the sun is in this region during the month of April.

planets—Small star-like objects in the sky that do not stay in a fixed position relative to the main body of stars. The modern model for the planets holds that they revolve in elliptical orbits around the sun, following the laws of Kepler and Newton. According to this model, the Earth is one of the planets.

Pluto—The most distant of the known planets. It was discovered in 1930 after a study of the effects it had on the planet Neptune. Its average distance from the sun is 39.4 astronomical units and its period is 246 years. Its orbit is quite eccentric, and it is expected to get closer to the sun than Neptune at its closest approach.

Ptolemaic system—The model for the motion of the planets suggested by Hipparchus, and set forth by Claudius Ptolemy in his book, the *Almagest* (2nd century A.D.). According to this model the planets moved in epicycles about their ideal positions, which in turn moved in uniform circular motion about the Earth. This model was commonly used in Europe for many centuries.

quarter moon—The times of month when the line from Earth to moon is at right angles to the line from Earth to sun. The tides are less pronounced at this time. Only half the face of the moon visible from earth is illuminated at quarter-moon time.

relativity—The idea that the laws of physics have the same form no matter how fast you are moving (in a straight line at uniform speed).

satellites—Objects such as moons, planets, comets, rockets, or rings (as of Saturn) that move in orbit about some central body.

Saturn—The most distant of the planets visible to the unaided eye. It circles the sun at a distance of 9.569 astronomical units every 29.5 years.

spring tide—The very pronounced tides, both higher and lower, that occur at the times of new moon and full moon. The effects of the sun and the moon in raising tides reinforce each other at these times.

sun—The nearest star; the massive central body about which the known planets, comets, asteroids, etc., revolve.

Telstar—An artificial satellite of Earth placed in orbit at such a distance that its period of revolution is exactly 24 hours. It will appear to stay fixed in the sky above some chosen spot on the earth's surface.

tides—The rise and fall of the oceans caused by the unequal pull of the moon (and to a lesser extent, the sun) on the opposite sides of the Earth. As explained by Newton, the tides rise and fall twice a day. The exact pattern of the tides at any given coast can be more complicated, depending on local geography.

Tycho Brahe (1546–1601)—Danish astronomer who set out to make accurate measurements of the motions of the planets, working without a telescope, but with great patience and ingenuity, Kepler used Tycho's work to develop his famous laws of planetary motion.

uniform circular motion—The principle of motion used in the Ptolemaic system and also by Copernicus. Since a circle was "perfect," it was reasoned, the heavenly bodies must move in circles, and at a constant speed. This principle was abandoned by Kepler in his first two laws, and eventually was superseded by Newton's principles of mechanics.

Uranus—The seventh planet from the sun. Not visible to the naked eye, it was discovered in 1785 by Sir William Herschel after he studied its effects on Saturn's orbit.

Venus—The second planet from the sun. It circles the sun, in the least eccentric orbit of all, at a distance of 0.7233 astronomical units every 224.7 days.

year—The time it takes Earth to make one revolution around the sun, 365.2564 days.

Questions

1. Is there a reason why spring begins when the sun passes a certain part of the sky? Is there a reason why tides are higher at new moon and full moon? Would you expect to find similar detectable effects from the positions of the planets?
2. When Galileo turned his telescope on Venus, he found that it had phases like our moon ("quarter Venus," "crescent Venus," etc.). What does this prove about the light by which we see Venus? Can we ever see a "full Venus" phase?
3. Show that the principle of inertia and the idea of relativity are related. Let us agree that an object at rest and subject to no forces will remain at rest. Let us further insist that this law must appear to hold true, no matter how fast our system is moving. What must we then conclude about the future course of an object in motion, if no forces act on it?
4. What was the *model* and the *principles* for the Ptolemaic system? What were they for Newton's theory?
5. Was the Ptolemaic system wrong? Why do you think so? Can you think of a better word than "wrong" to describe it?
6. Compare Copernicus' theory to the Ptolemaic system, on the basis of the criteria for acceptance of a new theory, as given in Chapter 1. Not all four criteria are met. Do you think this fact caused some of the delay in gaining acceptance for Copernicus' scheme?
7. What is friction? Name three ways in which friction can be useful.
8. Give three ways in which an airplane pilot can measure his ground speed. Show that each of these ways depends on referring to things outside the airplane. This includes the air outside the plane, that is not moving along with it.
9. Tycho's scheme had the sun and moon revolving around the Earth, and all the other planets revolving about the sun. Would this scheme give a prediction for the *relative* positions of the planets different from that of Copernicus? Why not?
10. Using string and thumbtacks, draw an ellipse that is almost a circle. Draw another that is highly eccentric.
11. Show that a comet in a very eccentric orbit will spend much more time in the far-out part of its orbit than in sweeping in close to the sun.
12. If the value given in Table 2-1 for Jupiter's distance and period were off by one unit in the last decimal place, would Kepler's ratio be closer to 1.0000?
13. Which planets and satellites have the highest speed, those closer in or those farther out?

14. Apply the four criteria of Chapter 1 to the case where Kepler's laws are the previously accepted theory, and Newton's scheme is the new theory.

15. Describe in your own words why the tides rise and fall twice a day.

Thought Provoker: In 1735 the French Academy of Sciences sent an expedition to Lapland, in the far north of Europe. They went there to measure the exact shape of the Earth. Newton had predicted that, because of its spinning motion, the Earth would bulge out near the equator and be flattened near the poles. When the expedition returned, reporting the predicted results, the philosopher Voltaire mocked them with the couplet:

> "*Vous trouvez en voyages dans pays pleins d'ennui*
> *Ce que Newton connût sans sortir de chez lui.*"
> ("*To distant and dangerous places you roam*
> *To discover what Newton knew staying at home.*")

Was Voltaire scientifically justified?
Can scientific theories be conceived solely from "armchair" thinking and debating, without being checked by experiments "in the field"?

3

the speed of light

Introduction and summary

The first example we gave of how a new theory came to replace a previously accepted model seems today to have been a rather one-sided case. Hardly anyone now takes the Ptolemaic model of the universe seriously. Looking back, a reader might think that the older theory was something of a "straw man," set up just to be knocked down.

In this chapter we will consider a second example, in which the "previously accepted theory" is one of the strongest and best verified structures of physics, the classical theory of electricity and magnetism. The result of the revolution in thinking that came about in the early 20th century was not to reject classical electromagnetic theory—we still use the same equations—but to change our interpretation of it.

One of the great areas of scientific advance in the 19th century was in the study of electricity and magnetism. These effects had been known since ancient times. But it was only after the spirit of scientific inquiry had been established, by such men as Galileo and Newton, that systematic study and interpretation became possible.

It had long been known that under certain circumstances, many kinds of objects can be "electrified," or in other cases, "magnetized." Such objects were known to attract or repel each other, even when they were not in direct contact. Such "action-at-a-distance" was much like the forces of gravitation, but the electrical and magnetic forces seemed to be very much stronger.

The effects of *electricity* and *magnetism* had been studied separately, and much written about them, before the year 1820. A series of discoveries in the following years showed that these two fields were related. Electric charges in motion, so-called electric currents, were shown to affect magnets. Magnets in motion could also give rise to electric forces. Electricity and magnetism were shown not to be separate effects, but different aspects of a single phenomenon, what we now call *electromagnetism.* Another great unification of fields of science had been achieved.

The story of unification was not over. In 1864 James Clerk Maxwell, a British scientist, wrote down some equations that seemed to summarize all that was known about electromagnetism. These equations had many solutions, most of which corresponded to the already well-known behavior of electric charges and currents. But there were some surprising new solutions. In certain cases, Maxwell found, electromagnetic signals could be produced and sent from place to place at a speed of 300,000 kilometers per second. This is almost exactly the speed of light! In fact, the new type of electromagnetic signal behaved in so many ways like light signals that he was led to conclude that light itself *is* an electromagnetic signal.

Another great unification had been achieved. The science of *optics*, until then an entirely separate field of study, now was studied as part of electromagnetic theory.

The unification of electricity, magnetism, and optics under the title of electromagnetism stood apart from the other great unification of physics, the union of such diverse fields as astronomy, heat, and sound under the edifice of Newtonian mechanics. Spurred by their success up to this point, scientists of the late 19th century tried to achieve the final unification of all branches of physics.

They did so mostly by trying to construct a mechanical model for electromagnetism. They thought of an elastic substance, called the "ether," that supposedly filled all of space. Electromagnetic effects were supposed to be the strains and vibrations of the ether.

There was a stumbling block that made unification of Newton's and Maxwell's theories difficult. Newton's laws, and all the consequences based on them, obey Galileo's principle of relativity. They are true, no matter how fast your reference system is moving.

Maxwell's equations, on the face of them, do not. There seemed to be one preferred frame, supposedly the frame in which the ether was at rest, in which they were true. In any other frame of reference, you would have to correct for your motion through the ether.

One or the other theory, that of Newton or of Maxwell, would have to be modified to achieve unification. The bias of the 19th century physicists was to give up on the principle of relativity. It was a nice idea, they felt, but it was valid for only a subset of physics. They thought that as soon as electromagnetism was accepted, relativity had to be dropped.

Further reasoning proposed that if there is no principle of relativity, then it ought to be possible to measure how fast we are moving "through space," or at least through the ether. In a famous experiment, Michelson and Morley attempted to do just that. To their surprise they found that the Earth is apparently not moving with respect to the ether. Even though we know that the Earth is circling the sun at a speed of 30 kilometers per second, and the experiment was sensitive enough to see such an effect, no motion could be detected.

Various explanations were offered. Either (1) just by accident the combined motion of the Earth and the solar system happened to match that of the ether, or (2) the Earth dragged part of the ether around with it, or (3) apparatus moving through the ether shrank in just the right proportion to give Michelson and Morley their null result. Explanations (1) and (2) were tested by experiment, and were found wrong. Explanation (3) is of such a nature that that it cannot be tested. If everything shrinks in the same proportion, there is no way to measure it. It does seem strange, however, that nature conspires to arrange things exactly so that we cannot measure our motion through the ether.

Albert Einstein proposed to take this natural conspiracy and make it into a basic principle. The laws of physics *do* appear the same in all reference frames,* he asserted. There is no way to measure how fast you are moving through space. The principle of relativity is still valid, he said.

There was a difference between Einstein's relativity and Galileo's. Einstein insisted that Maxwell's equations and their consequences—the speed of light, the null Michelson-Morley result—were true in *every* uniformly moving reference system, without correction. But for this idea to be valid, some other accepted notion had to be given up. What we had to abandon, Einstein suggested, was the idea that the size of an object, or the time interval between events, was the same no matter who did the measuring.

Einstein showed that, according to his new principle of relativity, the size of any object shrinks and that time appears to pass more slowly for an observer in rapid motion, as compared to another observer "at rest." He proved that there was no inconsistency in this idea. There really wasn't any way to make a direct comparison of rulers and clocks. Two observers moving with respect to each other, he demonstrated, could not even agree on whether two events happened at the same time or not. If they adopt reasonable procedures for trying to synchronize their clocks, he explained, each of them will soon conclude that the other's clock is running slow.

These were heady ideas. They go against all our senses. Perhaps we might be willing to accept the idea that solid objects shrink when they are going very fast. But our sense of time?

It is not easy to accept an idea that disagrees with all our experience. The truth is that we do not have much experience at travelling at very high speeds. When we do put Einstein's predictions to the test of experiment, they pass every trial. For example, some very accurate atomic clocks were recently put in an airplane and flown around the world. It turned out that they did run slower than identical clocks left on the ground, by precisely the amount predicted.

Einstein's theory of special relativity meets all the criteria for acceptance of a new theory. It agrees precisely with the previous theories, those of Newton and Maxwell, when we deal with speed very much slower than that of light. It explains effects like the Michelson-Morley experiment, that could be accounted for only very awkwardly, if at all, under the old ideas. It predicts

* Moving in a straight line at constant speed.

effects, such as the slowing down of clocks, that can be tested by experiment. And it is above all such a simple and elegant explanation of nature that it commended itself to the artistic soul of many scientists even before the experimental tests were done.

Electric charge

Demonstration: Attraction and Repulsion Two plastic spheres hang from separate threads. A third sphere, also plastic, is held at the end of a wooden stick in the lecturer's hand. At first the spheres are shown to have no visible effect on each other, even when they are brought quite close together.

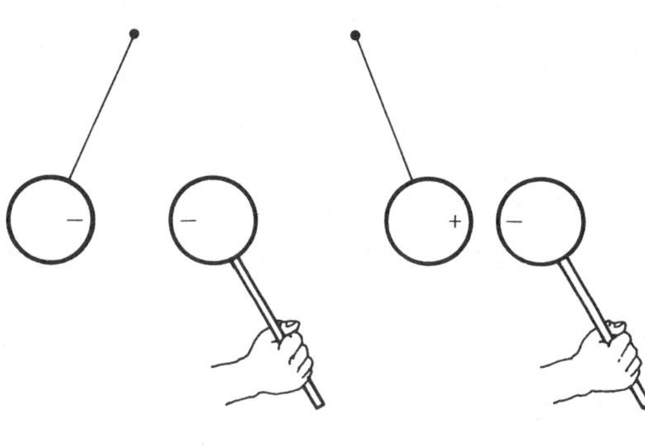

Figure 3-1

One of the hanging spheres is rubbed with cat's fur. The second hanging sphere is rubbed with silk. The hand-held sphere is also rubbed with cat's fur. Then the hand-held sphere is brought close to each of the hanging spheres. It is found to *repel* the first one, which, like itself, was rubbed with cat's fur. It *attracts* the sphere that was rubbed with silk. The two hanging spheres, when brought together, are also seen to attract one another.

The spheres are seen to exert forces on each other even when they are not actually touching. The forces can be either attractive or repulsive in different cases. They are obviously much stronger forces than the gravitational attractions between these objects, which would be too weak to measure easily. Objects that exert such forces on each other are said to possess *electric charge*.

Benjamin Franklin (1706–1790), the American statesman and also the first American physicist, showed that there must be two kinds of electric charge, which he named *positive* and *negative*. Positive electric charge is the kind the plastic ball had after it was rubbed with silk. Negative electric charge is the kind received by the sphere that was rubbed with cat's fur.

The rule for whether electric forces are attractive or repulsive is simple. Objects with the same kind of electric charge, positive or negative, repel each other. Objects with opposite electric charges—one positive, the other negative—*attract* each other. An object with no net electric charge, with an equal amount of negative and positive charge, is said to be electrically *neutral.* Electrically charged objects exert no forces, neither attraction nor repulsion, on neutral objects.

The modern picture is that all matter—cat's fur or silk, plastic or air, liquid or metal—is made of large numbers of very tiny *molecules* that in turn are composed of *atoms*. This will be discussed in more detail in later chapters. The atoms themselves have electrically charged parts. The parts with negative electric charge are called *electrons.*

An atom in its natural state has an equal amount of positive and negative electric charge. It is thus electrically neutral. But electrons can be detached from atoms in some circumstances. An atom that has lost one or more electrons has a net positive charge. Atoms can also pick up extra electrons. Such an atom has a net negative charge.

Cat's fur is a material from whose atoms electrons are easily detached. So the plastic balls rubbed with cat's fur obtained a negative charge. The atoms of silk have a tendency to attract extra electrons. So the sphere rubbed with silk lost some of its electrons and remained with a net positive charge.

Charles Augustin de Coulomb (1736–1806) in 1785 showed that the forces between electric charges have some features in common with gravitational forces. Whereas gravitational forces increase in strength in proportion to the *mass* of the objects involved, electric forces are proportional to the electric *charges* of the objects. The greater the net electric charge, the stronger the electric force.

Electric forces also become weaker the farther objects are apart. Coulomb showed that, similar to the case of gravitational forces, electric forces decline in proportion to the square of the distance the objects are apart. If objects are moved twice as far apart, the strength of the electric forces between them falls to one quarter of what it had been.

Despite these similarities between electric forces and gravitation, there are several important differences. Gravitation is a universal force, between every pair of objects, whereas electric forces affect only objects that have electric charge. Gravitation is always an attractive force, whereas electric forces can be either attractive or repulsive. Gravitation is a rather weak force, its effects easily observable only when one of the attracting bodies is at least as large as a planet or a moon. Electric forces between even small objects are strong enough to be observable in a classroom demonstration.

The forces that electrically charged objects exert on each other, as discussed in this section, are known as *electrostatic* forces. The electrostatic forces between two objects are the same whether they are at rest or in motion. There are *additional* forces between electric charges when they are in motion, as we shall study in the following sections.

Magnetism

Other kinds of objects, not electrically charged, have long been known to exert forces on each other, even when not touching. Such objects include bar magnets, compass needles, and the Earth itself. The forces that such objects exert on each other may be termed *magnetic* forces.

Up to a point, we can describe magnetic forces in ways we used to describe electrostatic forces. Just as there are electric charges, there seem to be magnetic *poles*, both positive ("north-seeking," or "N") poles and negative ("south-seeking," or "S") poles. Poles of the same kind repel each other. Opposite poles attract. There even seems to be an inverse-square law that governs the way the forces between magnetic poles decrease with distance apart.

Every bar magnet has both a North and South pole.
If we cut the bar magnet in half . . .

N S N S

. . . each half develops a new pole, opposite to the one it had.

Figure 3-2

The important difference is that it is not possible to separate positive and negative magnetic poles completely from each other. Up to the present writing (1978) *there has been no verified case of any object possessing a nonzero net magnetic pole strength.* Every object that seems to have a positive magnetic pole in one part of it has an equal negative pole somewhere else.

Suppose we have a long bar magnet that seems to have a positive (N) magnetic pole near one end and a negative (S) pole near the other. If we cut the bar magnet in half near its midpoint, we do *not* obtain isolated N and S poles in the two halves. Instead, we find that on the half that contained the original N pole there is now a *new* S pole near the cut. There is also a new N pole near the cut on the other half. Each of the halves thus keeps zero net magnetic pole strength.

The clue to the real nature of magnetic forces came after scientists learned how to set up electric *currents.* An electric current is a set of electric charges in motion together. The material, often a metal wire, through which an electric current can flow, is known as an electrical *conductor*.

The discoveries of how to make electric current flow through conductors were made by two Italian scientists, Luigi Galvani (1737–1798) and Alessandro Volta (1745–1827).

It is possible to have a conductor carry an electric current and still be electrically neutral. This is in fact the usual situation.

In a typical metal conductor, some of the electrons are free to move between the atoms. The current consists of the group motion of these electrons. The atoms from which the electrons have come, have a net positive charge. These atoms are not free to move.

There is an equal amount of positive charge on the fixed atoms as there is negative charge on the moving electrons. So the net charge on the conductor as a whole is zero. But the negative charges, the electrons, are in motion, while the positive charges are not. So there is a nonzero electric current in the wire.

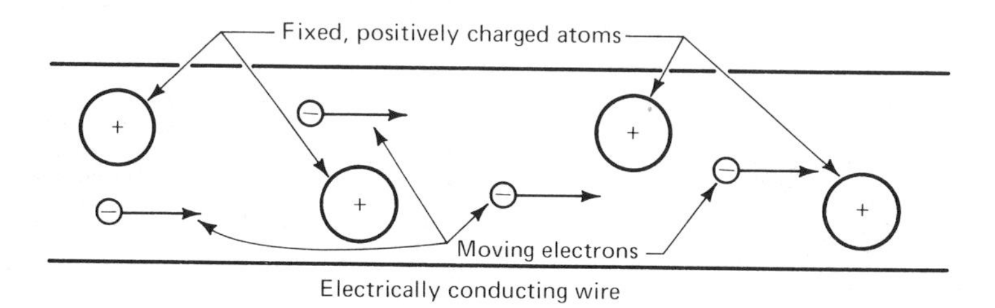

Figure 3-3 Electrically conducting wire

Hans Christian Oersted (1777–1851), a Danish schoolteacher, in 1820 made the important discovery that an electric current in a wire can deflect a magnetic compass needle. This discovery was followed rapidly by others, notably by André Marie Ampère (1775–1836) of Paris, and Michael Faraday (1791–1867) of London.

Ampère showed that two wires carrying electric current exert forces on each other. He pointed out that a coil of wire carrying electric current could behave exactly like a bar magnet or a compass needle. He went so far as to assert that *every* magnet behaved as it did because of electric currents that circulated among the atoms.

Thus the sciences of electricity and magnetism had been unified. *The forces that had been called "magnetic" were seen to be forces between electric currents.* These currents could be either measurable currents flowing in a wire, or internal currents flowing among the atoms of an iron magnet.

Demonstration: Forces between Currents Two long copper wires are set up parallel to each other. The wires are flexible enough to bend in response to forces. A battery and connectors are provided so that the electric currents in the two wires can be made to flow in the same direction, or in opposite directions.

When the currents flow in the same direction the wires are seen to attract each other. When the currents flow in opposite directions, the wires repel one another.

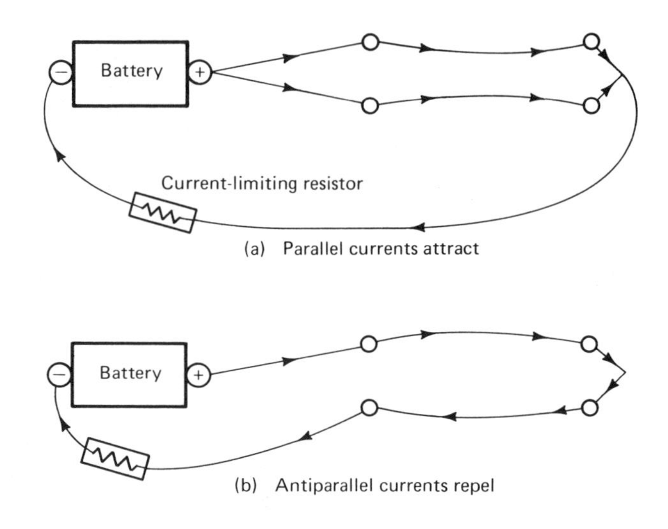

Figure 3-4 (*a*) Parallel currents attract; (*b*) Antiparallel currents repel

Thus we have demonstrated that current-carrying wires exert forces on each other. One of the laws governing such forces is that parallel currents attract each other, and anti-parallel currents repel each other.

Electromagnetic theory

Faraday pursued the connection between electricity and magnetism further. He had the intuition that if electric currents could produce magnetic effects, perhaps magnets could *induce* electric currents in wires. He was able to show that, by moving a magnet around in the vicinity of a conducting wire, electric current could be made to flow in the wire. The same effect also could be achieved simply by increasing or decreasing the strength of the magnet, without moving it at all. Faraday's discoveries lie at the base of most modern electrical machinery.

James Clerk Maxwell (1831–1879), a British mathematician, in 1873 published "Treatise on Electricity and Magnetism" (based on earlier work he had done) in which he summarized the discoveries of Coulomb, Oersted, Ampere, Faraday, *et. al.*, by a series of mathematical equations. *Maxwell's equations* are still used as the basic statement of *electromagnetic theory*.

Maxwell made one more addition to the already discovered laws. Faraday had shown that a changing magnetic configuration could induce electrostatic effects. For the sake of symmetry, and also to make his equations consistent with one another, Maxwell guessed that a changing electrostatic configuration might likewise lead to magnetic effects.

Maxwell then found that, if both these effects happened, it was possible to send an electromagnetic signal across space without any charges or wires being present at all. He began by setting up a changing electrostatic field. This then induced a magnetic field that also changed with time. The changing magnetic field induced more electrostatic fields, and so forth. The signal spread out from its source at a speed characteristic of Maxwell's theory, namely, 300,000 kilometers per second.

Figure 3-5 James Clerk Maxwell 1831–1879 THE BETTMAN ARCHIVE, INC.

Light as an electromagnetic signal This speed is familiar to us as the speed at which light signals travel. In fact, Maxwell's electromagnetic signals resembled light signals in many other ways. He concluded that *light signals are electromagnetic signals* of this nature. Another great unification of fields of physics was thus achieved.

How can electromagnetic signals be generated to begin with? Maxwell showed that such signals are always sent out whenever an electrically charged object is accelerated. How can such signals be detected? One way is to place a conducting wire in the path of such a signal. The electrons in the wire will move in response to the electrostatic fields of the signal. This motion of the electrons makes up a current in the wire that can be detected.

Light signals turn out to have such rapid variations in them that no man-made apparatus can yet generate them directly. They arise from the rapid motions of electrons within atoms. But it should be possible, Maxwell said, to make signals similar to light, but with much slower oscillations. Such signals are what we now call radio waves. These signals also travel at the speed of light.

In 1885 Heinrich Hertz showed that Maxwell was right. It was possible to send and detect electromagnetic signals that travel across space just at the speed of light. Within another decade Guglielmo Marconi had put the discovery to practical use by sending messages over long distance by means of radio signals.

Electromagnetic theory and Galilean relativity The unification of all physics, it seemed, was almost at hand. Already the studies of heat and sound, astronomy and dynamics had been gathered together under the principles of Newton's mechanics. Now it also had been shown that the fields of electricity, magnetism, and optics follow Maxwell's equations. All that remained was to join these two great edifices together, and physics would be complete.

There was one important stumbling block. Newton's Principles had Galileo's principle of relativity built into them. If they were true in one frame of reference, they should be equally true in any frame that was moving at constant velocity with respect to the first. No experiment, based on any of the fields of physics that obeyed Newton's principles, could ever determine how fast a given laboratory was moving "through absolute space."

Maxwell's equations do not follow Galileo's principle of relativity.

Consider, for example, the forces between two electrically charged plastic balls at rest in the laboratory. If they are at rest, they should exert electrostatic, but not magnetic, forces on each other. However, the Earth is moving at 30 kilometers per second around the sun. So the two spheres are not "really" at rest. They are in parallel motion through the solar system, at 30 km/sec. Charges in motion constitute electric currents. Currents exert magnetic forces on each other. We should be able to tell how fast the laboratory is moving *absolutely*, by measuring the strength of the magnetic attraction between the two charged balls.

Consider a second example. Maxwell's equations say that a light signal travels at 300,000 km/sec. But suppose the Earth, sun, and galaxy, are all speeding through space at, say, 10,000 km/sec. Would it not seem likely that a light signal that meets us head-on would appear to be coming at us faster than normally? At 310,000 km/sec, perhaps? And wouldn't a signal approaching us from behind seem to be closing the gap at only 290,000 km/sec? So can't we tell which way the earth is moving, and how fast, by measuring whether light signals approach us at different speeds when they come from different directions?

CHARGES AND CURRENTS

A plastic sphere that has been rubbed with silk has an electric *charge*. If it is not in motion, it does not have an electric *current*. It will react to *electrostatic* forces. It will not be affected by *magnetic* forces that act only on currents.

A metal wire can conduct an electric current, yet have no net electric charge. The negative charges in motion within the metal are balanced by positive charges that are at rest. Such a wire is electrostatically *neutral*. It will not be attracted nor repelled by electrically charged objects that are not in motion outside the wire. The wire carrying current will, of course, respond to magnetic forces.

It is possible to have both current and charge present at the same time. Any electrically charged object in motion constitutes an electric current as well as a charge. Such an object can give rise to, and be affected by, *both* electrostatic and magnetic forces. The *total* force on such an object is the *combined effect* of both types of forces.

An example of a situation where charges and currents are both present is a *beam* of electrons. Such a beam can be maintained in a vacuum. You have certainly seen many examples of electron beams inside evacuated glass tubes. Television tubes, oscilloscopes, *x*-ray generators, vacuum-tube amplifiers, all have electron beams inside them. The beams are electrically charged because each electron carries a negative electric charge. They constitute a current because the electrons are in motion.

The beam can be deflected electrostatically by putting stationary electric charges near it. This is commonly done in oscilloscopes. Or it can be deflected by magnetic forces by activating an electromagnet close to the beam's path.

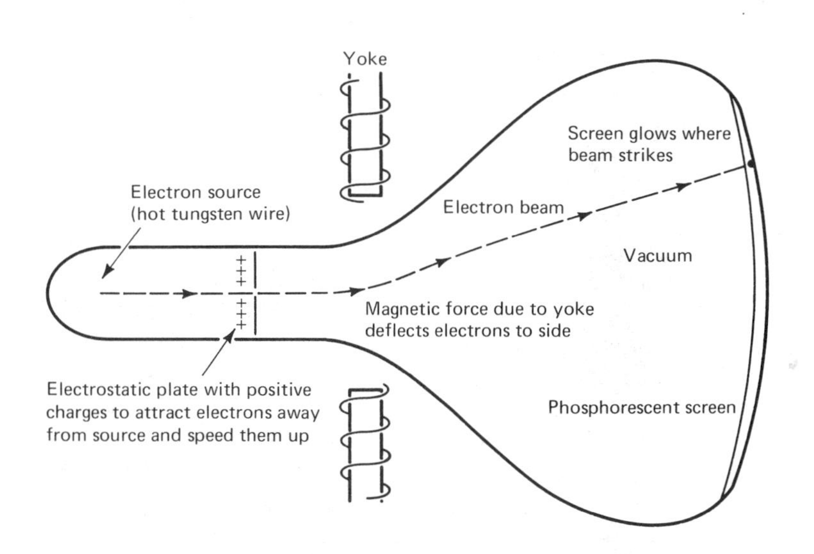

Figure 3-6 A television tube

This is the function of the *yoke* in a television set. Both kinds of forces can act at the same time. For example, a television tube uses electrostatic plates to speed up the electron beam at the same time that the magnetic forces, due to the yoke, are deflecting the beam from side to side.

It is possible, although not so common, to build a tube with *two* parallel electron beams. In such a setup the two beams will exert both electrostatic and magnetic forces on each other.

All the electrons have a negative electric charge. Since like charges repel each other, the beams have a tendency to blow each other away.

But the beams also constitute a current. So they will also exert magnetic forces on each other. The rule is that parallel electric currents attract each other. So the tendency of the magnetic forces is to pull the two beams together.

Which of these two types of forces will prevail? Apparently it depends on how fast the electrons are moving. If they are going very slowly, there is very little current. In that case, there is practically no magnetic force. So for very slow beams the electrostatic repulsion is the dominant effect.

If the electrons in the beam are moving faster, there is more current. The faster the charges move, the more current there is. The more current, the stronger the magnetic forces. The electrostatic repulsion, on the other hand, does not depend on whether the electrons are moving fast or slow, or are at rest.

The net effect is that the force of repulsion between the beams is *diminished* as the electron speed increases. The electrostatic repulsion still prevails, at least at low to moderate electron speeds, but the magnetic attraction cancels out part of it.

Clearly, magnetic forces depend on the speed of *both* beams. If the electrons

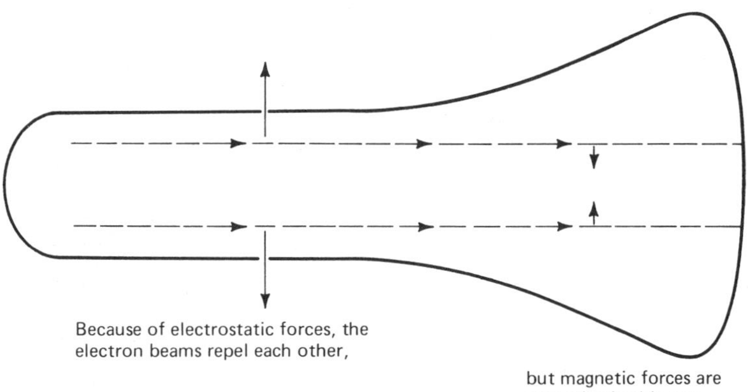

Figure 3-7 A tube with 2 electron beams

in one beam are moving rapidly and those in the other are not, there will be little current in the second beam. Therefore, it will not respond strongly to magnetic forces. It seems that the magnetic forces must depend on the *product* of the speeds of the electrons in the two beams.

If both beams are set up to have the same electron speed, the magnetic force should depend on the *square* of that speed. That is, the product of that speed (for the first beam) times the same speed (for the second beam).

The total force between beams is the difference between the electrostatic repulsion (which does not depend on the speed of the beams) minus the magnetic attraction (which increases in proportion to the product of the two speeds).

We can test this idea by experiment. We can run the electrons at different speeds and measure how much the beams repel each other. For the moderate electron speeds we can easily obtain, the data follow the predicted curve.

Thus, an interesting speculation arises. At high enough speed, it appears that the magnetic forces would just balance the electrostatic forces, and there would be no net force at all between the two beams. At even higher beam speeds, presumably the magnetic attraction would dominate.

Such speeds are not attainable, but their value can be calculated from the data. The speed at which the magnetic attraction between two parallel beams of charged objects just cancels their electrostatic repulsion, depends only on the laws of electricity and magnetism. It does not depend on what kind of particles are used, or how far apart the beams are, or any other details of the particular experimental setup.

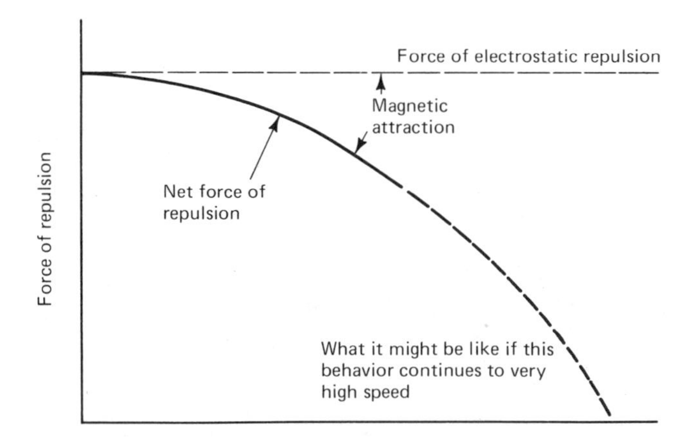

Figure 3-8

The speed at which this exact cancellation would take place turns out to be 300,000 kilometers per second.

This speed is a familiar one in physics. It is equal to the speed of light through empty space. Is it a remarkable coincidence that this speed is also built into the laws of electricity and magnetism?

Perhaps not. Perhaps the explanation lies in the fact that light signals are also electromagnetic in nature. The speed at which they travel is the speed that is built into electromagnetic theory, the speed known as c, 300,000 kilometers per second.

The complete cancellation between the electrostatic repulsion and magnetic attraction between two parallel electron beams never really happens. As we shall see in this chapter, we never can accelerate electrons, or any other material objects, up to quite the speed of light.

Maxwell's equations appeared to be absolutely correct in only one frame of reference. In any other frame, it seemed, correction would have to be made for the speed at which that frame was moving.

The ether The scientists of the late 19th century tried to close the gap between Newton's and Maxwell's laws by thinking up a *mechanical model* for electricity and magnetism.

They compared light signals to signals of sound. Sound travels through air (and also through water, solids, and other media). An observer at rest in the air hears sound signals coming at him at the same speed from all directions. If he is moving through the air, he observes signals coming at him at different speeds from different directions. This effect is not due to his absolute motion through space. It arises from his very real motion through the air. This motion can be detected by such other effects as the wind blowing in his face.

In a similar way, 19th century physicists argued that light signals travel through a medium they called *ether*. Unlike air that ends a few hundred kilometers above us, the ether was believed to fill all of otherwise empty space. They said it must extend at least as far as the most distant stars, for we certainly can see light signals from them.

We live in a vast bowl of jelly, they argued. When the jelly shakes, we see light signals. When it is strained or distorted, we observe electric and magnetic effects. The jelly must be very slippery to electrically neutral objects. Such objects, be they even as large as the moon or the planets, can apparently slide right through the ether without ever slowing down in the least. But an electrically charged object seemed to have "barbs" that hooked onto the ether-jelly, pulled it, and was pulled by it. Electric and magnetic forces arose, they claimed, when one charged object distorted the ether, which then pulled or pushed along other charged objects.

The mechanical ether was a nice model in many ways. It gave hopes of uniting Maxwell's theory with the older branches of physics. It provided the frame in which Maxwell's equations would be true without correction. And, like any useful explanation of nature, it could be tested by experiment. Unfortunately for the model-makers (Maxwell himself was one of their leaders), the model of ether failed the experimental test.

The Michelson-Morley experiment It seemed a reasonable test to measure just how fast the Earth was moving through the ether. So the American physicist, Albert Michelson (1852–1931), and the chemist, Edward Morley (1838–1923), set out to measure the apparent speed at which light travels in various directions, and from these differences tell how fast the Earth is moving absolutely.

The perplexing result they got was that the Earth does not appear to be moving at all! To the limit of their accuracy, which was very good, light signals travelled at exactly the same speed in all directions within their laboratory.

There was no shortage of explanations proposed to account for this result, once it was announced.

Perhaps, just by accident, the motion of the Earth around the sun was exactly opposite to the motion of the solar system as a whole. The two motions would then have cancelled out each other. At the date that Michelson and Morley did their experiment, the Earth might have been at rest, or nearly so, in the ether.

Michelson and Morley repeated the experiment six months later. The Earth was then moving in the opposite direction around the sun, as always at about 30 kilometers per second. Their apparatus was sensitive enough to see the difference. But the result was the same as before. The speed of light was still the same in all directions. The Earth did not appear to be moving through the ether.

Another explanation was that the Earth dragged part of the ether along with it, the way a fish-net drags along some of the surrounding water. This explanation failed, however. If it had been true, we should have seen certain shifts in the apparent positions of distant stars. Such effects were not seen. For this kind of explanation to work, the part of the ether that the Earth dragged along with it would have to reach all the way to the stars. This seemed too much to demand.

A third explanation was offered by H. A. Lorentz (1852–1928) and G. F. Fitzgerald (1851–1901). They pointed out that all matter is made of atoms that have electrically charged parts. It seemed reasonable to suppose that as the atoms move through the ether, they might become compressed along the direction of motion. If the atoms become flattened, all objects composed of atoms (in other words, everything) would also shrink. We would have no way of noticing this, because all our measuring rods are made of atoms, too. They would contract in the same proportion.

Lorentz and Fitzgerald calculated that if all atoms contracted by the right amount, the Michelson-Morley result would be explained. Michelson and Morley could not detect differences in the speed of light, they reasoned, because a speed is a distance divided by a time interval. The "fixed" distance over which they timed the light signals was, let us say, the distance between the ends of a meter stick. But if the stick changed length when it was turned in different directions, the experimenters might be deceived into calculating an incorrect value for the speed of light.

The Lorentz-Fitzgerald idea explained the results of the Michelson-Morley experiment, but it could not be tested in any other way. It required that *all* atoms flatten out along their direction of motion. It said that the amount of contraction depends only on the speed, not on the kind of atom. The effect could be observed only in experiments like that of Michelson and Morley. The size of the effect had to be exactly such as to give that experiment its null result. It was as if there was a conspiracy of nature to keep scientists from ever measuring the speed of the Earth through the ether.

The theory of relativity

Albert Einstein (1879–1955), a young man then working as an inspector for the Swiss patent office, proposed in 1905 that we turn this conspiracy of nature into a basic principle of physics. It appeared to him that it is not possible to measure the speed of the Earth, or any other system, through absolute space. Only its motion *relative* to other material bodies can be detected. Let us take it as a principle, he suggested, that no experiment can ever tell us how fast we are moving, except in relation to other objects. Let us assert, with Galileo, that *the laws of physics appear to have the same form whether we are standing still or moving at a constant velocity.*

But what about electromagnetic theory? We have seen how it did not seem to fit Galileo's idea of relativity. Must this theory, successful in so many ways, be abandoned?

No, said Einstein. Maxwell's equations are among the laws of nature. If they are correct, they must appear to be correct in every uniformly moving system. In particular, one very important conclusion from these laws, the numerical value of the speed of light, must appear the same, no matter how fast we are moving.

The difference, said Einstein, lies in the *interpretation* we give to Maxwell's laws, and to the speed of light. The mechanical model of electricity and magnetism would have to be given up. There is no ether. At least, there is no way to detect its presence. So by the scientific standards we set forth in Chapter 1 the concept of the ether is a very *unphysical* idea.

Figure 3-9 Albert Einstein 1879–1955 THE BETTMAN ARCHIVE, INC.

Simultaneity There need be no contradiction, Einstein pointed out, if we are very careful to understand how measurements are made. To measure the speed of light, we need to know how long it takes for the signal to get from one place to another. To do the measurement well, we need to have good clocks at both places. The problem of *synchronizing* the clocks is not as easy as it might seem.

It won't do simply to take a clock and move it from one place to another. Because of effects similar to the Lorentz-Fitzgerald contraction, a moving clock might not keep the same time as a clock at rest. We might send signals from one place to another, to get the clocks timed in. But we know of no signals that travel faster than light. So we are forced to use light (or radio) signals themselves to do the synchronizing.

This is in fact how such things are done in practice. There are government time stations near Washington and London that send out radio signals, as accurately as they know how, exactly on the stroke of each minute. Suppose the London station finds that the Washington signals always arrive 10 milliseconds after the minute in London. And suppose that the Washington station finds that the signals from London are arriving 20 milliseconds after the minute, Washington time. The two stations would conclude that the signals take 15 milliseconds to cross the ocean, and the British clock is 5 milliseconds slow compared to the American one. Appropriate adjustments in one or both clocks could then be made.

Einstein pointed out that, while this is a perfectly rational way of synchronizing clocks, certain untestable assumptions go into it. Most particularly it is assumed that the speed of light is the same in both directions. This is the conventional assumption to make, but there is no independent way to tell whether it is correct. We might have made a different convention. For example, we might have corrected for the Earth's motion around the sun.

This conventional approach leads, however, to some very strange conclusions. One of these is that events which seem simultaneous to one observer may not seem so to a second observer who is moving with respect to the first.

The example Einstein gave is that of a train struck by lightning at the front and back ends. The center of the train is a well-marked spot. The conductor is standing there, in position to watch both ends of the train. There is a track worker standing on the ground alongside the train.

A brief time after the conductor has passed him, the track-worker sees lightning flashes striking both the front and the back ends of the train. Both flashes reach his eyes at the same time. The conductor (who has very good time-discriminating ability) reports that he saw the bolt hit the front of the train slightly before he saw the flash from the back of the train. Everyone is agreed on what the two men saw. But the interpretation of the events may be different.

"I can explain it all," says the track worker. "The lightning hit both ends of the train at the same time, just at the instant the midpoint of the train was passing me. So both light signals had the same distance to travel to reach my eyes. It was only natural that I, standing still on the ground, saw both flashes at the same time. The conductor was moving toward the front end of the train, and away from the spot where the lightning hit the back end. By the time the signals reached his eyes, he had moved a bit. He actually saw the flash from the front end before I did. He saw the rear-end flash a bit after me, since that signal had farther to travel. But it is clear that the two flashes really were simultaneous."

"Why should I be the second-class citizen because I ride the train?" asks the conductor. "The train was going westbound at night, so I was more nearly at rest in the solar system than the man on solid earth. I saw the front end flash before I saw the back end flash. I was standing right at the midpoint of the train. That means to me that the front-end flash *did* occur first. The track-worker must make a correction for his motion around the sun."

Which of the two observers is correct? Einstein argued that both men have equal claims to validity. There is no basis for saying that the train is "really"

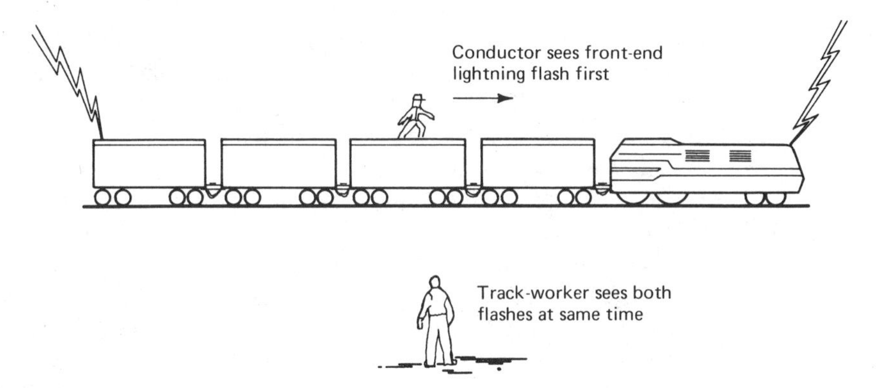

Figure 3-10

moving and that the track-worker is "at rest." From the point of view of relativity there is no preference between the two.

The accompanying box about the Mars explorer gives another illustration of how observers with different points of view can reach different conclusions about the simultaneity of events. This is an inescapable conclusion from the theory of relativity. If we accept the premise that the speed of light is the same in all directions and for all observers, we must expect that events that appear to occur at the same time in one system may not appear simultaneous to another.

WHICH CAME FIRST? A CASE OF IRRECONCILABLE DIFFERENCES

A landing vehicle on the planet Mars communicates with Earth by sending back television signals. These signals are analyzed by a computer. (Rapid decisions must sometimes be made and cannot be entrusted to the slow reactions of humans!) The computer can digest the information and send back an answering signal in less than a *millisecond*, one thousandth of a second.

However, the Earth and Mars are 100 million kilometers apart at this time. It takes about $5\frac{1}{2}$ minutes for the television signal, even at the speed of light, to get from Mars to Earth. It takes just as long for the answering signal to get back to Mars.

One of the television signals shows that the landing vehicle is about to crash into a large rock. The computer on Earth sends back a warning message to the landing vehicle to stop. Alas, the message arrives too late! The crash has already occurred.

In fact, reports the landing vehicle, the crash took place exactly halfway between the time the television signal was sent out and the time the answering message was received. How ironic, that the crash was occurring in precisely that millisecond that the Earth computer was composing the warning!

"Not at all," argues the engineer at the space center. "The distance between Earth and Mars is now decreasing at the rate of 6 kilometers per second. In $5\frac{1}{2}$ minutes, that amounts to 2000 kilometers. The return signal had to travel 2000 kilometers less than the original television signal from Mars. So the original signal had 6 or 7 milliseconds longer to travel than the return signal. The crash occurred halfway between your sending a signal and your receiving an answer. It must have been happening several milliseconds *before we* received the television signal."

"Maybe, and maybe not," commented a scientist who heard the story. "After all, the Earth is moving around the sun at 30 kilometers per second, and Mars is going at 24 kilometers per second. By the time the first signal reached Earth, the Earth had moved 10,000 kilometers closer to the source than when that signal was first emitted. By the time the answering signal got back to Mars, that planet had moved 18,000 kilometers from where it had been when the first signal went out. The return signal had 18,000 kilometers longer to travel than the first signal. At the speed of light, the return signal took 60 milliseconds *longer* than the first. The warning message must have been sent out well before the crash."

Which of these observers is correct? All of them, from their own points of view. The theory of relativity insists that it is impossible to decide which of several frames of reference is at rest, and which is "really" moving. Therefore, it is impossible to decide about events that take place far apart, whether they were simultaneous or whether one or the other came first. So it is a moot

Figure 3-11 A Mars landing vehicle (Courtesy NASA)

question whether the crash on Mars came before or after the signal was processed by the computer on Earth.

What all three observers can agree on is that, since no signal can travel faster than light, there was no way any warning message from Earth could have reached Mars in time to prevent the crash.

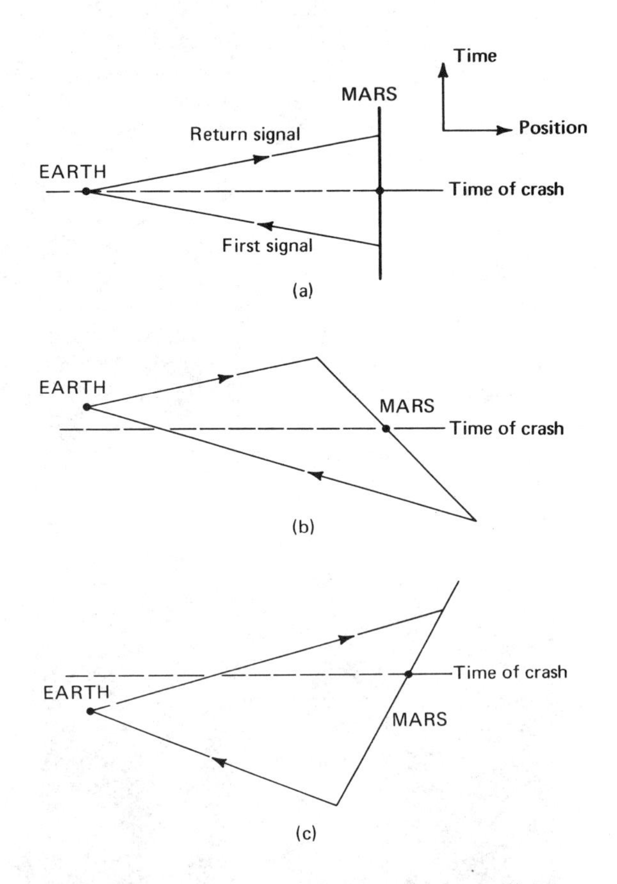

Figure 3-12 (*a*) From the point of view of Mars; (*b*) From the point of view of Earth; (*c*) From the point of view of the solar system

"If only," mused the track worker, "we had a signal that could go faster than light. Then we could use it to synchronize our watches, and decide once and for all who is right."

And there we are led to another conclusion of the theory of relativity. *There can be no signal that travels faster than the speed of light.*

If there were such a signal, the whole theory would immediately fall apart. We could use the faster signal to time in our clocks. Then we could measure the speed of light absolutely. It might be different in different directions. That would show how fast we are moving through space. But this is what the theory says we cannot do.

Many other contradictions would also arise. Thus, if the theory is to make any sense at all, faster-than-light signals cannot be allowed.

Material objects, like stones or electrons, can be used to carry signals. We can throw a stone, or send off a burst of electrons. The detected arrival of these objects is as good a signal as a radio pulse. The theory requires that *no* signal travel faster than light. This means that *no material object can move at faster-than-light speeds.*

No signal of *any kind* can go faster than light. If we push at one end of a meter stick, the side nearest our finger begins to move right away. But the far end of the stick may not change immediately. The motion of the far end can be used as a signal. Until there has been time for a light signal to travel the length of the stick, the far end must stay as it was. During this brief interval the rod is distorted. Its near end is moving, but its far end is at rest. This argument applies not only to meter sticks, but to any solid object. The theory of relativity requires that all objects become distorted when we try to change their speed. *There is no such thing as an absolutely rigid object.*

Time and space

Einstein's statements about relativity were very radical. They almost appear contradictory. They need not be, Einstein pointed out, but we must relax our ideas about time and space.

We have already seen that the simultaneity of events can depend on your point of view. We shall now see that other concepts, like the passage of time or the length of a meter stick, can also depend on who measures them.

We can define *one minute* as the time it takes the second hand of our watch to make one complete sweep. We try to make the timepiece as accurate as we can. If necessary, we can compare it with standard clocks, such as the signals sent out by the government radio time station.

We can make two watches so they are exactly the same, and keep the same time. If one of these watches is in my hand on the ground, and the other is in your hand in a moving automobile, we expect both of them to keep good time. If for any reason, they should later disagree, we would have no reason to suppose that one or the other is the more accurate.

As your car passes me, we can touch hands and agree that that is "time zero." You continue to drive and I to stand on the ground. We are both to report when exactly one minute has passed. By the time this has happened, we are no longer at the same place. Will the two events, the two watches completing a one-minute sweep, occur at the same time?

Not from every point of view. We know that simultaneity of remote events depends on the observer. It may turn out that, from my point of view, my watch completes its cycle before your watch does. You might reach the opposite conclusion. We both conclude that the other's watch is slow.

This in fact turns out to be a general conclusion of the theory of relativity. *A clock in motion* (from a given point of view) *appears to keep slower time than an identical clock at rest.*

The term "clock" can be applied to any natural process that can be used to tell time. Not only wristwatches or atomic standard clocks, but pulse-beats and biological rhythms that, however crudely, can be used to mark the passage of time must also fall under this category. If there were any such "clock" that did not appear to slow down when in motion, we could use it to calibrate the others. And this would contradict the assumption that you cannot tell by yourself whether you are moving or not. The conclusion is: from every observer's point of view, all other systems in motion appear to be "living slow."

This is a new prediction of the theory that can be tested by experiment. It has been shown many times that certain unstable subatomic objects appear to have longer lifetimes when they are in rapid motion than when they are standing still. It is as if their internal clock, which tells them when they should disintegrate, is running slower when they are in motion.

More recently some highly accurate atomic clocks were taken aloft in airliners and flown around the world. They were then compared with several clocks that had been left behind on the ground. Sure enough, it was found that all the clocks in motion ran slower, by the predicted amounts, than the clocks at rest. This remarkable prediction of Einstein's theory turned out to be accurate in all respects.

Our concept of distance turns out also to be relative. The length of a rod can be defined as the distance from one end of the rod to a point at the other end *at the same time*. Since observers in relative motion cannot agree on simultaneity, it should come as no surprise that they may not agree on where the far end of a rod is *at the same time* that the near end is at zero on their measuring stick.

Suppose the two observers set out to measure the speed of a radio signal travelling from Washington to London. The observer in the fast plane might conclude that the time elapsed between sending and receiving of the signal was a shorter interval than the observer on the ground would measure. But we expect, by the assumptions of the theory, that he will measure the same speed for the radio signal that the ground observer does.

Speed is calculated by dividing distance by time. So, from the point of view of the flying observer, the distance from Washington to London also shrinks. It contracts by the same factor that the time interval decreases. This turns out to be the same factor that Lorentz and Fitzgerald had calculated to explain the Michelson-Morley experiment.

So we see that an effect that had once been proposed solely to explain a single experiment now becomes a natural consequence of a simply stated and beautiful general theory.

How relativity meets the scientific criteria

Einstein's special[1] theory offers a classic illustration of how a new scientific theory came to supersede a previously accepted one. The previously accepted theory is in this case Maxwell's theory of electricity and magnetism.

The many well-tested predictions of Maxwell's theory continue to be valid under the new theory. It had previously been thought that Maxwell's equations would have to be corrected before they were applied to frames moving rapidly through the ether. Einstein now insisted that Maxwell's equations should be equally correct in all uniformly moving frames.

We have really not had much experience with rapidly moving frames. The fastest large[2] objects that we know about—rocket ships, comets, meteors, the planet Mercury—do not achieve speeds of one-thousandth the speed of light. At such "low" speeds the slowing-down of clocks, or the Lorentz-Fitzgerald contraction, do not amount to one part in a million. At speeds

[1] There is also a more complex general theory of relativity that applies the ideas of relativity to gravitation. More will be said in Chapter 14.

[2] Today we know of subatomic particles that travel at close to the speed of light. These had not been discovered before Einstein's time.

much less than the speed of light there is hardly any difference between the predictions of the older theory and those of Einstein's theory of relativity.

Einstein's theory accounted for facts that could not be explained, or were explained clumsily, by the previously accepted theory. The Michelson-Morley experiment could hardly be understood under the mechanical ether model of electromagnetism. Only by invoking what seemed like a special effect, the Lorentz-Fitzgerald contraction, which could only be observed in this one kind of experiment, could the older theory explain the Michelson-Morley results at all. Whereas the theory of relativity made these results naturally obvious.

The theory of relativity makes new predictions that can and have been tested experimentally. It predicts that "clocks" (by which we mean any process that has a natural time scale) slow down when in rapid motion. It predicts that no object, no matter how energetic, can travel faster than the speed of light. It predicts, as we shall examine in a later chapter, that objects increase in mass as they get more energy. All these predictions can be tested. The experiments to date all confirm Einstein's theory of relativity.

The theory is particularly simple and elegant. It can be summarized in a single sentence: *The laws of physics*, in particular Maxwell's equations and the speed of light, *must appear to have the same form in all uniformly moving frames of reference*. No experiment can ever, even in principle, tell which frame is moving and which frame is standing still. This is a statement about a basic symmetry of nature. No equations or other mathematical apparatus are needed to *state* the theory.

But the consequences that can be derived from the theory are wide-ranging. We were led to conclude that such intuitive concepts as the simultaneity of events, the passage of time, the distance between points, could be different for observers with different points of view. Yet many experimental results, some of them done before Einstein's time, some of them predicted by him, could be explained so simply and so naturally by the new theory. What was getting to be a complicated and unwieldy model, that of the mechanical ether, could now be replaced by a single sentence.

It should come as no surprise that scientists quickly, and virtually unanimously, came to accept Einstein's theory of relativity.

SOME OTHER CONSEQUENCES OF RELATIVITY

Einstein's theory is full of surprises. Again and again we draw conclusions from it that go against our "common sense." Can these results possibly be true? The only way to answer that question is to do the experiments.

Every experiment that has so far been done to test Einstein's ideas on relativity has come out exactly the way he would have predicted. Of course, all these tests can't ever prove him right. All it takes is one failing prediction to discredit the whole theory. All we can say is that, so far, that hasn't happened.

In this section we will show how some of the better-known, but still "surprising," results follow naturally from the theory of relativity.

1. No material object can travel as fast as light. We have already concluded that no signal can travel faster than c. Since material objects can carry information with them, this limit applies to them as well as to other kinds of signals. Now we shall prove an even stricter limitation. A material object must always travel *slower* than the speed of light.

We have to say what we mean by a material object. Let us define it as something that can sit still. A light beam is not a material object in this sense. It must always be moving, at its special speed, or it would not be a light beam.

Let us consider some material object, a bullet, say, or a rocket, or an atom. Even though it may be moving very fast from our point of view, there is always some frame in which it is standing still. This frame would be the one that is moving along with the object. From the point of view of someone sitting *in* a rocket ship, the ship itself is not moving.

Suppose there is a light beam about to strike the object. In the moving frame we would observe the object itself at rest. The light beam approaches it, at the speed c, and eventually hits it.

Now let us look at the same system from a frame in which the object is moving rapidly away from the light beam. According to our principle, the light beam is still a light beam and still is moving at the speed c.

We still see the same event happen. The light beam eventually strikes the object. From the new point of view we would say that the light beam is chasing the object and eventually catches it.

If a light beam, traveling at speed c, can overtake a material object, that object must be moving at a slower speed. This is true no matter from which frame we look at it.

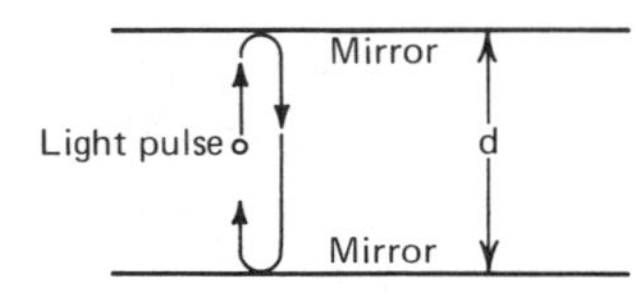

A "clock" using a bouncing light pulse for "ticks"

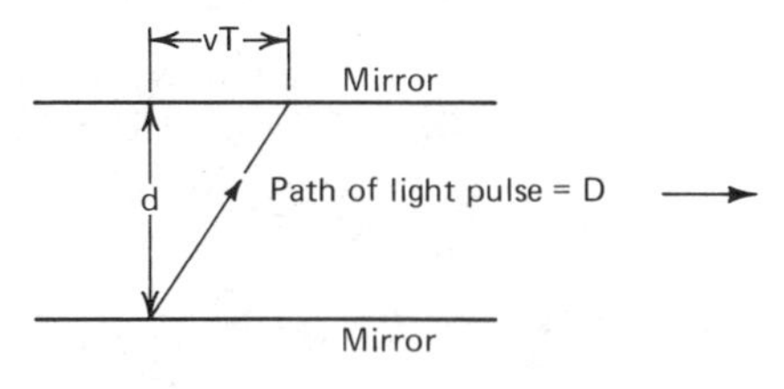

Figure 3-13

We conclude: An object that is at rest in some frame will have a speed less than c in every other uniformly moving frame.

This result has been tested. In some of our highest-energy accelerators (see Chapter 12), electrons have been speeded up until they are travelling at 99.999% of the speed of light. No amount of energy supplied can make them go quite as fast as c.

2. *A clock observed to be in motion will seem to be running slow.* Let us devise a "clock" that consists of a pulse of light bouncing back and forth between two parallel mirrors. The "ticks" of our clock occur every time the light pulse strikes one of the mirrors. If the mirrors are 30 centimeters apart (almost one "foot"), the speed of light is such that the time between ticks is 1 *nanosecond* (one-billionth of a second).

Let us call the distance between mirrors small d and the time between ticks small t.

Let us now mount this clock on a train moving at a velocity v parallel to the planes of the mirrors. That is, the light pulse bounces from side to side on the train.

From the point of view of observers on the ground, the time between clock ticks is capital T. This may not be the same as the time, small t, measured by observers on the train.

We cannot even take for granted that the distance, small d, between mirrors will be the same for both observers. This does turn out to be the case. We ask the observers on the ground to set up a passage that is just wide enough for the mirrors to pass through. If the width of the passage is not exactly 30 centimeters, we ask them to recalibrate all their meter sticks so that the numbers agree. In this way we make sure that observers in both systems are using the same length meter sticks. Notice that distances *along* the direction of motion cannot be calibrated this way.

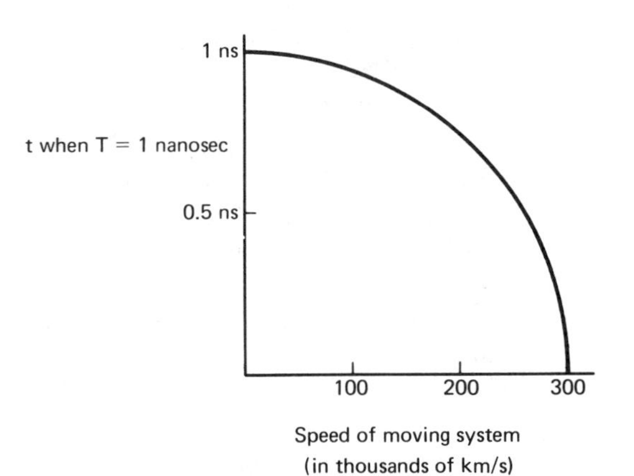

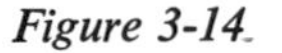
Figure 3-14.

From the point of view of the observers on the ground the light pulse follows a diagonal path. The distance travelled by the light pulse, from their point of view, is capital *D*.

It is obvious from Fig. 3-13 that capital *D* is longer than small *d*. What are the observers on the ground to conclude? Not that the light beam is travelling at a different speed. That would contradict the principle of relativity.

The only consistent conclusion is that capital *T* is a longer time than small *t*. The time between ticks, as observed from the ground, is longer than what the people on the train think it is. The people on the ground conclude that the moving clock is running slow.

The ratio of small *t* to capital *T* is in fact the same as the ratio of small *d* to capital *D*. A student with only a little skill in geometry and algebra can quickly calculate exactly how much the time-shrinkage factor must be. Figure 3-14 gives this ratio for different speeds of the train.

Let us point out that for slow train speeds the effect is very small. The third leg of the triangle will then be very short, and *D* will be hardly any longer than *d*. Only when v is a sizable fraction of *c*, like 10%, do the effects become noticeable.

This explains why the conclusions seem so strange. Such high speeds are completely outside our experience. The fastest large objects we know of—rocket ships, comets, the planet Mercury—are slower than one-thousandth the speed of light. The slowdown of a clock at such "low" speeds is less than one part in a million.

To test the prediction of time shrinkage we must use either very fast subatomic particles or very accurate clocks Both kinds of tests have been done, and the results were as predicted.

Glossary

Ampère, André Marie (1775–1836)—Showed that wires carrying electric current exerted forces on each other. He reasoned that since the behavior of any magnet could be imitated by a suitably designed electromagnet (which see), there were no "real" magnets at all, but only loops of electric current.

Atoms—The extremely small particles that can be put together to build all the molecules of normal matter. Atoms have electrically charged parts, including negatively charged electrons. Electrons can be removed from an atom, or extra ones added to it, but in its most common state an atom is electrically neutral.

conductor—A material through which electric current can flow easily. Most metals are good conductors because some of the electrons can move freely from atom to atom.

Coulomb, Charles Augustin (1736–1806)—Discovered that the forces between electric charges decrease in proportion to the square of the distance they are apart.

current—A flow of material. Any electric charges in motion constitute an electric current.

Einstein, Albert (1879–1955)—Famous for his theories of relativity. He also explained the photoelectric effect (see Chapter 11). His explanation of Brownian motion, the random motion of small objects being buffeted by nearby atoms, gave the first indication of what the size of an atom really was. He also made contributions to kinetic theory and the study of heat.

electric charge—An object that is attracted to or repelled by other objects known to be electrically charged, is said to possess an electric charge itself. Electric charge comes in two varieties, called positive and negative. We can tell them apart by the rule that similar charges repel each other and opposite charges attract. To complete the definition we say that a plastic ball rubbed with cat's fur acquires a negative electric charge.

electricity—The various effects produced by or on objects that have or can obtain an electric charge.

electromagnet—A magnet made by causing electric currents to flow through conducting wires. An electromagnet can be designed to behave exactly like any specific "permanent" magnet. It has the additional features that its magnetism can be "turned off" by shutting off the current, or made stronger by increasing the current.

electromagnetic forces—The forces produced on, and by, objects in which electric currents are flowing. Ampère showed that all effects previously thought to be magnetic in nature could be mimicked by a suitable arrangement of electric currents.

electromagnetic signal—A signal that can be sent across otherwise empty space by means of varying electric and magnetic fields. Maxwell predicted that such signals exist, and Hertz confirmed this by discovering what we now call radio waves. Light is also thought to be an electromagnetic signal.

electromagnetism—The combined study of electricity and magnetism, united by Maxwell on the basis of the many discoveries of the early 19th century.

electrons—The low-mass, highly mobile, negatively charged particles that are found in every atom of matter. Electrons can be detached from a neutral atom, leaving a positively charged atom behind. Some types of matter can attract extra electrons and acquire a negative charge. In conducting materials, including most metals, electrons are free to move from atom to atom through the whole conductor. A beam of free electrons can be set up in an evacuated space, such as inside a television tube.

electrostatic forces—The forces of attraction and repulsion between electrically charged objects that exist whether or not they are in motion.

ether—The mechanical medium suggested by 19th century physicists to try to unite electromagnetism with Newtonian mechanics. Electromagnetic signals were supposed to be ripples in the ether. The laws of electromagnetism (i.e., Maxwell's equations) were supposed to be exactly true only if things were at rest in the ether. In other frames corrections would have to be made for motion through it. The major difficulty was that experiments, such as the Michelson-Morley experiment, designed to measure how fast things were moving through the ether always gave a null result. Einstein suggested that since the ether was undetectable, we should leave it out of any physical theory.

Faraday, Michael (1791–1867)—English physicist who made many discoveries in electricity and chemistry. One of his most important discoveries was the law of electromagnetic induction: a changing magnetic field—whether caused by moving a bar magnet or by switching an electromagnet on and off—can force charged objects to begin moving and make electric currents flow in conductors.

Galvani, Luigi (1737–1798)—Italian physician who first studied the effects of electric currents. He was interested in how animal nerves and muscles responded to these currents.

Hertz, Heinrich (1857–1894)—German physicist who first made and detected what we now call radio waves. This experiment confirmed Maxwell's prediction that electromagnetic signals, similar in many ways to light, could be sent and received in this way.

light—A signal detectable by our eyes that brings us most of the information we receive about the world around us. Light signals travel through empty space at almost 300,000 kilometers per second, and at lower speeds through materials like glass, water, and air. On the basis of Maxwell's theory of electromagnetism we consider light to be one form of electromagnetic signal, similar to radio waves in many respects.

Lorentz-Fitzgerald contraction—An effect proposed to help explain the Michelson-Morley experiment. The original argument was that all atoms, and hence all matter made of atoms, since they have electrical parts, might get shorter when they move through the ether. This contraction would not be detectable since everything shrinks by the same fraction. Einstein's special theory of relativity gives a natural explanation of the Lorentz-Fitzgerald contraction without referring to the ether.

magnetism—The forces that magnets exert on each other and on magnetizable materials like iron. Wires carrying electric current are also affected by magnetic forces. Such wires in turn exert magnetic forces on magnets and on each other.

Marconi, Guglielmo (1874–1937)—Italian inventor who made use of the radio waves discovered by Hertz to send wireless signals over long distances.

Maxwell, James Clerk (pronounced Clark) (1831–1879)—British physicist and mathematician. He formulated the laws of electromagnetism, predicting that electromagnetic signals, such as radio waves, would exist and travel at the speed of light. He also made important contributions to the kinetic theory of gases.

Maxwell's equations—Four equations of calculus that express the discoveries of Coulomb, Ampère, and Faraday in mathematical terms. They form the principles for the theories of electromagnetism and optics.

Michelson-Morley experiment—An experiment, first done in 1887, to measure how fast the Earth is moving through the ether. The idea was to see whether the speed at which light approaches us is different in different directions. The surprising result was that there was no difference. It was as if the Earth was always standing still in the ether. Various attempts to explain this result led finally to Einstein's special theory of relativity.

molecule—The smallest particle of a material that can exist alone and still behave like that material. All normal matter is thought to consist of large numbers of very small molecules. Molecules themselves are made up of one or more atoms bound together.

negative electric charge—The kind of electric charge that a plastic ball acquires when rubbed with cat's fur. Any electrically charged object that is repelled by a negatively charged object can be said to have negative charge itself. Electrons have negative charge. Most large objects with negative charge are thought to have that charge because they hold an excess of electrons.

neutral—An object is electrically neutral if it is neither attracted nor repelled by electrically charged objects. Atoms and larger objects that are electrically neutral are thought to contain an equal amount of positive and negative electric charge.

Oersted, Hans Christian (1777–1851)—A Danish schoolteacher who first noticed that electric currents flowing in wires can deflect magnetic compasses, and other magnetized objects. This discovery was the first step in finding the connection between electricity and magnetism.

optics—The study of the nature and propagation of light. After Maxwell showed that light could be explained as an electromagnetic signal, optics became a branch of electromagnetism.

poles—Magnetic poles once played the same role in magnetism that electric charges play in electrostatics. Magnetic N poles are attracted to the Earth's north magnetic pole (see question 4), and magnetic S poles are repelled by it. Like magnetic poles repel each other, and unlike poles attract. The difficulty is that no object has ever been found that does not contain equal strength of N and S poles within itself. Ampère insisted that there are no real magnetic poles, that all magnetic effects can be explained by the forces between coils of currents. Thus, as every coil of wire must have two ends, every magnet must have both N and S poles.

positive electric charge—The kind of electric charge that a plastic ball obtains when rubbed with silk. Any electrically charged object that is repelled by a positively charged object is said to have positive charge itself. The nucleus of an atom has positive charge. Most large objects that have positive charge are thought to have that charge because they have lost some electrons.

relativity—Einstein's Special Theory of Relativity holds that (1) the laws of physics have the same form no matter how fast you are moving (in a straight line, at uniform speed); (2) among these laws are the value c for the speed of light, as deduced from Maxwell's equations. From these assumptions we derive such "strange" conclusions as that no signal can travel faster than light, or that clocks perceived in motion appear to run slow. These conclusions can be tested and have been verified many times.

simultaneity—The perception that two events happened at exactly the same time. According to Einstein, two different observers may not agree whether two events occurred at the same time. Neither is right; neither is wrong. It is just that the perception of simultaneity depends on the frame of reference from which the events are viewed.

Volta, Alessandro (1745–1827)—Italian physicist who showed how to maintain steady electric currents in wires, using a battery he invented. This invention made possible the study of the effects of such currents, including the discoveries of Oersted, Ampère, and Faraday.

yoke—The electromagnet that flanks the neck of a television tube, by means of which the current of the electron beam within the tube can be deflected from side to side.

Questions

1. In the light of our modern picture of atoms and electrons, would it have been better for Benjamin Franklin to have named positive and negative charge in the opposite sense from what he did?
2. What do we mean when we say that an object has an *electric charge*?
3. What is an electric *current*?
4. According to the definition of "poles" in the glossary, should the Earth's north magnetic pole be considered an N-pole or an S-pole?
5. Why should the magnetic forces between two beams of electrons depend on the *product* of the speeds of the two beams?
6. Give an example in which electrostatic forces, but not magnetic forces, are present. Give an example in which magnetic forces, but not elec-

trostatic forces, are present. Give an example in which both are acting at the same time. Which force is stronger in the last example you gave?

7. Suppose some unconventional scientist insisted today that the mechanical ether really does exist and that Einstein's theory is not needed. Do you think he could account for all experiments? What objections might you raise to his ideas?
8. Why does the conductor, in Einstein's train example, see one flash of lightning before the other? Did one flash *really* occur before the other?
9. Which of the points of view about the Mars landing vehicle is the "correct" one? (None of them? All of them?) Why?
10. How fast are you moving right now? Think: What meaning does this question have?
11. In what respects does Einstein's theory of relativity differ from Galileo's idea of relativity? What do the two have in common?
12. Show how Einstein's theory of relativity meets the criteria, stated in Chapter 1, for a new theory of nature to supersede a previously accepted theory.
13. The instructor stands in the middle of the room and asks students at opposite ends of the room to synchronize their watches to a light flash that the instructor sends out. Should the students correct for the fact that the room is moving, with the Earth, at 30 kilometers per second around the sun?
14. Suppose a faster-than-light signal existed that could have been used to prevent the crash of the Mars landing vehicle. Show that, from at least one point of view, such a message would have arrived before it was sent.
15. Speculate on the possibility of prolonging your life by flying about at high speeds, thus slowing down all your biological processes.

Summary question: Give three examples of experiments or observations that confirmed the predictions of some scientific theory. In each case, name the theory being tested, describe the measurements that were made, and tell how the results were different from what might have been expected if the theory were wrong.

Avoid "predictions" of results that were already known at the time the theory was proposed, or of behavior that was already known and could be expected to repeat itself.

section two

THE IDEA OF ENERGY

4

energy

A BASIC LAW OF PHYSICS

Introduction

Up to now we have been trying to show the way in which some of the laws of physics came to be accepted. We have chosen examples from several different parts of science. If any of these ideas interested you, maybe you will find time to read more about them.

Now we are going to talk about a specific idea of physics, the idea of energy. Energy can be defined in terms of what it can be used for. It takes energy to lift a weight, to wind up a spring, to set a car in motion, to boil a cup of water. Any property of any system that can be converted to one of these uses must be a form of energy.

The most important thing about energy is that the total amount of it does not change. Energy is not created out of nothing, nor does it disappear, though it often changes its form. When you add up all the forms of energy after some process, they add up to exactly the same total as before the process began. This is the law of conservation of energy.

The idea of an essence that never changes, despite the appearance of constant change in the world, came from Plato. To some thinkers it is a comfort to know that, despite everything that happens, there are some things that never change. One of these constants is, to the best of our knowledge, the total amount of energy in the world.

Businessmen have long made use of a similar principle, that account books must balance. The label on a certain sum of money may change from "inventory" to "accounts receivable" to "cash on hand." But when the books are closed each day, "debits" must always equal "credits." In a similar way the chemical energy in a liter of gasoline turns into kinetic energy when it is used to speed up an automobile, then into gravitational energy when the car climbs a hill, and finally into heat energy when the brakes are applied. There is a global system of bookkeeping that makes sure that, in one form or another, the energy is always there.

There also is a moral to the law of conservation of energy. We get out only as much as we put in. If a job takes energy, then it will not be done until the required amount of energy is supplied. If energy is once put into a system, it must all be accounted for. You cannot stop a speeding train without absorbing the energy that went into speeding it. In the world of energy you get what you pay for, and you pay for what you get. Nobody has yet found a way to cheat the system.

Conservation laws

Some of the most useful and important laws of physics take the form of conservation laws. Suppose there is something called X that obeys a conservation law. Such a law will say that whenever X increases in one form or place or object, there must be an equal decrease of X in some other place at the same time. The total amount of X stays always the same. X is said to be *conserved*.

In this chapter we will talk mostly about the law of conservation of energy. Energy is not the only thing in physics that is thought to be conserved. There are, for example, the laws of conservation of electrical charge and of momentum. However, because energy is so much a part of every known process, and because it has so many different forms, it probably is the most important of these conserved substances.

A conservation law does not tell the whole story about any process. It does not tell whether an allowed process will actually happen. It does not say how fast the change will take place. The most a conservation law can say about any process is whether or not such a change is possible. Other laws of science are needed to describe all the details.

"CONSERVATION OF WATER": A "LAW" THAT DOESN'T QUITE WORK

To give an example of how a conservation law works, let us pretend that we have found a new law of nature, the "law of conservation of water." We might have noted that we can measure out a liter of water, and pour it from one container to another. We can separate it into several smaller cups, and then pour them all back together again. If we have been careful not to lose any, we should have as much left as we started with.

But suppose we have a critic among us, who does not believe in this law. "Put the water in a cold enough place," says this spoilsport, "and it will freeze. It will no longer pour easily like the water we know, but become a cold, solid substance that we call ice." "Yes," we answer, "but if we weigh the ice, it will be just as heavy as the water it replaced. If we melt the ice, we can get back all the water we started with. So let's consider that ice is another form of water. Then the sum of the weight of water plus ice should always stay the same."

"All right," says the critic, "but you know very well that if you let water stand open in a warm place, it will gradually disappear." This is harder to answer. We make a guess that there is another form of water called water vapor. This is an invisible gas that mixes with air. We can test this idea by putting a small amount of water in a large glass bottle that is tightly closed with a stopper. We can weigh the bottle with the water in it. Then we gently heat it until much of the water has evaporated. The water vapor will still be trapped in the bottle, so the whole thing should weigh the same. As a further test we can let the system cool down again. The water vapor will then condense into liquid water again, the same amount that we started with. So now our conservation law must include the weight of water vapor plus ice plus liquid water.

The critic, however, has some more problems for us. Mix the water with

One liter of water

Pour it into smaller containers.

Pour them back together. Still one liter!

Weigh the water.

Freeze it.

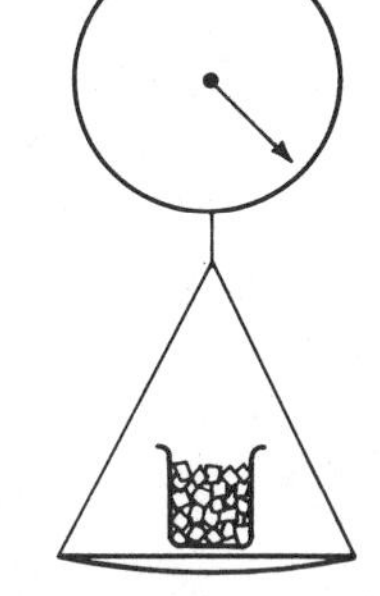

Still weighs the same.

Water in a closed container.

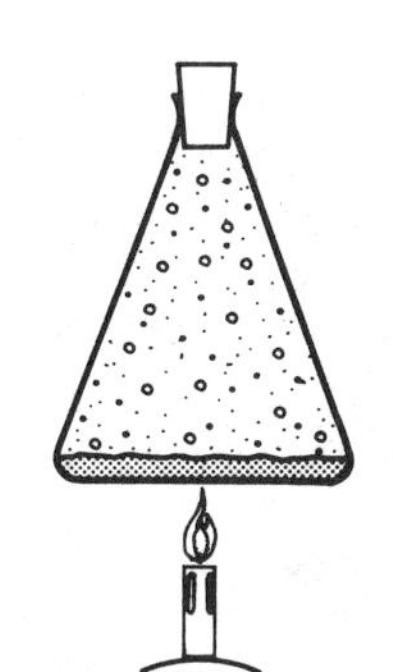

Heat it. Some of it turns into water vapor.

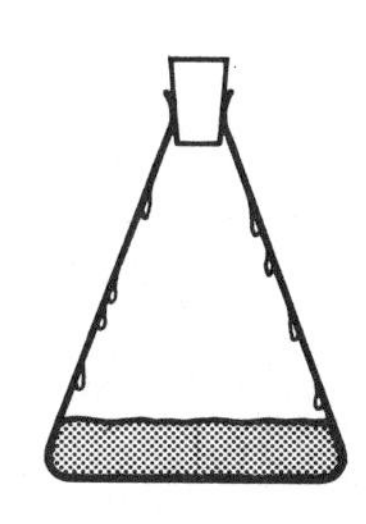

Cool it down. The vapor condenses into water again.

Figure 4-1

cement powder, he suggests. The whole thing will turn solid. Yes, we agree, but perhaps the water is trapped in the crystals of solid cement. If we heat it enough, maybe the cement will crumble and release the water as vapor. Then we could condense it and recover it all. This is a more difficult experiment than before, but in principle it could be done.

But now the critic has a devastating argument. Pass an electric current through the water, he urges. The water will separate into hydrogen and oxygen gases in a process called electrolysis. These gases are very different from water. They will not condense at all except at extremely cold conditions. If they are kept separate, you cannot get back water from either gas alone. So this experiment proves that water is not conserved in every process.

It is true that if you mix hydrogen and oxygen together under the right conditions, they will combine explosively to form water. So perhaps it is possible to define a new form of "water" that would allow us to keep using the conservation law. This would not be very useful, however, because it would soon force us to define any chemical that contains hydrogen or oxygen to be a form of water.

So our critic wins the day. We are not able to invent enough forms of water to explain every experiment. Reluctantly we conclude that there is no law in physics of the conservation of water.

The fact that a conservation law does not depend on the details can be very useful. Sometimes the details are very complicated. When a hard elastic ball bounces off a hard flat surface, many things happen. The ball changes shape, shock waves travel back and forth across it. Using the conservation laws of physics, we can figure out how high the ball will bounce, without knowing exactly what went on at the moment of collision.

Sometimes we do not yet have a good enough theory to make all the calculations. When two subatomic particles collide, we do not know in advance what reaction will occur. But we can use the law of conservation of energy to help us tell whether any of the final products of the reaction have escaped our detectors.

Energy, as we have said, has many different forms. Perhaps once in a generation scientists have found a process in which the total of all the *known* forms of energy was not being conserved. In each case they were able to define a new form of energy that made it come out right. These discoveries marked some of the greatest advances in our understanding of nature.

Can we go on inventing new forms of energy without end? There are some restrictions on how energy may be defined, as we shall see in the next section. If there were too many different forms of energy, say, thousands of them, the idea would not be very useful. So there are limits to how far we can extend the notion of energy.

Energy is not something, like water, that we can easily recognize for what

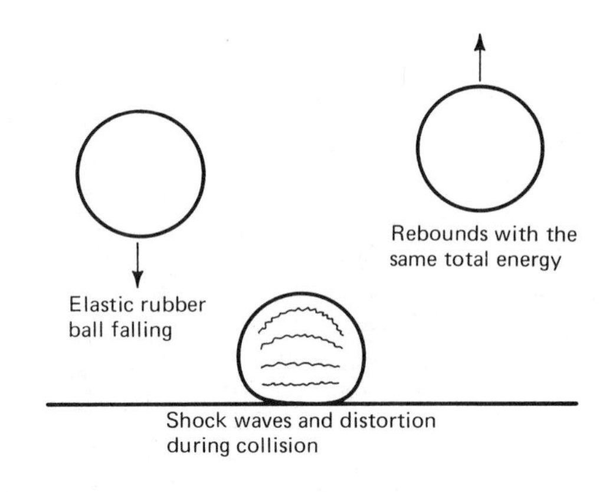

Figure 4-2

it is. The most important feature about energy is that it obeys a conservation law. If energy were not conserved, we would have a hard time defining it, and it would not be a very useful idea even if we could. We should not be asking: "Is energy really conserved?" A better question, given the restrictions on how it may be defined, would be: "Is there such a thing as energy at all?"

The answer, like all scientific explanations, can be only a tentative one, possibly to be changed as our understanding increases. To the best of our present knowledge, there are no exceptions to the law of the conservation of energy.

Restrictions on what we may call energy

From time to time we may have to define a new form of energy, but we are not completely free in the way we do this. It is not enough to say, "The wind carried some energy away," or "The sun supplied us with some energy."

If the idea of energy is to have any meaning, we should be able to find some measurable change that goes with the gain or loss of energy. Did the air get a little cooler after the wind passed? Does the sun lose a small amount of its mass when it sends out its light? Whenever energy seems to be appearing or vanishing, we ought to be able to track it down. We ought to be able to find the changes that the gain or loss of energy produces.

Measurability. We must be able to figure out the energy of a set of things from measurements we can make on them. There are many different kinds of measurement that can be made. Nearly all of them are important in determining some form of energy.

The *speed* of an object, how fast it is moving, relates to its kinetic energy. This is the first form we shall study.

The *position* of an object, how high it is off the ground, or how far from other objects, gives a measure of its *gravitational potential energy.*

The shape and size of an object, such as a spring that is stretched out of its normal shape, determines its *elastic potential energy.*

These are only a few examples of the sort of measurements that can be made. (In one of the questions at the end of this chapter, you are asked to suggest some others.)

Once we have made the measurements, we can get the energy by some known calculation. Sometimes we have to multiply two measurements together. In other cases we can read the energy from a chart or table. For every form of energy there is a set of measurements to be taken, and a way of combining the results to tell how much energy is present in that form.

The fact that we can measure the energy leads to an important result. *The energy does not depend on how things got the way they are.* We can always figure out the energy from measurements on the present state of things. The past history makes no difference.

The energy of a car parked on top of a hill is the same, no matter how it got up there. It makes no difference whether it was lifted there by a crane, or whether it was driven there under its own power. It does not matter

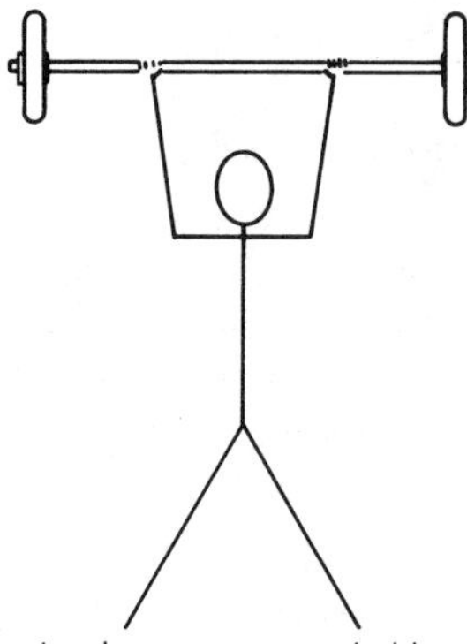

It takes no energy to hold a weight perfectly steady!

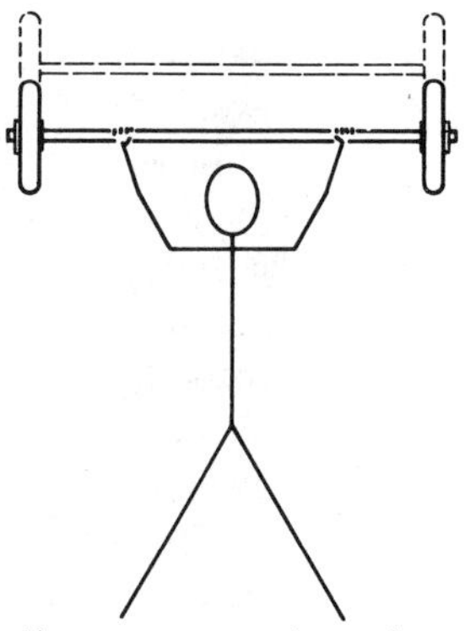

However, our muscles tend to tremble. The up-and-down motion costs us energy.

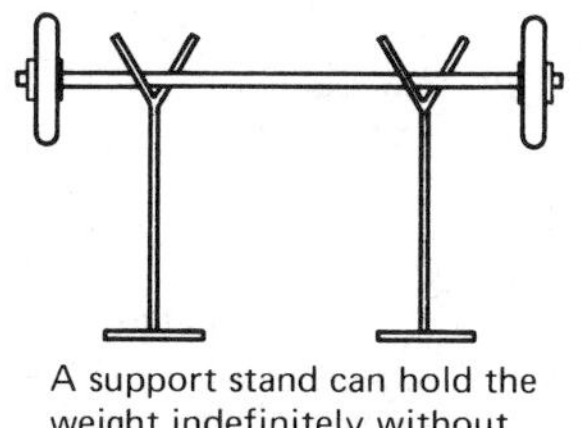

A support stand can hold the weight indefinitely without any energy input.

Figure 4-3

whether it went up slowly or fast, by a short-cut or the long way around. (Of course, by some of these methods, some energy may have been transferred to other forms: heating up the tires or the air, or making noises, for example.) The amount of energy stored in the car, the energy that could be retrieved by letting the car coast back down the hill, is the same in every case.

If no changes take place, there is no transfer of energy. It takes no *energy* to hold a weight perfectly still over your head. You may get tired, but this is because you are not really holding it steady. Human muscles tend to *tremble*, relaxing and then tensing again. The resulting up-and-down motion consumes energy. If you put the weight on a support stand, the stand can hold the weight steady as long as desired without any input of energy.

Additivity. The total energy of a system, made up of separable parts, is the sum of the energies of the parts. If several different forms of energy are present, the total energy is the sum of the energies in each of the forms.

This seems like a trivial thing to say. If we have more than one kind of energy present, or we have several objects, each with its own amount of energy, we get the total by adding all the amounts. The whole is equal to the sum of all its parts. It is important to understand that we are making this assumption, however natural it seems. There are some consequences that are not so trivial.

If we have one piece of steel moving at a certain speed at a certain height off the ground, that represents a certain amount of energy. Suppose we have a second piece of steel, exactly the same size as the first, moving at the same speed and located at the same height. Since all the measurements on the second piece would come out the same as for the first piece, its energy must also be the same. So if we put the two pieces together we have twice as much energy as either piece alone.

Thus we can add more identical pieces to our system, all with the same speed and height. The more we add, the more energy there is in the system. This is a general feature of most forms of energy. In other words, if all other measurements are equal, the amount of energy in a system increases as more material is added.

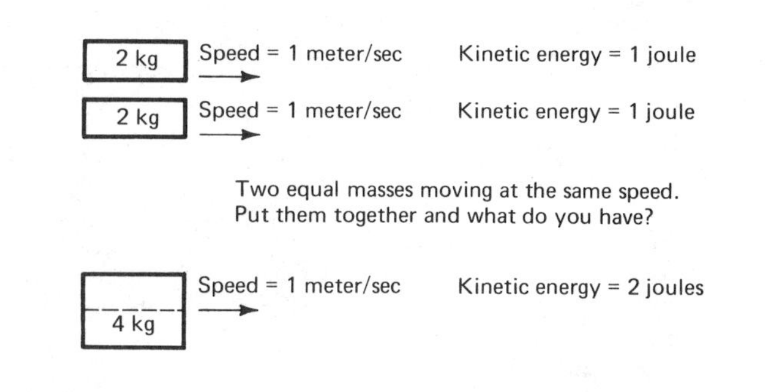

Figure 4-4

If all the objects we have to think about were made of the same material, say, steel, it might not be very hard to figure how much material each contained. We could measure its size and shape, and calculate its volume. If the shape is too complicated, we might plunge it into water and see how much water it displaces. Or we might weigh it.

Weighing an object seems to be a natural way of measuring how much material it contains. The more material, the more it weighs. When it comes to comparing the masses of two objects made of different materials, almost the only practical method of comparison is to weigh them together. Two objects that weigh the same at the same place are presumed to contain the same amount of material. This we shall call the *mass* of the objects.

Since energy is additive, we expect that, other things being equal, the more mass something has, the more energy it carries. We shall find that the formulas for many important forms of energy have the mass of the objects in them explicitly as a multiplying factor. This shows that these forms of energy increase directly in proportion to the amount of matter involved.

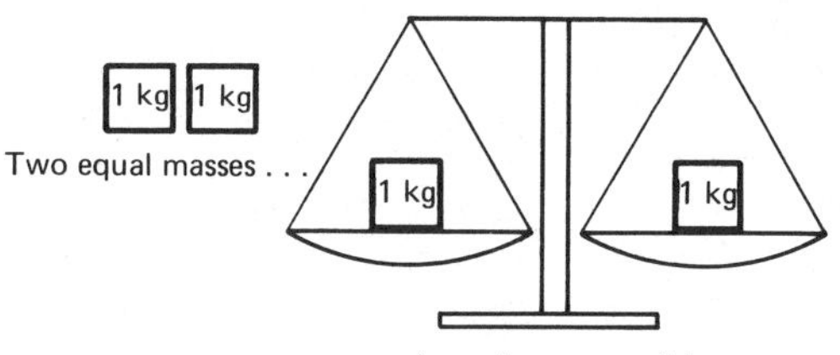

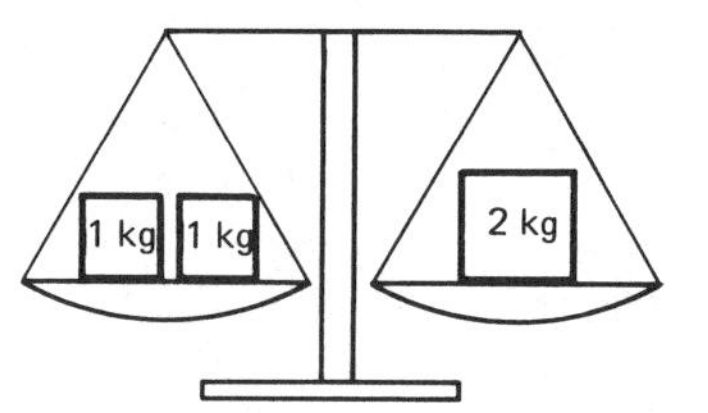

Figure 4-5

WEIGHT AND MASS

The *weight* of an object is the force of attraction that Earth (or other bodies) exert on it. The same piece of steel will have a different weight in different places. It will weigh less on the moon than on Earth. Far out in space, it will have practically no weight at all.

On the other hand, we recognize that there is something the same about the piece of steel, no matter where we weigh it. We have been calling this the "amount of material." In physics it is called the *mass*.

Two identical pieces of steel obviously have the same mass. They will also have the same weight, no matter where we weigh them. If we cut up one of the pieces and reshape it without adding or taking away any material, it will still have the same total mass. We can then verify that it weighs the same as the unchanged piece. If we put the two pieces together, there will be twice as much *mass* as either alone. The combination will also *weigh* twice as much. If you cut one piece into two equal halves, each half will have half the mass and half the weight of the original.

The more mass a piece of steel contains, the more it weighs. We can compare the masses of any two pieces of steel by weighing them. The weighing must be done at the same place to avoid the problem of varying forces of gravity.

The above discussion is fine, so long as we are comparing pieces of the same material. How do we compare the mass of a piece of steel with that of a block of wood? We can guess that the way to do it is to weigh them both. We can say that if two objects made of different materials have the same weight at a given place, they also have the same mass.

This is the way masses are compared in practice. There is a standard mass, a cylinder of platinum-iridium, kept in the International Bureau of Standards in Sèvres, France, that is 1 *kilogram.* Any other object with the same weight as the standard, at a given place, is said to have a mass of 1 kilogram. Fractions and multiples of a kilogram can be defined by the process of halving and doubling, as we did with pieces of steel.

The mass defined by weighing something is sometimes called the *gravitational mass*. The mass that must be used in the formula for kinetic energy is called the inertial mass. Are the two exactly the same? This is something that can be tested.

To the best that anyone has been able to measure, and some experiments were sensitive to a difference of less than one part in a billion, there is no difference between the two ways of defining mass. If two objects, no matter what they are made of, have the same weight at a given place, they also have the same inertial mass.

Conservation. Whenever the amount of energy in some place or form increases, there must be a corresponding decrease in some other place or form of energy, and vice versa. This is, as we have said before, the most important fact about energy. Without it, the very idea of energy would not exist.

The conservation law is also important because we can use it to find out *how much* energy it takes to make a given amount of a new form of energy. Whenever we notice that the amount of known forms of energy seems to be disappearing, we expect that energy will show up in some new form. We can recognize the new form by changes that take place. The object goes up in the air, for example, or it gets warmer. But how much energy does it take to make it go up 1 meter, or to get 1 degree warmer? Is it always the same? We can answer these questions by letting a known amount of energy change into one of the new forms. This procedure is called the calibration of the new form of energy.

The law of conservation of energy, in a practical sense, is really a set of rules for calibrating a new form of energy. Whenever we suspect that such a new form exists, we follow certain steps to define it. First, we list the changes by which we can recognize the new form. Then we measure how much energy, taken from previously known forms, is needed to make a certain change. Then we must show that it always takes the same amount of energy to make the same change, no matter when or how it takes place. Only when all these steps have been followed can we say that the new form is well defined. And only then can we say that the law of conservation of energy is verified.

Kinetic energy

We have to start somewhere by defining one form of energy. We recognize that it takes energy to make things begin to move. It takes energy to make a moving object go faster. The energy associated with the motion of an object is called its *kinetic energy*, abbreviated KE.

Let us make the following primitive definition:

The amount of energy required to set a 2-kilogram mass, previously at rest, into motion at a speed of 1 meter per second, shall be known as 1 *joule.*

All other forms of energy, including different amounts of kinetic energy itself, can be measured in terms of the joule.

The way we made the definition implies several facts about kinetic energy. It says that the only measurements we need are mass and speed. The kinetic energy of an object does not depend on the material it is made of, nor on its size and shape. It does not depend on where the object is, whether on Earth or on the moon or in outer space. These assumptions can be tested. If they are wrong, we get inconsistent results in trying to define other forms of energy.

The additivity of energy implies that the more mass an object has, the more kinetic energy it has, other things being equal. In this case the "other things" can be only the speed. So we must conclude that two objects moving at the same speed have kinetic energy directly in proportion to how much mass they contain. Put another way, the kinetic energy must be equal to the mass multiplied by a factor that depends only on the speed.

This factor must be zero if the speed is zero. An object at rest has no kinetic energy. The factor must be 1/2 if the speed is 1 meter per second, to satisfy our definition of the joule. What the factor is at other speeds is something that can be measured.

One way that we can do this is to use a slingshot that can be stretched to store the same amount of energy each time. Then we can load the slingshot with balls of different masses. We measure the speed of each shot just as it leaves the sling, when all the energy stored in the sling has been converted into kinetic energy of the ball.

Different projectiles will leave the sling at different speeds. They will all have the same kinetic energy, namely, the energy that was stored in the slingshot. Their speed should depend only on the ratio of energy divided by mass. The relation between the speed and the energy per unit mass is what we wish to measure.

Experiment: How KE Depends on the Speed. We will use an air-track on which small projectiles can coast horizontally with very little friction to slow them down. A timing device measures the speed of each projectile. The mass of each one is measured by using a balance.

A spring at one end of the air-track is set up with a reproducible trigger, so that it can give each projectile the same amount of energy. We do not know in advance how much energy this will be. Let us call it 1 "springload."

Figure 4-6 The air-track apparatus for measuring the relation between kinetic energy and speed (Photo by Barrie Rokeach)

Use projectiles of several different masses, differing by at least a factor of four. For each projectile, record the mass and the time it takes to travel a known distance. Calculate its energy-per-unit-mass (in "springloads per gram") and its speed.

Make a graph of energy-per-unit-mass versus speed. Is the graph straight or curved? Make a second plot of energy-per-unit-mass versus the square of the speed. Do the points fall on a straight line now?

Repeat the experiment with a different springload.

Many experiments like this one have shown that, for the range of speeds that we can usually measure, the kinetic energy per unit mass goes up with the square of the speed of the projectile.

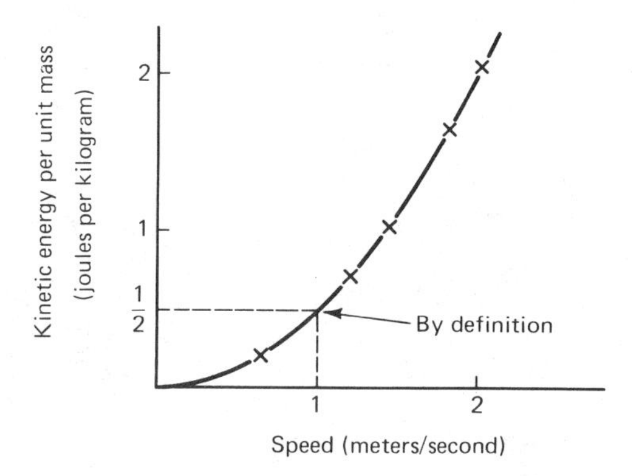

Figure 4-7

Putting all this information together, we can say that *the kinetic energy (in joules) of an object is equal* to one-half its mass (in kilograms) times the square of its speed (in meters per second).

The factor of one-half appears for historical reasons, to agree with a definition of kinetic energy based on Newton's laws of motion.[1]

Kinetic energy is always a positive quantity when it is not zero. It is zero only when everything is at rest. If there is any motion at all, there is more kinetic energy than if all is still. So energy must always be supplied from some other form to make things begin to move.

Motion can be started only if some other form of energy can decrease. There are many situations in which the other forms of energy have their lowest possible value. A stone lying at the bottom of a pit, or a watchspring that is completely unwound, are examples. Nothing will begin to move

[1] The energy required to accelerate an object of a given mass from rest to a given final speed is equal to the product of the mass times the final velocity times the average speed during the acceleration. If the speed goes evenly from zero to the final velocity, its average value during that time is one-half the final speed.

because the other forms of energy cannot decrease any more. Such a situation is said to be *stable.*

In a stable situation the sum of all the forms of energy, except kinetic, has its lowest possible value. If the kinetic energy is also zero, nothing more will happen, at least until new energy is supplied from outside the original system.

Examples of stability are all around you. The books on the shelf, the chairs on the floor, the pencil on the table, do not suddenly jump out of place. Any motion requires kinetic energy. There is no immediate source of energy on which these objects can draw.

Clearly, kinetic energy by itself is not conserved. We constantly see things start to move or come to rest, speed up or slow down. There are a few interactions, such as the collision of billiard balls, in which kinetic energy is the only form of energy involved. For the most part, the things that we see happen every day require exchanges of energy between kinetic energy and other forms.

CALIBRATION: FIGURING THE CONVERSION FACTOR

Suppose we have a water bucket and we want to find out how much water it holds. Suppose also that we have a wine bottle that is known to hold 1 liter. We can pour one wine bottle full of water at a time into the bucket, keeping count of how many it takes to fill it up. Or we could fill the bucket first, then empty it by drawing off 1 liter at a time in the wine bottle. Once we have done this, by either method, we can use the bucket itself to measure other water volumes. We are said to have calibrated the bucket as a volume measurer.

In a similar way we can calibrate new forms of energy. Suppose we know how to measure kinetic energy, but do not yet know how to measure gravitational potential energy. So we set out to convert a known amount of kinetic energy into the latter form. We can do this by throwing a ball straight up and seeing how high it rises.

We can weigh the ball beforehand, and thus learn its mass. We can use a motion picture camera to help find out how fast it is moving when it leaves our hand. From these measurements we can figure how much kinetic energy we started with.

At the top of its climb the ball is momentarily at rest. It has no more kinetic energy. If energy is conserved, it must all have been converted into the new form. So we conclude that the gravitational potential energy at the top of the climb must equal the kinetic energy it started with.

It is easy to see that the faster we throw the ball, the higher it will rise. So height is one measure of gravitational potential energy. The additivity of energy suggests that the mass of the ball also must be considered. The more mass we raise to a given height, the more energy it takes.

We can throw the ball up at different speeds and see how high it rises. We can plot a graph of height reached against kinetic energy at the start. The

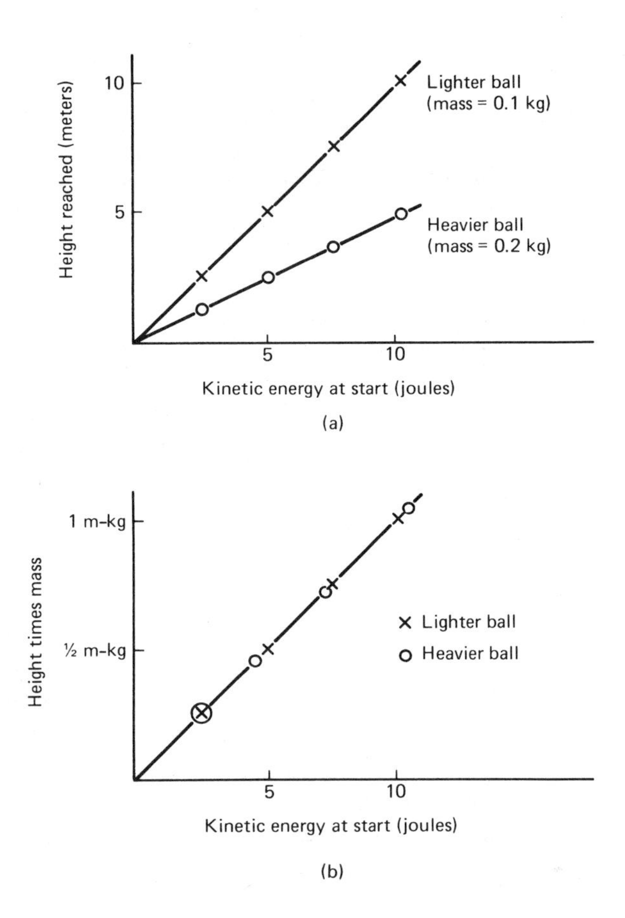

Figure 4-8

graph seems to be a straight line. But when we use a heavier ball we find that it does not rise as high for the same energy input. However, if we plot not the height alone, but the product of height times mass against kinetic energy input, we find that all the points lie on the *same* straight line. So we conclude that the gravitational potential energy is equal to some number—call it g—times the product of height times mass.

Once we know the value of g, we know completely how to measure gravitational energy. We call g the "gravitational field strength." For our purposes it can be called the *conversion* factor for this form of energy. It is the number you have to multiply the measurements by to get the amount of energy present.

We shall see that there is a conversion factor for each new form of energy. Finding its value is a necessary part of defining the energy form. Until the conversion can be determined, and shown to be the same in many different experiments, we cannot be sure that the new form of energy is well defined.

Gravitational potential energy

If we let go of a pencil, it falls. A falling body continues to increase in speed all during its fall, up to the moment it hits the ground. If you throw a ball up in the air, it will rise to a certain height, slowing down as it rises. The rule for inanimate objects in free flight seems to be, the higher it goes, the slower it moves. Conversely, the lower it falls, the faster it goes.

The form of energy associated with the height of an object above the earth (or distance from any gravitating body) is known as the gravitational potential energy.

Experiment: Calibrating GPE. Use the same spring that we have used before, one that can be cocked repeatedly with the same amount of energy. Fire balls of different masses straight up in the air. Measure how high each ball rises. Calculate the product of mass times height in each case. Are these products the same for all the balls?

Experiment shows that, for distances that we can conveniently measure, the gravitational energy is proportional to the product of mass times height. We would like to express this form of energy in joules. To do this we must multiply the product of mass times height by a *conversion* factor, which is

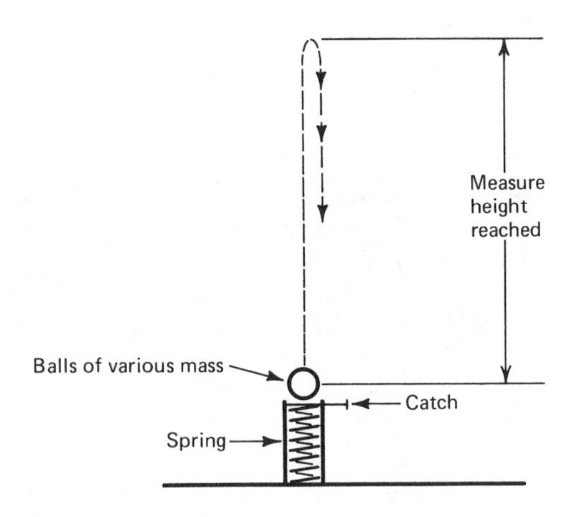

Figure 4-9 Calibrating gravitational potential energy

usually called g. We can measure the value of g by converting a known amount of kinetic energy into gravitational potential energy.

The gravitational potential energy of an object is equal to the product of its mass times its height times the gravitational conversion factor, g.

The factor g is not exactly the same everywhere. Its value is close to 9.8 joules per kilogram-meter at most places on Earth. It is slightly higher where there are heavy mineral deposits underground, slightly lower at high altitudes. At distances of thousands of kilometers from the Earth g is quite a bit less than 9.8. Its value is only 1.6 (in these units) on the surface of the moon. It is calculated that g is about 26 joules per kilogram-meter at the surface of Jupiter, and ten times that at the surface of the sun. Far out in space, g may be very much smaller.

To make the numbers in our sample calculations come out simpler, we will often "round off" the value of g to 10 joules per kilogram-meter. From now on, we will abbreviate gravitational potential energy as GPE. Kilograms will be abbreviated as kg, meters as m, and joules as J. So the value of g on Earth is approximately 10 J/kg-m.

The height that appears in the formula for GPE is, strictly speaking, the height above sea level. However, since only increases or decreases in GPE lead to changes in other forms of energy, it makes no difference what level we take as the "zero" of height.

Rapid City, S.D., is nearly 1000 meters above sea level. We can get the *relative* GPE of any object by measuring its height above ground level. If we insist on having the "sea-level" GPE, we can easily get that by adding to the relative GPE the product of the object's mass times 1000 meters times g, that is, 10,000 times the mass in kilograms, if g is 10. If several objects are involved, we can get the "sea-level" GPE by adding 10,000 times the total mass to the relative GPE. After some change has taken place, the relative GPE may have changed, but the total mass will still be the same. So the difference between the relative G.P.E. and the "sea-level" GPE is still 10,000 times the total mass. The change in the "sea-level" GPE, the amount of energy converted into other forms, is the same as the change in the relative GPE. So there is no need to carry the large term, 10,000 times the total mass, which never changes anyway, in all our calculations.

We can take as our zero level of height the ground level, or the level of the floor in our room, or the level of the table top, or any other convenient level. We only need to be consistent. In a given situation, all measurements of GPE should be referred to the *same* zero level of height.

The law of conservation of energy holds at every stage of a process, not just at the beginning and the end. Let us take as an example the flight of a ball thrown straight upward by a baseball player. We can take motion pictures of the entire flight. From the pictures we can measure the height and the speed of the ball at each point of its motion. From these measurements, and knowing the mass of the ball, we can calculate the kinetic energy and the G.P.E. at each stage.

Let the mass of the ball be 1/10 of a kilogram. Suppose the player gives it an initial speed of 40 meters per second; that is about as fast as the best players can throw a ball. The kinetic energy when the ball leaves his hand is

$$\frac{1}{2} \times (1/10 \text{ kg}) \times (40 \text{ m/sec}) \times (40 \text{ m/sec}) = 80 \text{ joules.}$$

For GPE we use as zero level the height of the ball as it is released. Then the GPE is zero when it leaves the player's hand, since the height is zero. The total energy is the sum of kinetic energy and GPE, 80 joules plus zero, or just 80 joules.

At the peak of its flight the speed of the ball drops momentarily to zero. At this point there is no kinetic energy. The total energy must still be 80 joules. So the height must have a value to give 80 joules of GPE. Since the mass is 1/10 kg, and we use g of 10 J/kg-m, the product of mass times g is just 1 joule per meter in this case. For this mass ball, the GPE in joules is the same number as its height in meters. The ball has a GPE of 80 joules when it has risen 80 meters above its launch point.

Using the motion picture film, we can prepare a graph of the GPE of the ball versus time. This is the same as the graph of the height of the ball against time, since the GPE and the height are numerically the same for this mass ball. For other masses, GPE and height would have different values, but the shape of the two curves is the same.

The kinetic energy begins at 80 joules and decreases to zero at the moment when the ball is at the top of its flight. Then the ball begins to speed up again as it falls toward the ground. When the ball has fallen back to its starting height the GPE is again zero. At this point the kinetic energy is back up to 80 joules. The ball has the same speed as it started with, 40 meters per second.

At an intermediate point in the flight, the ball has both kinetic energy and GPE. Take the moment when the ball is at a height of 60 meters. The GPE is then 60 joules. The total energy is always 80 joules, so that leaves 20 joules of kinetic energy. The ball will have this much kinetic energy if its speed at that moment is 20 meters per second.

$$\frac{1}{2} \times (1/10 \text{ kg}) \times (20 \text{ m/sec}) \times (20 \text{ m/sec}) = 20 \text{ joules}$$

We can make the same kind of calculation at every point of the ball's trajectory.

What does a graph of total energy versus time look like? At every point in the ball's motion, add up kinetic energy and GPE. The total is always 80 joules. So the graph of total energy against time is just a horizontal straight line. This is the meaning of the law of conservation of energy. The total energy does not change with time.

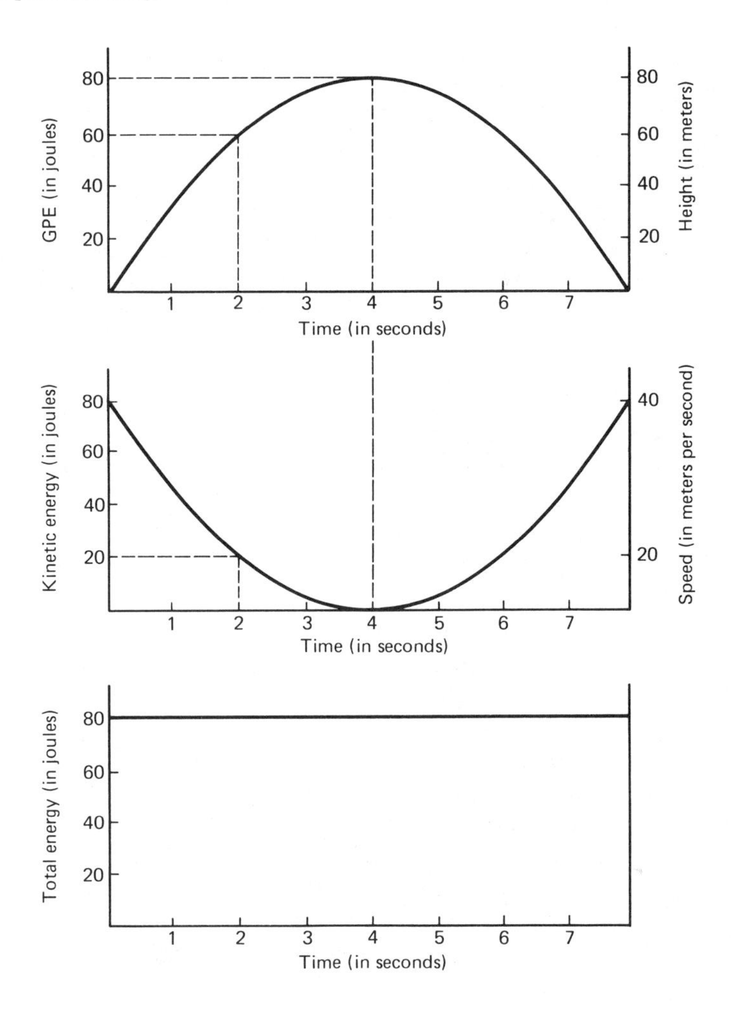

Figure 4-10

Elastic potential energy

All solid objects have a natural size and shape that they take on when they are not under stress. Stress can be applied in many ways. Objects can be squeezed or stretched, bent or twisted. It takes energy to do this. If the stress is too great, a permanent change can be made. The object can be broken or crushed or permanently bent. But if we are careful not to go too far, the object will tend to return to its natural size and shape when the stress is removed.

We can usually define some number to measure how far an object is out of its natural shape. In the case of twisting stress, for example, this measure would be the angle of twist. Let us confine ourselves here to a spring which can be stretched or compressed from its unstressed length. We call x, the *displacement*, the difference between its length under stress and its natural length. The displacement can be positive when the spring is stretched, or negative when it is compressed.

The lowest state of energy of such a spring is when it is in its natural state and x is zero. We define this state to have zero *elastic potential energy*. Any other value of x, whether positive or negative, requires an input of energy to the spring. This suggests that the elastic potential energy of a spring depends on the square of x, which is always positive, no matter whether x is positive or negative. We guess that the formula for elastic potential energy, which we shall abbreviate as EPE, is some number, k, times the square of the displacement.

$$\text{EPE} = kx^2$$

The number, k, the *elastic conversion factor*, is different for every spring. The stiffer the spring, the more energy it takes to stretch it by a given displacement; therefore, the larger k is.

The formula for EPE is an approximation that is expected to work only for small displacements. Certainly, it is not valid under stresses large enough to break the spring. How well the formula works depends on the spring.

Each different spring, or any other elastic object, must have its EPE calibrated separately. This can be done in the usual way. A known amount of energy is converted into E.P.E., and the resulting displacement is measured. On the other hand, we can start with a certain displacement in the spring, convert it all into kinetic energy and/or G.P.E., and see how much is released.

The curve of EPE against displacement can thus be measured for every elastic object. The exact formula may be rather complicated and is different for every spring. The important fact is that it always takes the same amount of energy to cause the same displacement. That is the meaning of conservation of energy as applied to elastic objects.

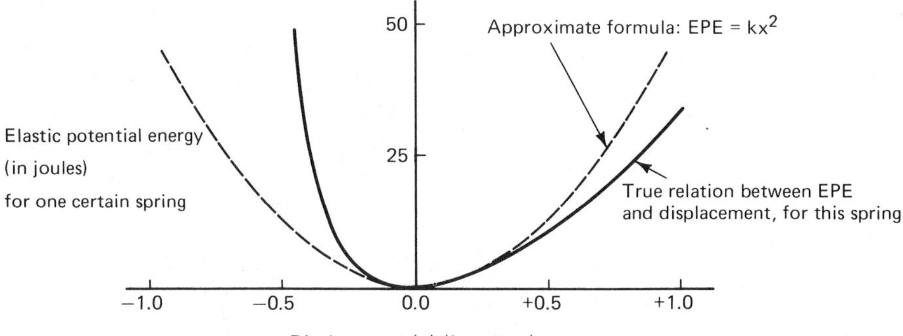

Figure 4-11

Mechanical energy

The three forms of energy described in this chapter, kinetic energy, GPE, and EPE, are known collectively as *mechanical energy*. They are the forms of energy that can be determined by measuring distances and times and masses. Other forms of energy require more subtle measurements.

In a later chapter, mechanical energy will be called *useful* energy. The different forms of mechanical energy can be easily and completely converted into each other. Each of the three forms is a way in which energy can be stored. Water can be pumped into a high mountain reservoir. A watch spring can be wound up tight. A massive flywheel can be set spinning. Each of these cases represents a storehouse of energy, ready to be used for any purpose as it is needed.

When the kinetic theory of matter was proposed in the last century, some scientists thought that maybe mechanical energy was the only kind of energy known. You will recall that this theory proposed that all matter is made up of large numbers of very tiny molecules. According to kinetic theory, *sound* energy is the organized vibrations of millions of molecules moving back and forth together. *Heat* energy is supposed to be the random disorganized motion of all the molecules. If only we could make measurements on a small enough scale, we could account for all these forms of energy as mechanical energy on a molecular scale.

This simplification turned out to have exceptions. It was learned that *electromagnetic* forms of energy cannot be explained as mechanical energy. *Nuclear* energy was discovered, but it did not fit into any of the known forms.

On the other hand, most elastic energy seems to arise from the attractions and repulsions among the molecules within the elastic object. This is also true of what we call *chemical* energy, such as the energy released by burning a chunk of coal. The forces between molecules, in turn, are thought to be electrical in nature. The different forms of energy are closely intertwined.

In later chapters we shall explore some of these other forms of energy.

Glossary

additivity—The property that the total energy of a system, made up of separate parts, is the sum of the energies of the parts.

calibration of energy—Determining how many joules it takes to produce a given amount of a new form of energy. Whenever a new form of energy is defined, we must calibrate it by converting a known amount of energy into (or out of) the new form. In this way we measure the conversion factor.

conservation law—A law of nature that says that the total amount of some physical quantity (such as energy) never changes. Though it may change its form, it is never created out of nothing, nor can it disappear completely. Whenever the amount of the conserved quantity increases in one place, an equal amount of it must decrease in some other part of the system. There are laws of conservation of electric charge and of momentum, as well as the law of conservation of energy.

conversion factor—The number of joules it takes to produce a certain change in a new form of energy. For example, it takes 9.8 joules to raise 1 kilogram 1 meter against the pull of Earth's gravity. We will later see that it takes 4.186 joules to make 1 calorie of heat energy, and that it takes 90 million billion joules to create 1 kilogram of matter.

displacement—The amount by which an elastic solid is distorted out of its natural shape. The elastic potential energy of the solid depends on the magnitude of its displacement.

elastic potential energy (EPE)—The energy that must be supplied to distort an elastic solid out of its natural shape and size, and which can be released when it returns to its undistorted condition. We say that the object possesses the EPE while it is distorted.

electrolysis—The separation of water into hydrogen and oxygen by passing an electric current through it.

energy—The measurable, additive property of a system, the amount of which can be converted into, or out of, forms which are known to require energy. These forms include the setting of massive objects into motion, the lifting of weights and the stretching of springs.

g—The conversion factor for gravitational potential energy, it expresses how many joules it takes to raise one kilogram one meter. On the surface of the earth, g has the value of 9.8 joules per kilogram-meter.

gravitational definition of mass—Two objects, of whatever material, are said to have equal masses if, at a given place, it takes the same amount of energy to raise either one a certain distance against the pull of gravity. Thus, any two objects that balance each other on a balance scale, are said to have the same mass, by this definition.

gravitational potential energy (GPE)—The energy that must be supplied to lift a massive object a given distance against the pull of gravity, and which can be released when the object falls to a lower position. We say that the object possesses the GPE while it is in the elevated position.

height—The position of some object above a level of reference. This reference can be sea level, but any other reference level will do. The height of an object is an important measurement in determining its GPE.

hydrogen—A very simple and lightweight chemical element. Pure hydrogen is an invisible gas. But hydrogen atoms can combine with many other kinds of atoms to form compounds. The most common such compound is water, each molecule of which contains atoms of hydrogen and oxygen.

ice—The solid form of water. Ice is made of the same kind of molecules as liquid water. When ice melts, the water that forms weighs the same as the ice did before melting. So we conclude that ice and water are simply different forms of the same basic substance.

inertial definition of mass—Two objects, of whatever material, are said to have the same mass, under this definition, if it takes the same amount of energy to bring them each from a state of rest up to a state of motion at a given speed. Experiments have shown that, to a very high precision, the inertial and the gravitational definitions of mass are equivalent.

joule—The unit in which energy is measured. It is the amount of kinetic energy possessed by a mass of 2 kilograms moving at a speed of 1 meter per second. Energy in any other form that can be converted into, or out of, this much kinetic energy, is equal to 1 joule.

kilogram—The unit in which mass is measured. It is the mass of a standard kilogram kept at the International Bureau of Weights and Measures, and of any other mass that would balance the standard kilogram. A liter of water—1,000 cubic centimeters—has a mass of close to a kilogram (in English units, 2.2 pounds).

kinetic energy (KE)—The energy that must be supplied to accelerate a massive object from a state of rest to a state of motion at a given speed, and which is released when the object slows down to rest again. We say that the object possesses the KE while it is in the state of motion.

mass—The amount of material in an object. The mass of two objects made of the same material can be compared by comparing their sizes. Objects made of different materials can be compared by using either the gravitational or the inertial definitions of mass (which see).

measurability—That property of energy that the energy of a system can be determined completely from measurements of the present state of the system. This means in particular that the energy does not depend on how the system got that way.

mechanical energy—The forms of energy measurable by the large-scale motions and positions of the objects in a system. Mechanical energy includes kinetic energy, gravitational potential energy, and elastic potential energy.

oxygen—A common chemical element. Pure oxygen is a gas; it is the part of the air that we breathe. Atoms of oxygen can combine with other atoms to form molecules of many kinds. One important kind of molecule that contains oxygen is water.

speed—How fast an object is moving; the distance covered divided by the time elapsed. Speed is an important measurement in determining the kinetic energy of a system.

stable—Not sensitive to small change. A system in a stable state is one in which *any* change requires energy to be supplied. It is likely to stay in the stable state until the extra energy is put in, and to return to the stable state if the extra energy is lost.

volume—The three-dimensional space occupied by a solid object or by a fluid.

water vapor—The gaseous form of water. At high temperatures it is also known as steam. The water vapor that forms when water boils, or evaporates, has the same mass as the original mass of the water.

weight—The gravitational force that Earth, or any other gravitating body, exerts on an object. The energy it takes to raise the object 1 meter against the pull of gravity.

Questions

1. What is the most important fact about energy?
2. What is the meaning of "conservation of energy"?
3. Can you invent another "form of water" so that the total amount of water is conserved in a process like electrolysis?
4. It is sometimes said that the total energy of a completely isolated system never changes. What is meant by a "completely isolated system"? (The inside of a turned-off refrigerator? The universe as a whole?) Is any real system ever 100% free of energy leaks?
5. The statements made in the "Weight and Mass" section about the equality, doubling, or halving of the *masses* of various pieces of steel amount to definitions of what we mean by the "same," "twice," or "half" the mass. Are the statements made about the *weight* of these pieces also true by definition, or are they subject to experimental test?
6. Can you calibrate GPE, not in the way described in this chapter, but by letting balls drop from various heights? Tell how you would go about doing this, what measurements to take, and what graphs to plot.
7. A *balance* is a device for comparing the weights of two objects. It is made in such a way that if one of the objects moves downward, the other moves upward exactly the same distance. Show that the GPE of the system is so arranged that, starting from rest, motion is possible

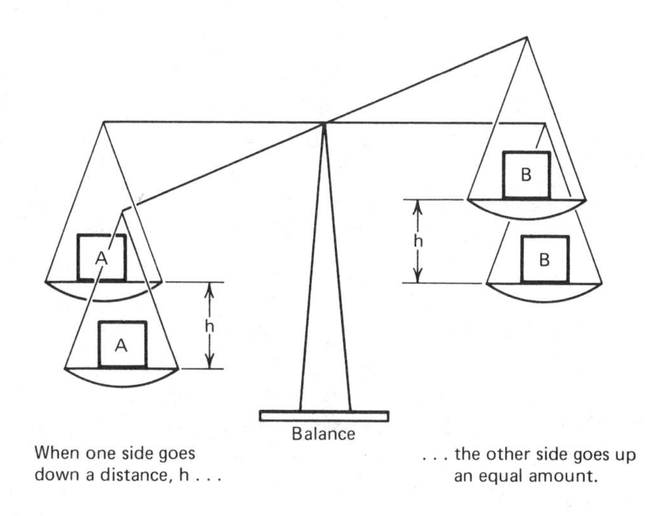

Figure 4-12

if, and only if, the heavier object goes down. What if the masses are exactly equal?

8. Give some examples of measurements that can be made—other than mass, speed, and position—to help determine the energy of something.
9. What is the shape of a graph of total energy versus time?
10. Suggest a way to measure the gravitational conversion factor, g, at different places on earth.
11. Suppose an object is tumbling or spinning as it moves, so that not all parts of it are moving at the same speed. How would you go about finding its total kinetic energy?
12. If an object is accelerated from rest to close to the speed of light, the usual kinetic energy formula doesn't work. Use the information in the footnote on page 101, and take account of the fact that the object spends a large fraction of the acceleration process moving at nearly the speed of light. What formula emerges in this case?
13. Give a way to measure the elastic conversion factor, k, for a given spring. How would you check how well the formula suggested for EPE works for any given spring?
14. Suppose you had a certain amount of energy stored in the form of a spinning, heavy wheel. What form of energy is this? How could you use that energy to lift a weight?
15. Suppose you have a certain amount of energy stored in the form of water behind a dam. What form of energy does this represent? How could you use that energy to make a mill wheel turn? Can you think of a way to use the energy to make a vehicle move?

5

more about energy

USING THE IDEA

Introduction

In the last chapter we defined what we meant by the idea of energy. Now we are going to show how the idea of energy can be used to explain the features of many kinds of motion. Our examples will range from the motion of a pole vaulter to that of an automobile, from a mass bobbing at the end of a spring to an evacuated can collapsing under the pressure of the air.

Some important quantities will be defined in terms of the *rate of transfer* of energy. *Power* is the rate of transfer of energy per unit *time*. *Force* is the rate of transfer per *distance moved*. *Pressure* is the energy required to make a unit decrease in *volume*. *Torque*, *surface tension*, and *voltage* are defined as the rate of energy transfer with, respectively, *angle*, *area*, and *electric charge*. Thus we see that many familiar terms in physics are directly related to the concept of energy.

We have the intuitive notion that the harder a task is to do, the more energy it takes. But some jobs that we find difficult to do right away, can be done if we take our time, or otherwise stretch out the job. We cannot lift an automobile straight up by ourselves. But we can get the car lifted up eventually, using only our own strength, by making use of a bumper jack.

Human muscles, or any kind of engine, given enough time and enough fuel, can transfer an indefinite amount of energy. But there are always limits on the rate at which the energy can be delivered. There are a maximum power and a maximum force that our muscles can exert. Motors have their limitations, too, though perhaps higher limits than ours.

From our understanding that energy is the same, no matter where it comes from, we can design devices to help us concentrate the energy we have, and do the hard jobs that face us.

Mass on a spring

Experiment: *Mass Hanging on a Spring*. Hang a spring from a support stand. Attach a weight to the end of the spring. A pointer attached to the assembly helps to measure the position of the weight with respect to a meter stick alongside.

Hold the weight up so that the spring is in its unstressed condition. Release the spring from rest. Observe the motion.

We can prepare a graph of the various forms of energy versus position of the pointer. The elastic potential energy is zero at the starting point where the spring is not stressed. The energy increases for any displacement, up or down, away from this position.

The gravitational potential energy can be graphed against pointer position in a similar way. The GPE *decreases* as the pointer goes down. Through most of the motion, then, the GPE is lower than it is at the starting point.

We can add the two forms of potential energy together. Notice that both of these forms depend only on the position of the pointer. The *potential* energy does not depend on how fast things are moving. Whenever the pointer is at a given mark, the potential energy is always the same.

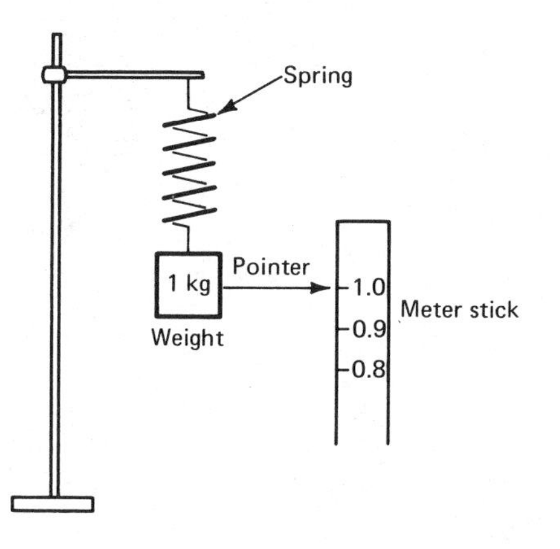

Figure 5-1

The kinetic energy at the starting point is zero, since the weight is released from rest. At other points in the motion, the kinetic energy will be greater than zero. If the total mechanical energy is constant, the kinetic energy must make up the difference between the total and the potential energy.

As the weight descends from the starting point, the potential energy drops from its initial value. So the kinetic energy increases by an equal amount. The speed continues to increase until the pointer passes the 0.90 meter mark, where the potential energy has its lowest value. At this point the kinetic energy is highest. Beyond this mark the kinetic energy decreases. The motion slows down.

Finally, the pointer reaches the 0.80 meter mark, where the potential energy is equal to its starting value. The kinetic energy must be zero at this point. The weight is momentarily at rest. The pointer cannot go lower than this unless more energy is supplied from outside the system.

The place where the potential energy equals the total mechanical energy is

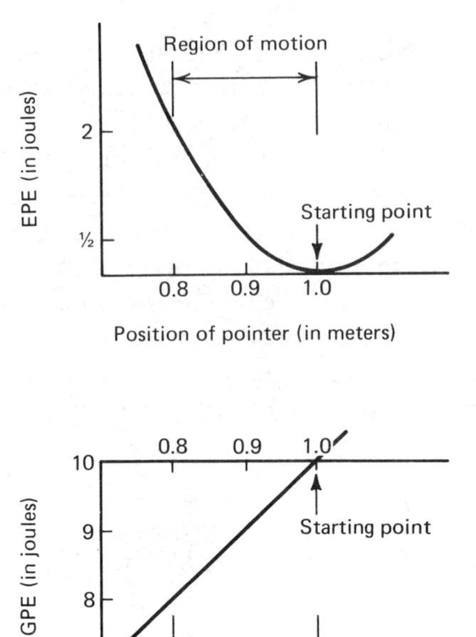

Figure 5-2

called the *turning point*. Here the system comes to rest for an instant. Since the motion can go no farther, the only thing it can do is to reverse its direction. That is what happens. The pointer retraces its steps. The kinetic energy increases again, as we pass positions of lower potential energy.

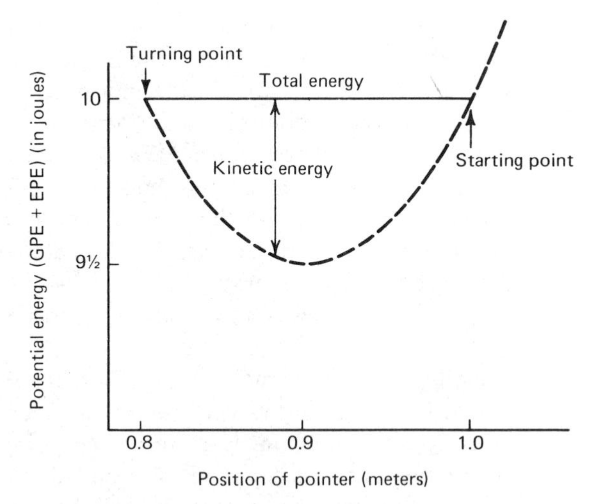

Figure 5-3

We pass the minimum potential energy point. After this the kinetic energy declines again. The pointer approaches the starting point. The starting point is really another turning point, for the potential energy curve crosses the total energy line. So once again the motion reverses itself.

The motion continues in this pattern indefinitely. The system swings back and forth between the turning points. It speeds up near the potential energy minimum. It slows down to a momentary stop at each end. Such a repeated back-and-forth behavior is called an *oscillation*.

Dissipation

If we follow the oscillations of the spring-and-mass system for several minutes, we might notice some changes. The pointer no longer returns all the way up to its original starting point. The other turning point is not as low as it once was. The maximum speed is less than it had been.

The reason for these changes is that some of the original mechanical energy is being converted into other forms. The spring becomes warmer. The air around the system gets stirred up. Creaking noises are emitted. The transfer of energy from mechanical (or electrical) energy to less useful forms is known as *dissipation*.

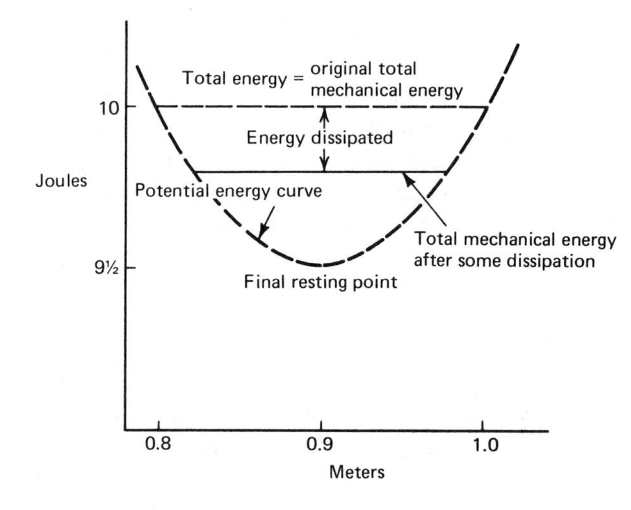

Figure 5-4

In the system we have been using, the dissipation is slight. We hardly notice a change over a period of a few swings. Only after many oscillations do the effects of dissipation become noticeable. So we are justified in drawing a horizontal straight line for the total mechanical energy. Over the period of one swing, this line will not change enough to make a noticeable difference.

After many oscillations, however, enough energy may have been dissipated that we can measure a total mechanical energy somewhat lower than we started with. We can draw a horizontal straight line that is lower than the original. This new line will account for the motion over the next few swings.

The turning points are closer together. The kinetic energy at any position is less than it used to be. But the total mechanical energy stays nearly constant at this reduced value over the period of a few oscillations.

As dissipation continues, the level of mechanical energy keeps dropping. Eventually it reaches the value of the minimum potential energy. When this happens, the oscillations come to a halt. No further motion takes place, because the kinetic energy is zero. The potential energy is as low as it can get. There is no more dissipation, since there is no mechanical energy left to dissipate. The system remains at rest at the stable position, where the potential energy has its lowest value.

The pole vaulter

A second example of the application of energy conservation is provided by the film of a pole vaulter in action.

The aim of the pole vaulter, in energy terms, is to provide as much GPE as possible to the system composed of himself plus the pole. In other words he wants to get his own mass up as high as he can. The key to his problem is the fact that there is only about 1 second between the time he leaves the ground and the time he crosses the bar. The amount of energy his body can provide in 1 second is quite limited. The energy for the vault has to be provided before the jump begins. The trick is to convert as much as possible of the kinetic energy from his running start into GPE at the crucial moment.[1]

The vaulter begins by running along the approach path as fast as he can. The kinetic energy he develops during this stage becomes the total store of energy upon which he draws during his later motion. From the time his pole strikes the pivot, and his feet leave the ground, until he hits the mat, the vaulter and his pole form an isolated system. The total energy of vaulter plus pole remains essentially constant for the rest of his jump.

As soon as the vaulter leaves the ground, several changes take place. The direction of his motion changes from horizontal to mainly upward. His speed slows down, as kinetic energy is converted into other forms. His GPE increases as he rises from the ground. Most important, the pole flexes sharply. Elastic potential energy is being stored in it. By the time he has reached half the height of the bar, the kinetic energy has dwindled to a small fraction of its starting value. The rest of the energy is by now divided about equally between GPE and EPE.

For the second half of the climb, the energy stored in the pole is converted into GPE. In a successful vault, as little energy as possible is left in the pole.

[1] The world track records correspond to speeds of 10 meters per second. This is a kinetic energy of 50 joules per kilogram, or 3500 J for a 70-kg man. The pole vault record is 5.5 meters. This corresponds to 55 J/kg, or 3750 joules for a 70-kg man. The near equality of the energy for the two events suggests the truth of our assumption.

Figure 5-5 Pole vaulter in action (Photograph by Ira Atkins)

The vaulter releases it unflexed (that is, with little EPE) and not quivering (i.e., no kinetic energy left in the pole).

Center of gravity There are some other tricks that pole vaulters know. One of these relates to the fact that the height that enters into the GPE formula is the *average* height of the whole body. This is sometimes called the height of the *center of gravity*. The center of gravity of a large object is the point such that, if all the mass of the object were concentrated at that point, the GPE would be the same as for the real distribution of mass.

A man standing upright has his center of gravity halfway up, in the vicinity of his hips. If he bends his body to touch his toes, the center of gravity is between his knees and his chest.

The pole vaulter can clear the bar in such a way that much of his mass, at first his legs, later his head, is always below the level of the bar. If he does this

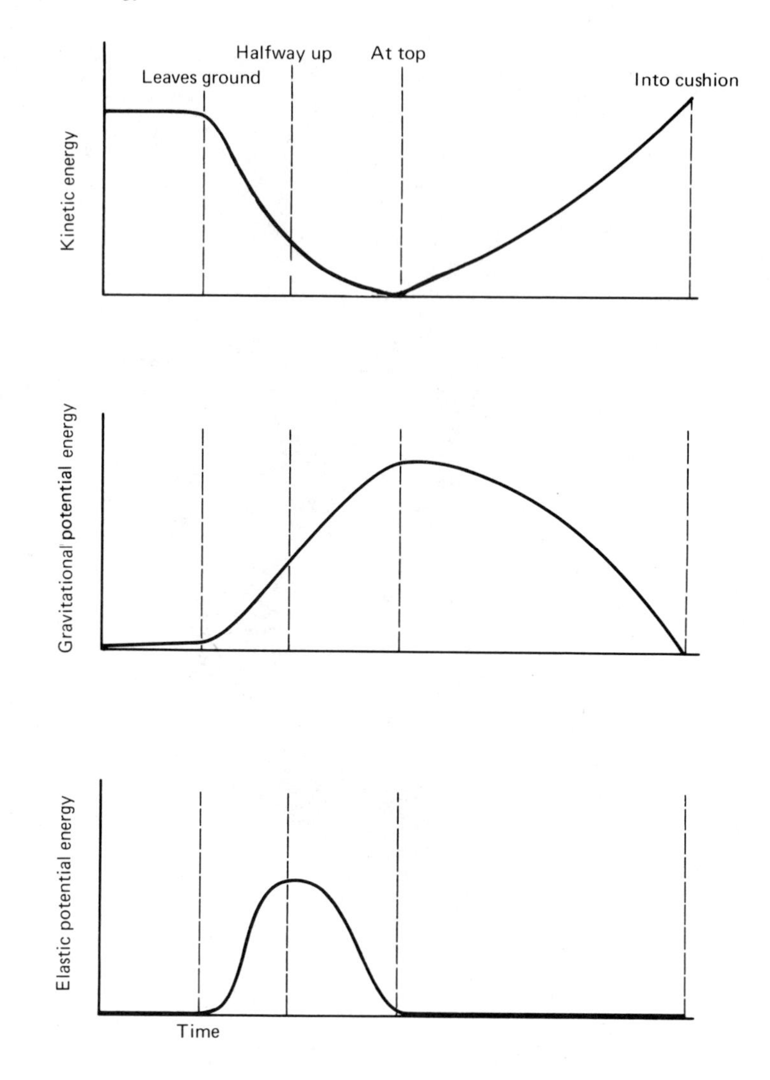

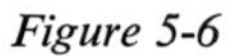
Figure 5-6

properly his center of gravity can stay always below the bar. This cuts down on the GPE needed to clear a given height.

The motion of the vaulter after he releases the pole, until he hits the cushion, is free fall. Kinetic energy replaces GPE as he drops, in the same manner as in a falling baseball.

Power

We have stated in the previous chapter that the energy of a system does not depend on how it got that way.

Consider for example the system of a man whose mass is 70 kilograms climbing a hill 100 meters high. Taking g as 10 J/kg-m, we find that the increase in GPE is 70 kg times 10 J/kg-m times 100 meters, which is 70,000 joules.

It takes this much energy to get the man to the top of the hill (assuming very little energy is wasted) no matter how he gets there. Whether he runs up the hill at top speed, or takes all day to make the climb; whether he climbs straight up a steep face or goes around by the gentle slope at the rear—in principle it makes no difference. The energy involved depends only on the difference between the final state (man at the top of the hill) and the initial state (man at the bottom of the hill). We grant that some climbing methods may be more wasteful than others, so that a complete description of the initial and final states should include measurements of the man's body temperature, the air, and so forth. We assert that for a reasonably steep climb accomplished in a reasonable time, the GPE is the predominant form of energy involved.

But, you claim, this violates all your senses. It is obviously harder to run up the hill than to walk. So something is not the same in the two cases. The thing that is different is not the energy, but the *rate* at which energy is supplied.

We define *power* as the rate at which energy is transferred per unit time.

$$\text{power} = \frac{\text{transfer of energy}}{\text{time}}$$

Given enough time a human being can supply very great quantities of energy. But the amount of energy he can supply *per unit time* is limited. Therefore, a bicycle rider going up a steep hill steps down his gears. So his bicycle moves more slowly. Thus, it takes longer for his bicycle to reach the top. Therefore, the denominator in the power formula is larger, and the power is reduced to the level that the bicycle rider is able to supply.

BICYCLE RIDING VS. RUNNING

The great advantage of a bicycle over a runner on a level or downhill course is that the bicycle wastes much less of the rider's energy. It takes very little energy to keep going on level ground at constant speed. Only the small losses to wheel friction and air resistance must be supplied.

Whereas a runner, on close examination, is undergoing a series of up-and-down partial stops with every footfall. A large fraction of the runner's kinetic energy is dissipated on every step and must be resupplied.

On an uphill course, where most of the energy supplied goes into GPE, the bike rider loses his advantage. It might be worth testing whether, on a steep enough hill, a runner can make just as good time as a bike rider.

If energy is measured in joules, then power is measured in joules per second. This unit is used so often that it has a name of its own; the *watt*.

watts = joules per second

The electric company supplies you with power, so many watts or kilowatts. But it bills you for energy, that is, power multiplied by the time you use it.

$$\begin{aligned}\text{one kilowatt-hour} &= (1000\ \text{watts}) \times (\text{one hour})\\ &= (1000\ \text{joules/second}) \times (3600\ \text{seconds})\\ &= 3{,}600{,}000\ \text{joules}\end{aligned}$$

Force

Another derivative of the energy is the force. We have mentioned force before, but we have not yet defined it. The following definition (not the same as in many other textbooks) is a valid one:

$$\text{force} = \frac{\text{energy transferred}}{\text{distance moved}}$$

The usual formula is to turn this equation around and say that:

$$\text{energy transferred} = \text{force} \times \text{distance moved}$$

Force can also be defined in the absence of motion. We can say that force is the energy that *would* be transferred per unit distance moved, if motion in a given direction were to take place.

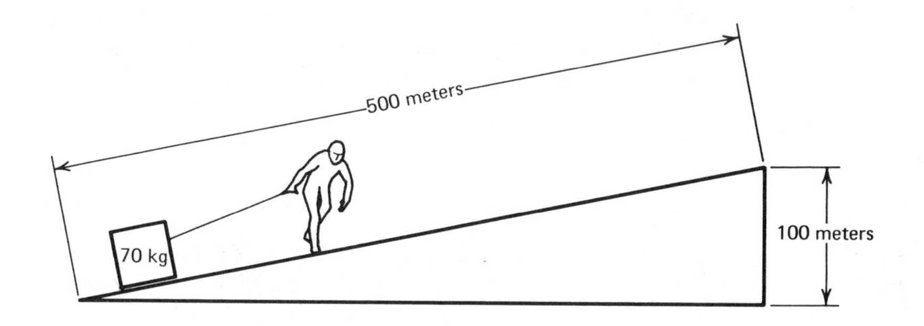

Figure 5-7

For example, consider a 70-kg weight being moved up a hill 100 meters high. The energy required is 70,000 joules, as we have calculated before.

If we try to pull the weight straight up the steep face, the force required is 70,000 joules divided by 100 meters, or 700 joules per meter. (A joule per meter is also called a *newton.*) If we pull it up the long slope, of length 500 meters, the force required is only 70,000 joules divided by 500 meters, or 140 joules per meter. This is obviously a much easier job.

A similar principle operates when you jack up an automobile. You do not have the strength to lift up the automobile directly. You use a jack, which has the property that it transmits energy with much smaller distances moved than the motion of the jack handle. You then make many moves through long sweeps, exerting a moderate force each time. The jack transmits the energy in such a way as to move a heavy weight (the car) a small distance. The product of force times distance, the energy transmitted, should be the same at each end.

Other derivatives of the energy

There are several other physical quantities that are related to the rate of transfer of energy. Power and force were defined as the rate of energy transfer with respect to time and distance. If we put other measurements in the denominator, we can define other quantities of interest.

Pressure The *pressure* exerted by a system is the energy required to compress it by a unit amount of volume.

$$\text{pressure} = \frac{\text{energy transferred}}{\text{volume decrease}}$$

The harder it is to compress something, the more energy it takes to make its volume decrease, the higher the pressure it is said to be exerting.

The pressure of the air at sea level is about 100 joules per liter. This means that it takes 100 joules to push all the air out of a volume the size of a 1-liter wine bottle. No wonder there is such a strong tendency for air to rush into any evacuated space.

If you pump the air out of a thin-walled can, the can will be crushed by the pressure of the air outside. A suction cup clings to a flat surface because it has been pressed so close that, to pull it away, you must create a small vacuum between the cup and the surface.

To keep your own body from being crushed by the pressure of the air, your body fluids, especially your blood, exert an equal pressure outward.

THE PRESSURE OF THE ATMOSPHERE

Place a flat rubber disk smoothly against the ceiling. It will resist falling, because to do so would require the creation of a small vacuum between disk and ceiling. Pushing the air out of any volume takes energy.

Let the area of the disk be 10 square centimeters. To pull it 1 centimeter away from the ceiling would create a vacuum of 10 cubic centimeters, or 1/100 of a liter. (A liter is 1000 cubic centimeters.) If the air pressure is 100 joules per liter, it takes an energy of 1 joule to evacuate such a volume.

Hang a 1-kilogram mass from a hook attached to the disk. If this weight were to drop 1 centimeter, its GPE would decrease by (1 kg times 10 J/kg-m times 1/100 meter) 1/10 of a joule. This is not enough to provide the energy needed for the evacuation. So the disk can support the weight. It can support any mass up to 10 kilograms.

Ten square centimeters of disk can support up to 10 kilograms hung from it. The larger the disk, the more weight it can support. The pressure of the air is such that it can support a mass of 1 kilogram per square centimeter of surface area. This is a unit of pressure often found on pressure gauges in Europe.

The pressure of the air can support a column of water about 10 meters high. The space just above the water column must be cleared of most of its air. Suppose the column has a cross-sectional area of 1 square meter. The volume of the column 10 meters high is 10 cubic meters, or 10,000 liters. The mass of such a volume of water is 10,000 kg.

If the whole column were to drop by 1 meter, the GPE would decrease by 10,000 kg times 10 J/kg-m times 1 meter, or 100,000 joules. But such a drop would create an additional cubic meter of vacuum above the water. That is, it would displace 1000 liters of air. Under an air pressure of 100 joules per liter, it takes 100,000 joules to do this. So the energy exchange just balances. If there were less water in the column, there would not be enough GPE to release. So the water level could not drop. If there were more water in the column, the level would drop until it was only 10 meters.

Such a high column of water is inconvenient to handle in the laboratory. Air pressure is often measured by using a column of liquid mercury. Mercury is 13.6 times as dense as water. The normal air pressure will support a column of mercury 0.76 meters high.

Why is the air pressure as high as it is? It is mainly because the air at sea level has to support the weight of all the air in the atmosphere above it. If the

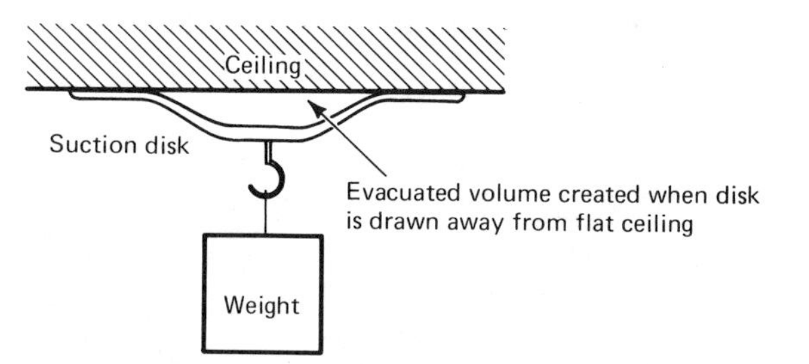

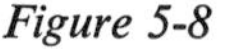
Figure 5-8

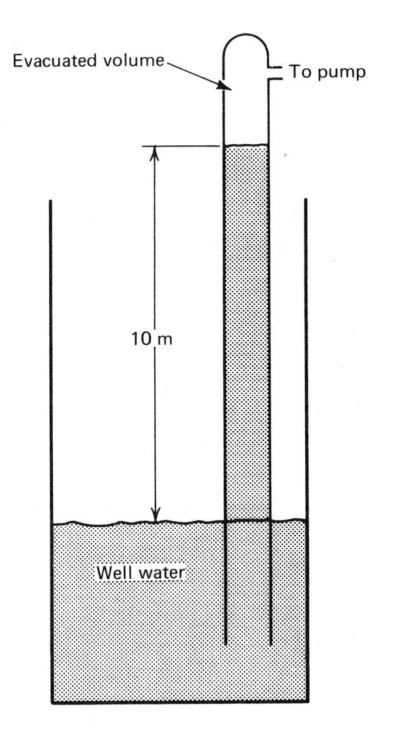

Figure 5-9

air did not resist compression, the whole atmosphere would collapse, and lie in a shallow layer close to the ground. When we try to evacuate a small volume at sea level, we are really displacing a column of air many kilometers high.

Surface tension All liquids have a tendency to form up into round drops. A sphere is the shape that has the smallest surface area for a given amount of liquid. It appears that there is a form of energy associated with any increase of the surface area of a liquid. The stable shape for a liquid, the shape with lowest energy, is the one with smallest surface area. We can define *surface tension* as

$$\text{surface tension} = \frac{\text{energy transferred}}{\text{surface area increase}}$$

Surface tension helps determine the size and shape of bubbles. It enables small insects to walk on the surface of water. The increase in surface energy required to make the insect sink into the water is greater than the decrease in GPE. So the insect cannot sink without outside energy input.

Torque When we are dealing with machinery that is rotating or twisting, the quantity of interest is often not the distance moved, but the angle turned. We define *torque* as

$$\text{torque} = \frac{\text{energy transferred}}{\text{angle turned}}$$

As in the case of force, it is not necessary for the energy actually to be transferred, for torque, pressure, or surface tension to exist. These quantities can all be defined as the energy that would be transferred if there were a unit small change in angle, volume, or area.

Electrical units The electrostatic potential, or *voltage*, can be defined as

$$\text{voltage} = \frac{\text{energy transferred}}{\text{electric charge moved}}$$

Electric charge is measured in units called *coulombs*. The unit of electrical potential is the volt. One volt is the electrical potential between two points such that it takes 1 joule of electrical energy to move 1 coulomb of charge from one point to the other.

1 volt = 1 joule per coulomb

Electric *current* is the rate of transfer of electric charge. If 1 coulomb of charge moves through a wire in 1 second, the wire is said to be carrying a current of 1 *ampere*.

1 ampere = 1 coulomb per second

If a volt is a joule per coulomb, and an ampere is a coulomb per second, we can multiply them together to get

volts × amperes = joules per second = watts

A typical home electrical power circuit delivers up to 30 amperes of current at 120 volts. You can draw 3600 watts maximum power from such a circuit.

The same amount of power can be delivered by sending 0.3 ampere at

12,000 volts. Such high voltage might not be safe in your home, but it is convenient to use it in *transmission lines* that carry energy between power stations. The losses in transmission lines increase rapidly with higher current. So it pays the electric company to use high voltage and low current in transmitting electrical energy. Just before reaching your home the power is *transformed* to a low-voltage house current.

The energy transmitted per second is the product of volts times amperes. This should be the same both before and after the transformer. The transformer has no extra energy to supply. If it is a good transformer, it does not dissipate very much energy.

ENERGY STORED IN A BATTERY

An *electrical storage battery* is a device for storing electrical energy. It makes use of the fact that the molecules that make up the material of the battery are hard to pull apart. It takes energy to break the molecules into smaller groups of atoms. The energy can be released by allowing the atoms to recombine into the original molecules. A storage battery is set up in such a way that the atoms cannot recombine unless an electric current flows outside the battery itself, from one terminal of the battery to the other.

A *lead storage cell*, the kind used commonly in automobiles, stores energy by breaking up *lead sulfate* molecules. When the battery is fully charged, there is no lead sulfate present. There is a negative plate made of lead and a positive plate coated with lead dioxide. Both plates are dipped into the same solution of sulfuric acid.

The solution is mostly water, with many *ions* moving around within it. Ions are atoms, or groups of atoms, that have an electric charge. There are two types of ions in sulfuric acid: positively charged hydrogen ions and negatively charged sulfate ions.

Energy is released when the ions move to the plates, combine with the lead, or lead dioxide, to form lead sulfate, which sticks to the plates. Some new water molecules are also formed at the positive plate. About 2000 joules of energy are released for each gram of lead that combines into lead sulfate.

There is some excess negative charge brought to the negative plate by the sulfate ions. The hydrogen ions bring an excess positive charge to the other plate. If this excess charge is allowed to build, it soon begins to repel the ions still in the water. The reaction stops, and no more energy is released.

The excess charge can be allowed to flow from one plate to the other through wires outside the cell. When this happens, the reaction in the battery will continue, releasing energy as it goes on, until the sulfate ions are depleted. The energy so released might be converted into kinetic energy of the electrons moving through the wire.

This energy can be extracted by suitable electric devices for various useful purposes. It can be used to turn the starter motor of the car or to drive the fan. It can light the headlights or power the radio. When energy from the cell

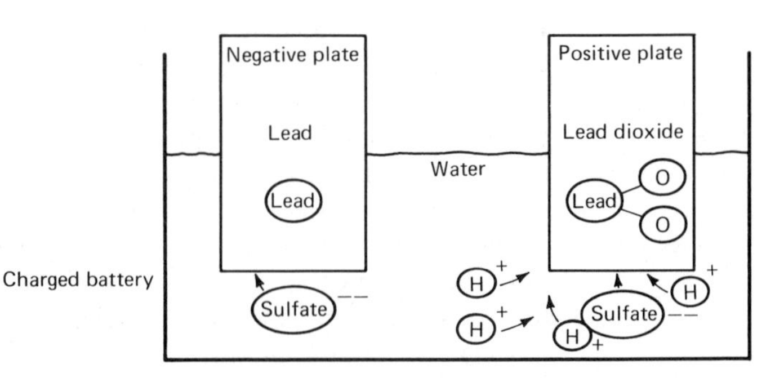

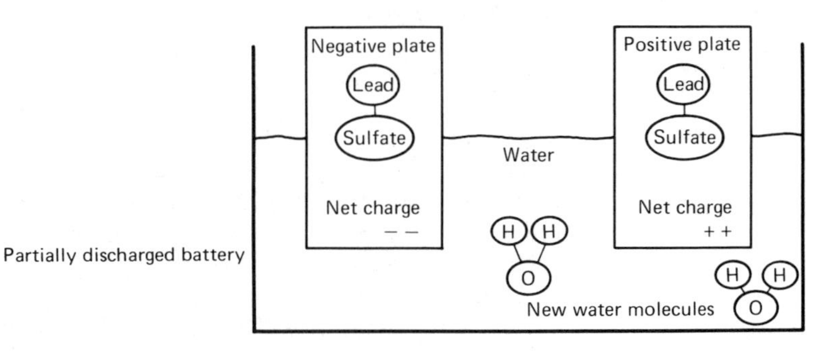

Figure 5-10

is not needed, the energy still remaining can be preserved by simply disconnecting the wire between the plates.

To put energy back in the cell, we force the current to flow in the opposite direction. This causes the lead sulfate molecules to break up again, restoring the battery to its starting condition. One device that can force the current to flow this way is a battery charger. It draws its energy from the local power company's lines. Another such device is the automobile alternator (in older models, the generator). Its source of energy is the gasoline burned in the car's engine.

How much energy can a battery store? Consider a typical 6-volt automobile battery. Such a battery is formed by connecting three lead storage cells in series. Each cell maintains a voltage of 2 volts between its plates. Suppose such a battery is given a charge of 10 ampere-hours through a battery charger. The total charge passed through the battery is

$$\begin{aligned}\text{10 ampere-hours} &= \text{10 coulombs/second} \times \text{3600 seconds}\\ &= \text{36,000 coulombs}\end{aligned}$$

The energy stored is

$$6 \text{ volts} \times 36{,}000 \text{ coulombs} = 216{,}000 \text{ joules}$$

(since a volt is 1 joule per coulomb).

What could be done with that much energy? Suppose the car's mass is 1000 kilograms, typical of a small European model. How high a hill can the car climb, driven by the battery alone?

Answer: 21.6 meters, about the height of a seven-story building, because,

$$\begin{aligned} \text{GPE} &= 1000 \text{ kg} \times 10 \text{ J/kg-m} \times 21.6 \text{ m} \\ &= 216{,}000 \text{ joules} \end{aligned}$$

How fast could you make the same car go on a level course, with this much energy?

Answer: A little more than 20 meters/second, because

$$\begin{aligned} \text{kinetic energy} &= \tfrac{1}{2} \times 1000 \text{ kg} \times (20 \text{ m/sec})^2 \\ &= 200{,}000 \text{ joules} \end{aligned}$$

Twenty meters per second, which is 72 kilometers per hour, is about as fast as you would want to drive on residential streets, but barely fast enough for the slowest lane on a superhighway. You can achieve this speed in a car driven by such a single 6-volt battery, only by completely using the energy stored in the battery.

Some designs for battery-driven cars provide for *regenerative braking*. This is a scheme by which the energy released when the car slows down, or coasts down hill, is used to recharge the battery. Nevertheless, the numbers illustrate some of the limitations of a battery-driven automobile.

Glossary

ampere—The unit of electric current; a flow of 1 coulomb of electric charge per second.

battery, electric storage—A device for storing electrical energy. An electric storage battery maintains a steady voltage between its two terminals. Energy is stored in the battery by forcing electric charge to flow from the negative to the positive terminal. This energy can be released by allowing the charge to flow back the other way.

center of gravity—The average position of the mass of an extended object. The height of the center of gravity is equal to the GPE, summed over all the parts of the object at their different heights, divided by the weight (mass times g) of the whole object.

coulomb—The unit of electric charge. In principle it should be defined in terms of a standard coulomb as the kilogram is. In practice, it is easier to define a standard ampere and to say that a coulomb is the amount of electric charge deposited by a current of 1 ampere flowing for 1 second. A coulomb is equal to the electric charge of 6 billion billion electrons.

dissipation—The transformation of mechanical energy into nonmechanical forms such as heat and sound. This loss of useful energy takes place through the action of friction ("rubbing"), air resistance, collisions, etc.

force—The energy that is transferred (or that would be transferred) to an object per unit distance moved.

lead storage cell—An electric storage battery, used in automobiles, that stores (and releases) energy by the chemical change of lead sulfate into (and out of) lead metal and lead oxide plus sulfuric acid.

newton—Unit of force. One newton is a force of such strength that it takes 1 joule of energy to move an object 1 meter against such a force.

oscillation—A back-and-forth motion that repeats itself at regular intervals.

power—Energy transferred per unit time.

pressure—The energy it takes, or would take, to decrease by 1 unit the volume of an object, or a fluid. Comparing the definitions of pressure and force, we can see that pressure is the force per unit surface area exerted on (and by) the substance.

regenerative braking—A means of avoiding dissipation when a battery-powered vehicle is brought to rest. With regenerative braking, the disappearing kinetic energy of the vehicle is used to charge the batteries. In ordinary cars this energy appears as heat in the brakes.

surface tension—The energy it takes, or would take, to increase the surface (usually of a liquid) by 1 unit of area.

torque—The energy that is transferred (or that would be transferred) to an object per unit of angle turned.

transformer—An electrical device that can change the high-voltage, low-current power in an electrical transmission line into the low-voltage, high-current power needed inside your house. The transformer supplies no power of its own. The product of voltage times current, which is the power delivered, is therefore the same on both sides of the transformer.

transmission line—Cables that carry electrical power from the generating plant to the user. Losses in a transmission line depend strongly on the current it carries. It is, therefore, an advantage to operate the lines at low current and high voltage.

volt—The unit of electrical potential (or voltage). It takes an energy of 1 joule to move an electric charge of 1 coulomb from one point to another against a voltage difference of 1 volt.

voltage—The energy that is transferred (or that would be transferred) to an electrically charged object, per unit of electric charge moved from one point to another.

watt—The unit of power. It is equal to an energy transfer of 1 joule per second.

Questions

1. Can the energy of a system be less than zero? What does it mean for the energy to be negative?
2. Why can a system without energy input not move beyond a turning point, where the potential energy curve crosses the line of total mechanical energy?
3. Using your own weight and the length of your own arms, figure how many joules it takes to "chin" yourself, that is, to lift your whole body's mass by a height equal to your arms' length. How many seconds does it take to chin yourself? How much power can your arm muscles develop?
4. A boy stretches a slingshot and uses it to propel a stone into the air. The stone rises to its maximum height, then falls into a nearby pond with a splash. Draw graphs of the following quantities versus time: EPE, GPE, kinetic energy, total mechanical energy, total energy.
5. An acrobat is bouncing on a trampoline. Draw graphs versus time of kinetic energy, GPE, EPE, and total energy. Cover the period from the peak of one bounce to the peak of the next bounce.
6. Time how fast you can run up a steep hill. Time yourself riding a bicycle up the same hill. How much power can your leg muscles develop?
7. How much force does it take to lift a 10-kilogram weight? (Think: How much energy does it take to raise it 1 meter?) What is the heaviest weight you can lift? How much force can your body exert?
8. If an animal is placed in an evacuated chamber, its blood vessels break open, and bubbles form within the body fluids. ("The blood boils.") Explain why this happens.
9. Air weighs about 1 gram (1/1000 of a kilogram) per liter. How high a column of air at this density can be supported by the air pressure at sea level?
10. If you spill some water onto a flat table, it forms into drops that are rounded on the edges, flat on the bottom. What would happen if there were no surface tension? What shape would the drops take if there were no GPE?
11. The performance of an automobile is often quoted in terms of "miles per gallon." The energy released in burning a gallon of gasoline is about

127 million joules. Show that the gasoline consumption of a car is directly related to the *force* required to make it go.

When the car is not accelerating or climbing hills, the main forces to overcome are (a) friction within the engine and other moving parts, and (b) air resistance. At high speeds the force of air resistance dominates. It increases rapidly with increasing speed. How does gasoline consumption behave as the car is driven at high speeds?

12. The *gear ratio* in an automobile relates the distance moved by the car itself to the distance moved by the engine parts. Explain why a high gear ratio helps cut down on energy wasted in engine friction. When extra force is needed, as in climbing a steep hill, why is a low gear ratio better?

13. About 1000 coulombs of electric charge flows between the plates of a lead storage cell for each gram of lead converted into lead sulfate. Knowing the energy released for each gram converted, 2000 joules, tell how many volts appear between the plates of a single such cell.

14. With the aid of a long crowbar, a man can lift an object with a mass of thousands of kilograms. Explain how this is possible. Hint: How far does the man's end of the crowbar move compared with how far the heavy object is lifted?

15. Compare the volume displaced by the point of a thumbtack to the volume swept out by the head of the tack. With respect to which physical quantity does the design of a tack increase our capabilities?

6

wave motion and sound

Introduction and summary

The idea of wave motion seems to show up in nearly every branch of physics. A wave is not a physical object, the way a ball or a stone is thought of. Rather, a wave is a shape or a form or a pattern that moves. We can lump these terms together into the word "disturbance." A wave is then defined as a disturbance that moves.

Waves can be classified in many ways. There are physical waves that can carry energy and information, and there are nonphysical or "phase waves." We shall mostly deal with physical waves in this chapter. Physical waves move because the wave disturbance acts on the less disturbed neighboring parts of the system. There is a cause-and-effect relation in the motion of a physical wave.

Waves can be compressional or transverse. Compressional waves, of which sound waves are an example, have their disturbances directed along the direction of motion. Transverse waves, such as surface waves on water, have disturbances at right angles to the motion.

A wave pattern that repeats itself at regular intervals is called a wave train. A wave train has a *period*, the time between repetitions. It has a *frequency*, the number of periods per second. It has a *wavelength*, the distance in space between repetitions. It is a basic relation for all wave trains that the product of wavelength times frequency is equal to the speed at which the wave moves.

The energy carried by a wave is not really a new form of energy. Usually it can be identified as a form of energy we already know. Kinetic energy, elastic, and gravitational potential energy can all be recognized in the parts of a system that are disturbed when a wave passes. In the case of waves of light, electromagnetic energy is known to be involved. The so-called wave energy is really just an organization of the more familiar forms of energy.

Just the same, the organization is so striking, and stays together so well, that we are tempted to give names to the various forms of wave energy. We talk about *sound energy* and *light energy*, as if they were really separate new forms.

In this chapter we will study some behavior that is typical of wave motion. The motion of a physical object like a ball or a bullet will not have these features.

The motion of a wave depends only on the kind of wave it is and the material through which it moves. It does not depend on the motion of the body that starts the wave. This fact leads directly to the *Doppler effect*. In this effect the frequency of a wave train becomes higher (or lower) than the frequency of the source when the waves are sent forward (or backward) from it.

When the source of the waves is moving faster than the wave speed, a shock wave can develop. This shock wave is known as a *bow wave* or a *sonic boom* or *Cerenkov light* for the various kinds of waves. It carries all the wave energy emitted over the whole path of the source's motion in a single sharp blast.

The most characteristic feature of wave motion is the ability of individual wavelets to reinforce each other or to cancel each other. This is called wave *interference*. It can lead to situations where each of two separate sources can be heard by itself, but the combination of the two is silence.

Diffraction is the process by which waves bend around obstacles. When the wavelength is short enough, the obstacles can cast sharp geometric shadows that would make you think the wave energy travels in a straight line. But when the obstacles are not large, compared to the wavelength, the shadows fuzz out. The waves then have no difficulty reaching the regions behind the obstacles that should be in shadow.

Wave motion is found in nearly every branch of physics. It is a simple idea whose application to many different situations speaks for the unity of science.

Demonstration: Waves in a "Slinky" Toy A long metal coil ("slinky" toy) is stretched across the front of the room. The instructor squeezes a handful of coil turns together, near one end of the long coil. Then he lets go.

The group of compressed coils seems to move like a ripple down the coil. It travels across the room to the far end of the long coil. There it seems to

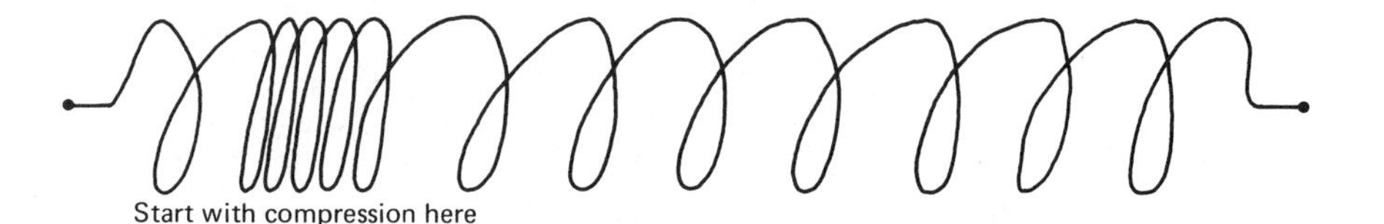

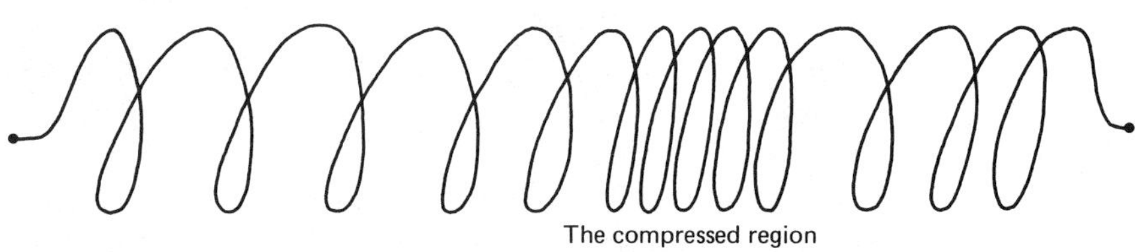

Figure 6-1 A compressional wave moving down a metal coil

bounce against the coil's end, and come travelling back to its starting point. It gets reflected again at this end, and continues to move back and forth across the room.

As the motion continues, the compressed region gets wider, and less sharply outlined. At last it becomes hard to recognize this region. The motion is spread out over the whole coil, just a general shaking and tumbling. Eventually even this disorganized motion ends, and the coil comes to rest.

Definition of a wave

Everybody who saw the demonstration must have felt that *something* was moving back and forth along the coil. Yet there was no single piece of metal that ever got more than a few centimeters away from its starting point. What was it that moved across the room?

What moved was the *pattern* of the coil turns, the region where the turns were closer together than usual. Any pattern or *shape* or *form* that moves can be called a *wave*.

We can be more precise and define a *disturbance* as any arrangement of the parts of a system that is different from its natural, resting condition. A physical wave is a disturbance that moves because of the effect that the parts of the system have on each other.

It is not hard to think of a wave that is *not* a physical wave. We can shine a flashlight on the moon. In a fraction of a second we can turn the flashlight so that the spot shifts thousands of kilometers from one end of the moon to the other. The light spot on the moon's surface fits the definition of a wave as a "moving pattern." But such a wave could not be used to send a signal directly from one base on the moon to another. Waves of this type, which carry no information or energy, are called *phase waves*.

A physical wave can be used to send *information*. Suppose we turn out the lights, with the instructor at one end of the coil and a student at the other. The instructor can send signals by pinching the coil turns at his end. The student can receive the signal by feeling when the compressed region reaches his end. After the motion dies down, another signal can be sent. A code, such as the Morse code, can be used to send long messages. In this way the entire lecture can be delivered.

Actual classroom lectures are usually delivered in a faster, but not basically different, way. Instead of compression waves in a "slinky" toy, we use sound waves in air. The instructor sends messages by talking, and the student receives them by listening. The code we use is known as the English language. We hope that some useful information gets transmitted in this way!

Physical waves can also carry *energy* from one place to another. Fasten a slip of paper to the far end of the metal coil. Before the wave is sent, the paper is at rest. When the disturbance passes the slip of paper, it flutters back and forth. Since the paper has mass, this motion indicates that some kinetic energy has been conveyed to the paper.

The energy carried by a wave has such a special character that we often treat it as a separate kind of energy. We will talk about sound energy, light energy, and other kinds of wave energy. We can see that the energy the metal coil carries is just the kinetic and elastic potential energy of the various pieces of metal within it. This sort of thing is true for other kinds of wave energy. Sound energy is mechanical energy of the molecules that transmit the wave. The energy of a light wave is electromagnetic energy. Though we shall often talk about sound energy, and other wave energies, we should remember that they are not really new forms of energy. They are basically well-organized forms of other, well-understood kinds of energy.

For the rest of this text we will use the word "waves" to mean physical waves. These are the waves that carry energy and information. Clearly these are the kinds of waves that play the most important roles in physics.

Types of waves

The wave we sent down the metal coil in our earlier demonstration was started by pinching some of the coils together. We can send a different kind of wave down the coil by plucking one or more turns to the side and releasing it. The sideward disturbance in the coil moves down the length of the coil and back. The two kinds of waves have many features in common. Some properties are different. For example, they probably travel at different speeds.

Waves in which the disturbance is along the direction of motion are known as longitudinal or *compression* waves. When the disturbance is at right angles to the direction of wave propagation, the wave is called *transverse*.

Sound waves are longitudinal in nature. Surface waves on water are mostly transverse. Light waves are transverse. The elastic waves that travel through the earth after an earthquake carry both longitudinal and transverse disturbances.

We will take *surface waves* on water and other liquids as our model for all wave motion. Everybody has seen these waves. They are the ripples that spread out when you drop a pebble in a pond. They are the swells that form

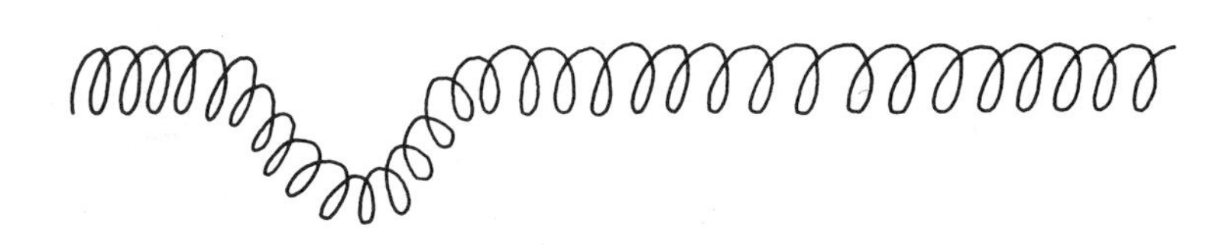

Figure 6-2 A transverse wave moving down a metal coil

when the wind blows over oceans and lakes. Nearly all of the features that are connected with wave motion can be seen in surface waves.

Sound is the type of wave motion we deal with most often. Sound waves are compression waves. You can start them off by "pinching" the air. When you snap your fingers, or clap your hands, you make a sound wave very much like the wave we made in the metal coil.

Sound waves travel through every kind of material. Solids, liquids, and gases are all conductors of sound. The speed of sound is different for different materials. It travels at about 350 meters per second in air, and at 1500 meters per second through water. It goes as fast as 5 kilometers per second in strong metals like iron and aluminum. In vulcanized rubber the speed of sound can be as slow as 50 meters per second.

The only "material" that will not transmit sound waves is an absolute vacuum. But sound waves do have trouble crossing the boundary between materials in which the speed of sound is very different. Waves tend to be *reflected* back from such surfaces. So a very good sound barrier can be made by forcing the waves to cross many such reflecting boundaries.

It is probably not very obvious that the sound we hear with our ears really is a form of wave motion. We cannot see the waves. The disturbance is really very slight. The loudest sound our ears can tolerate corresponds to a variation in air pressure of only 0.03%. There are now instruments sensitive enough to detect these small changes. But the idea that sound is a wave motion was accepted long before such instruments were ever made.

We can make the *analogy* between surface waves and sound. Surface waves are easy to see. We can observe their behavior and can show that they behave in certain ways. Then we use the behavior of these surface waves as a *model* for sound waves. We can then show that sound behaves in just the way we would expect of waves. The more effects we can explain in this way, the more successful the model is. This is the method of argument that convinced scientists of the wave nature of sound. It is the method we will use in this chapter.

Frequency, period, and wavelength

Up to now we have talked about waves whose initial disturbance had a short extent, and was not repeated. The compression wave in the metal coil, the pebble in the pond, the hand-clap and finger-snap, all had the character of "once briefly and nevermore." Such a short-burst type of wave is known as a wave *pulse*.

When the disturbance that starts the wave is repeated at regular time intervals, we have a *train* of waves. You can make a train of surface waves by moving a stick up and down repeatedly at one spot in the water. You can make a train of sound waves by singing or whistling or playing a single musical note.

Wave trains differ from each other in how rapidly the pattern is repeated. You can bob the stick up and down in the water vigorously, making a train of closely-spaced ripples. Or you can do it more slowly. The ripples will then

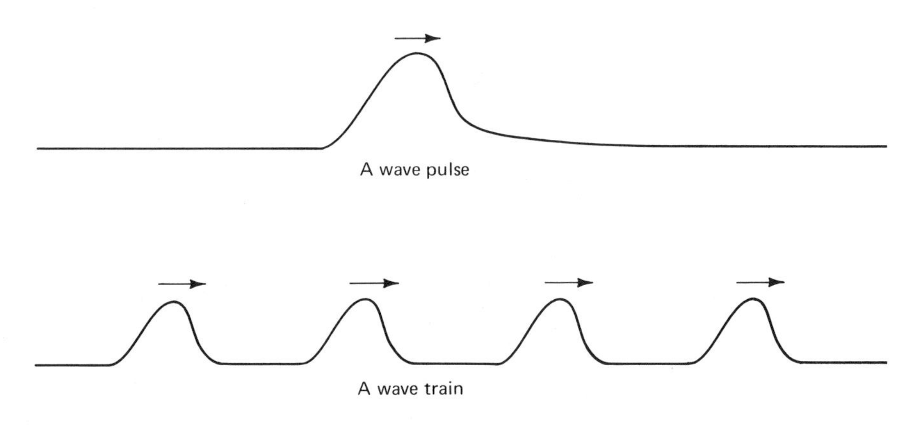

Figure 6-3

be farther apart. You can whistle a high-pitched musical note. Or you can hum a low note.

The time between repetitions of the pattern is called the *period* of a wave train. The number of repetitions each second is called the *frequency*.

The frequency and the period of a regular wave train have a simple relation to each other. The product of the frequency times the period is always the number one. If the period is one-tenth of a second, the frequency is 10 cycles per second. If the frequency is 1000 cycles per second, the period is one-thousandth of a second (or 1 millisecond).

The unit of frequency, "cycles per second," is often abbreviated as a *hertz*, or Hz. In the usual manner of the metric system, then, a *kilohertz* (or *kHz*) is 1000 cycles per second, and a *megahertz* (MHz) is 1 million cycles per second.

One way to make a train of sound waves is to clap your hands rhythmically. Each clap sends out a sound pulse. The repeated clapping sets up a wave train. If you clap once each second the period of the train will be 1 second. The frequency will be 1 hertz.

By the time you make the second hand-clap, the sound pulse from the first clap will have travelled considerable distance. If the speed of sound in air is 350 meters per second, the first pulse will have a 350-meter "head start" on the next one. If you keep on clapping once per second, each sound pulse will be 350 meters ahead of the next one. You will thus set up a train of sound pulses, each pulse spaced 350 meters from the next one.

The distance over which the pattern of a wave train repeats itself is called the *wavelength* of the train. The sound wave train you set up by clapping once per second has a wavelength of 350 meters.

You can clap your hands faster than once per second. Suppose you clap twice per second. The claps are then just half a second apart. The first pulse travels only 175 meters in the half second before the next clap. So the wavelength of this train is 175 meters.

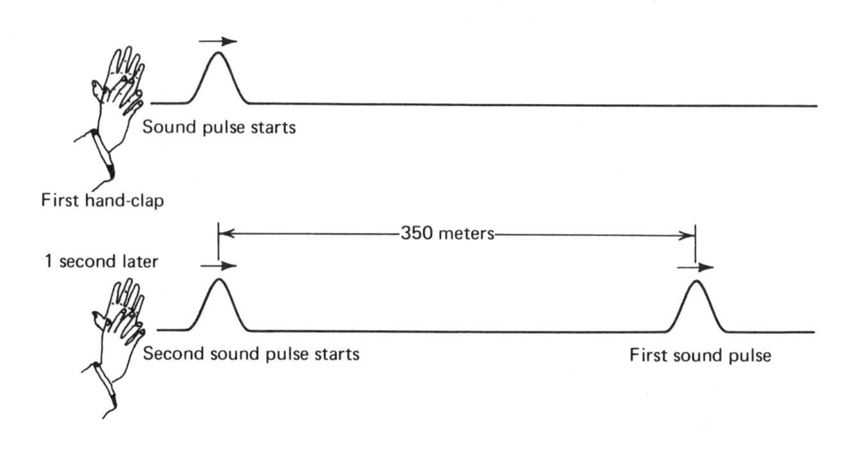

Figure 6-4

We see that the wavelength of a train is the distance a pulse can travel during one period. But the speed of the wave is just this distance divided by this time interval. So we reach the important conclusion that the *wavelength divided by the period* of a wave train is equal to the *speed* of the wave motion.

We can use the relation between period and frequency to state this a slightly different way. The *product of the wavelength and the frequency* of any wave train is equal to the speed of propagation of the wave.

Suppose a musical note has a frequency of 350 cycles per second (350 hertz). The wavelength of the sound wave train it produces in air will be 1 meter. The product of the wavelength, 1 meter, times the frequency, 350 hertz, is just the speed of sound in air, 350 meters per second.

If the note has a frequency of 10,000 hertz, the wavelength of its sound wave must be 0.035 meters (or 35 millimeters). Again, the product of 0.035 meters times 10,000 hertz comes out 350 meters per second.

The Doppler effect You have certainly had the experience of standing by the side of a highway as fast automobiles or motorcycles sped by. You must be familiar with the "rrrr-mmmm" change in sound as each vehicle passed.

MUSICAL INSTRUMENTS

An easy way to set up a train of waves is to use the way waves get reflected at boundaries.

Suppose we pluck one string of a guitar. An elastic wave travels down to the end of the string. It bounces back and comes travelling back down the full length of the string. The wave gets reflected again at the other end. Thus it returns to its starting point, going in the same direction as it started.

How long did it take the wave to get back to its starting point? Just the time

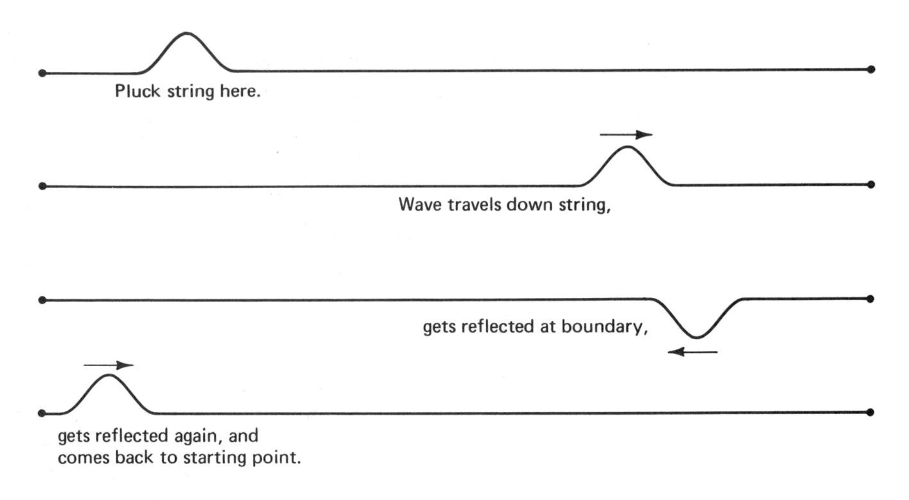

Figure 6-5

it takes a wave, moving at normal speed, to travel up and down the length of the string. The action will now repeat itself over and over again, with a repetition period just equal to this length of time.

A stringed instrument like a guitar has a *sounding board* to help make sound waves out of the vibrations of the string. The sounding board is often the hollow wooden body of the instrument itself. When the string vibrates, the sounding board shakes at the same frequency. The air inside and around the sounding board also begins to move back and forth at this rate. This leads to a sound wave we can hear. In this way the energy of the elastic wave in the string gets transformed into the sound energy of a musical note in the air.

In a wind instrument, such as an organ pipe or a flute, the original wave train is already a sound wave. A pulse of sound energy can be made by blowing or pumping air into the pipe of the instrument. This pulse then bounces back and forth inside the pipe. The repeating period of the system is determined by how long it takes the sound pulse to make one round trip. Openings at the end of the instrument allow some of the sound waves inside to escape into the air of the room and be heard.

There are two ways that a musician can change the frequency (that is, the "pitch") of the notes he plays. One way is to change the length of the string or the wind-pipe. A guitar player presses the string against the *fret* to effectively shorten the string. A trombone player changes the length of the wind-pipe. The round-trip time for the original wave will be less if the length it must travel is shorter. So we can produce higher frequencies (more rapid repetitions) by shortening the instrument. Conversely, we can make lower-pitched notes by lengthening it.

The second way to change the frequency is to change the speed at which the waves travel. This is easily done in a stringed instrument, simply by tightening or loosening the strings. This procedure is known as *tuning* the instrument. It is less easy to tune a wind instrument. The speed of sound does change with changes in the temperature or the pressure. So the tuning of a wind instrument can depend on the weather.

A more dramatic change can be made by changing the nature of the gas in the wind-pipe. The speed of sound in helium is nearly three times as fast as in air. So any wind instrument filled with helium will have a much higher pitch than when it is filled with air.

Most of the instruments in a modern orchestra produce their notes from either a vibrating string or sound pulses in a windpipe. Both of these types of systems are long and thin. They have the property that a wave bouncing back and forth within them cannot avoid passing by its starting point.

A different class of instruments make up the percussion section. These instruments, such as drums and cymbals, have broad areas or volumes over which the original wave can travel. A wave pulse, once started on a flat drum-head, can bounce back and forth from the edges many times before it returns to its starting place, if it ever does. Such instruments produce a different kind of sound than wind or string instruments. It is not usual to speak of a musical "note" produced by a drum.

The human voice can also be considered a musical instrument. Your wind-pipe is located at the rear of your throat, in a region called the "voice box" or the *larynx*. Muscles in the larynx can change the length (tension) of the vocal cords to produce different pitches. Other muscles are effective in making pulses of air that then bounce back and forth and produce the sounds of your voice.

The sound you hear arises from the vibrations of the engines. This is hardly a musical note in the ordinary sense, but it does have a definite repetition rate. The sound waves thus produced register in your ear and brain as a humming noise.

The change in pitch of the engine hum, as the auto passes you, is quite noticeable. You know very well that every driver does not suddenly decrease his engine speed just as he passes you. The pitch change takes place because of a difference in the way the soundwaves reach your ear. This shift in frequency goes by the name of the Doppler effect.

The Doppler effect is a change in the frequency of a wave because of the motion of the source emitting the waves, or of the observer receiving the wave energy.

Consider an automobile moving at a speed of 35 meters per second, whose driver honks the horn once each second. The speed of sound in air is 350 meters per second. By the time the driver makes his second horn honk, the wave pulse from the first honk has travelled 350 meters from the place where it was emitted. But the car has also in that time moved 35 meters from that spot.

The sound wave pulse from the second honk of the horn is only 315 meters behind the forward-going pulse from the first honk. The two wave pulses will stay 315 meters apart from then on. If the car keeps moving at the same speed, and honking at the same rate, a train of sound waves will be emitted forward from the car, each pulse 315 meters ahead of the next.

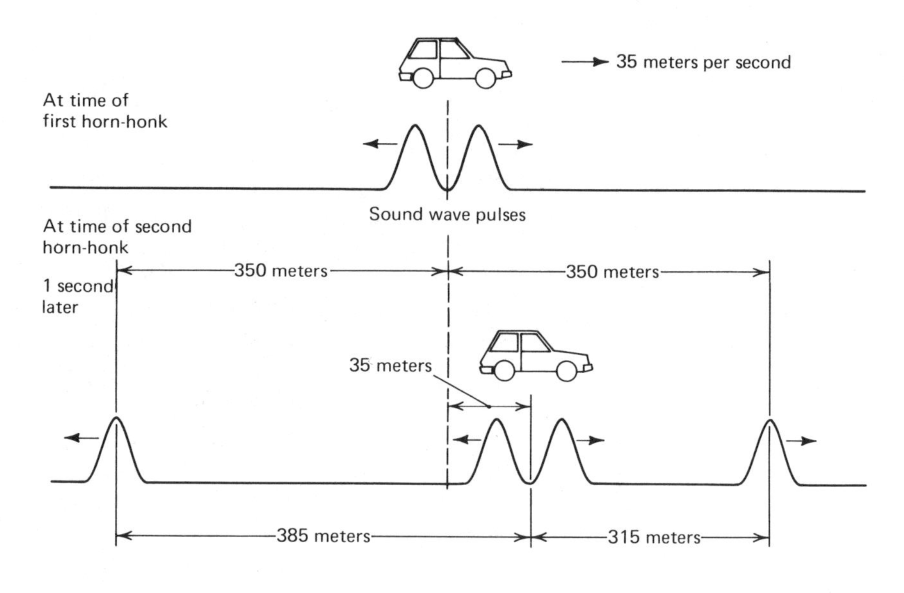

Figure 6-6

To a listener on the ground, far in front of the car, the sound pulses will be arriving faster than once per second. The pulses are 315 meters apart in the air, and travelling at 350 meters per second. The time between the listener hearing one sound pulse, until he hears the next one is 315/350, or 0.9 second. If the honking continues long enough, he can count 10 honks in 9 seconds. This is a general result. *A listener stationed ahead of a moving sound source will hear the sound pulses at a higher frequency than the source emits them.*

There is a different story for the sound waves emitted *backward* from the car. By the time the car emits its second horn-honk, it is 385 meters away from the backward-going pulse from the first honk. So the backward wave train has pulses 385 meters apart. It will take more than 1 second for a listener behind the car to hear successive sound pulses. In fact it will take 385/350, or 1.1 second between pulses. It will take the listener behind the car 11 seconds to count 10 honks.

Again this is a general result. *The frequency of a wave train emitted by a source moving away from the observer will have a lower frequency than that of the source*. The wavelength of the train will be longer in this case than if the source were standing still.

And so the shift in pitch of the engine hum, as a fast car moves past you, is explained. The "rrrr" sound that you hear as the car approaches you is a higher frequency than the actual engine rate. The "mmmm" sound you hear after the car passes is a lower pitch. Both sounds are Doppler-shifted, one up, the other down, from the sound you would hear if you were riding in the car.

The Doppler effect depends on the fact that the speed of a wave is the same no matter how fast the source that emits it is moving. Once a sound wave leaves its source, it travels at the speed of sound, in air or whatever medium it passes through. There is no Doppler shift for objects, like baseballs or rifle bullets, fired from a moving platform. These objects, unlike wave pulses, move at speeds that depend on how fast the platform was moving. (See Question 10 at the end of this chapter.)

The fact that Doppler shifts are observed for optical signals is taken as evidence that (1) light is a form of wave motion, and (2) the speed of light waves is independent of the motion of the source.

Do waves travel in straight lines?

If you stand in an open field (to avoid echoes) and listen to someone call from a distance, you can usually tell which direction the sound is coming from. You can argue that this proves that the sound pulse of a voice comes directly to you in a more-or-less straight line. If it did not, you would hear the sound coming from many directions, or from the wrong direction, and would not be able to locate the speaker.

On the other hand, if someone walks out of the room you are in, and calls to you from the hallway, you can still hear the voice. There may be a wall between you that is a strong baffle to all sound waves. The sound can still make its way through the open door and reach your ears by a path that is far from a straight line. This common experience shows that sound waves can bend around obstacles.

The process in which waves bend around obstacles is known as *diffraction*. The amount of diffraction that takes place in any given case depends on the relation between the size of the obstacles and the wavelength of the wave.

A good way to study this feature of wave motion is to look at surface waves in water.

In Fig. 6-7(a) we have an example of a short-wavelength wave passing through a wide hole in a barrier. The waves to the right of the barrier are strong directly in front of the hole. There is very little wave motion outside this region.

You can draw straight lines outward from the edge of the hole. Between the lines the wave activity is strong. Outside the lines there is practically no wave present. It is as if the barrier was casting a sharp *shadow*.

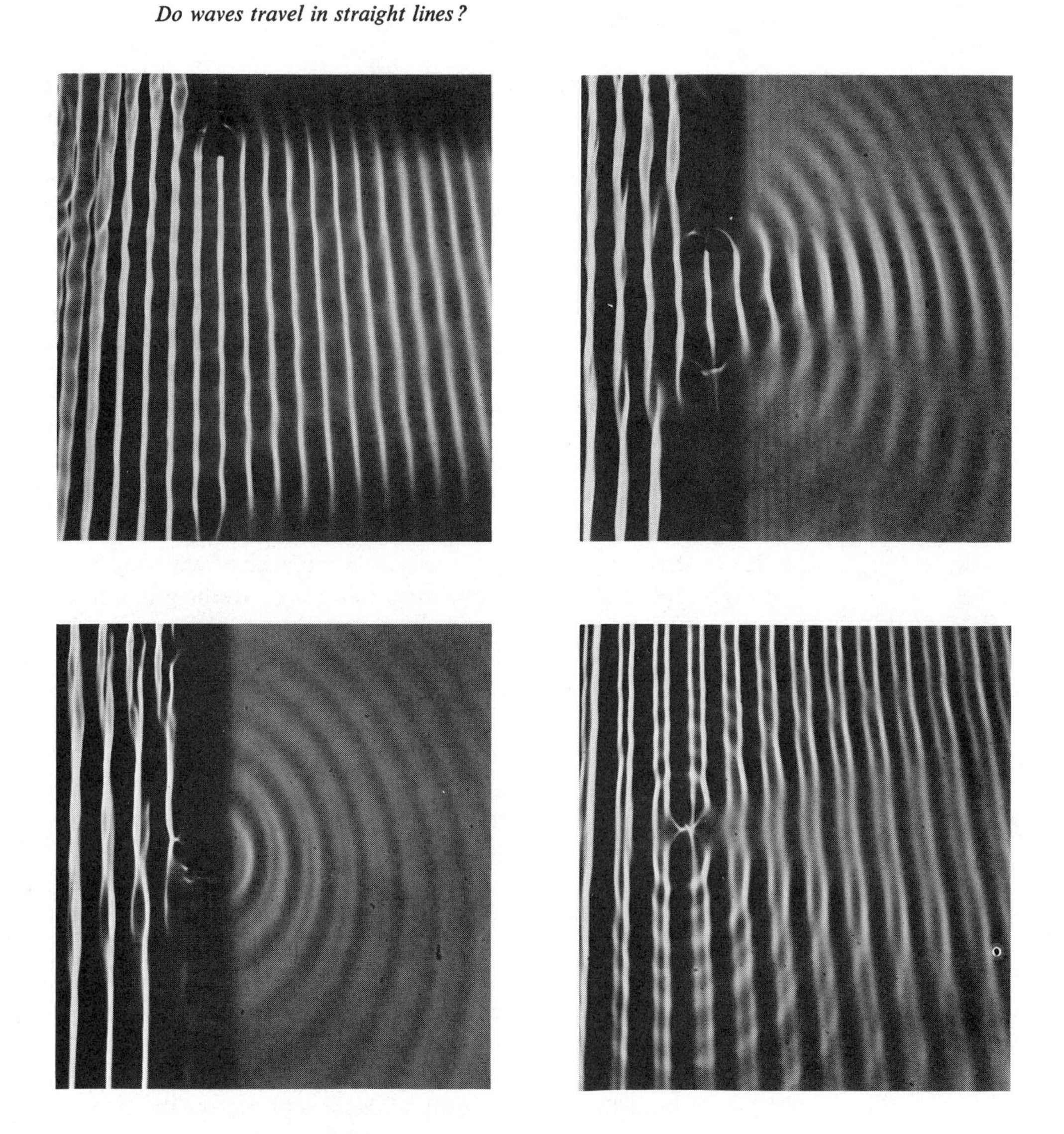

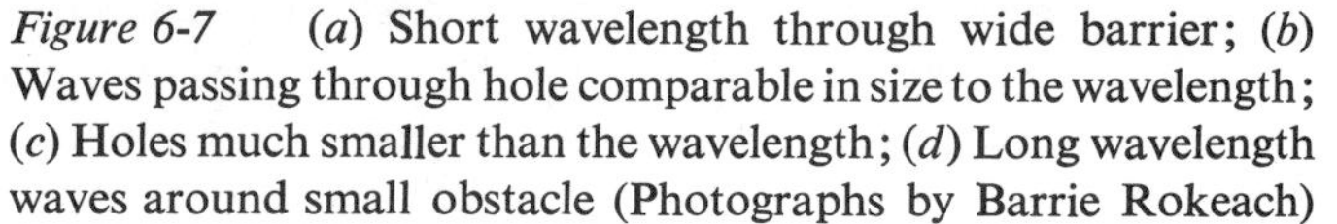

Figure 6-7 (*a*) Short wavelength through wide barrier; (*b*) Waves passing through hole comparable in size to the wavelength; (*c*) Holes much smaller than the wavelength; (*d*) Long wavelength waves around small obstacle (Photographs by Barrie Rokeach)

What do we mean when we say that the wave appears to travel in straight lines? Don't we mean that if we put a barrier between source and receiver, it will stop the wave? If I put a screen between you and me, you can't see me. If I put a sound barrier between us, you will have difficulty hearing me.

The ability of barriers to cast a sharp shadow is often taken as evidence that the wave energy travels in straight lines. Our experience with water waves shows that this is nearly the case when the wavelength is short and the obstacles are large. *We conclude that waves whose wavelength is very short compared to the obstacles they meet behave very much as if they travelled in straight lines.*

Figure 6-7(b) shows water waves passing through a hole whose size is comparable to the wavelength. The wave pattern is still strongest directly in front of the hole. But there is also quite a bit of wave activity along either side of this region. There is no longer a sharply defined shadow. Some wave energy does penetrate into the shadow region. The wave energy is able to bend around the obstacle.

In Fig. 6-7(c) the hole has been made much smaller than the wavelength. The wave pattern emerging to the right of the hole spreads out more-or-less uniformly in all directions. There is no information in the wave pattern of the size or shape of the hole. There is no semblance of any shadow.

In Fig. 6-7(d) we have the wave pattern of long-wavelength light encountering a small obstacle. There is some disturbance of the pattern close to the obstacle. Several wavelengths beyond it, however, the waves have formed back into their originally smooth pattern. You would hardly know, from looking at the waves far downstream, that there had been any obstacle at all. Long wavelength waves cannot be used to detect small objects.

We conclude that *when waves encounter obstacles that are the same size or smaller than the wavelength, the waves will tend to bend around the obstacle.*

Interference

Demonstration: Interference between Two Loudspeakers A pair of loudspeakers, both playing a musical note from the same source, are mounted on a pivot.

There are certain directions in the class such that the sound is very faint. When the speaker system is rotated about the pivot, every student in the class can hear the sound grow faint and loud again as the system goes through different positions.

We now hold the system steady and concentrate on the line of students who say the sound is faint. When we disconnect one of the speakers, those students can hear the sound from the other speaker quite well. When the first speaker is reconnected, and the second disconnected, they can again hear the first speaker well. But when both play together, the sound is faint!

We have the remarkable situation that the sound waves from two speakers, either one of them loud enough by itself, can *cancel* each other when they are both on at the same time.

The ability of waves to cancel each other is one of their most notable features. The process by which waves can cancel, or reinforce, each other is called *interference.*

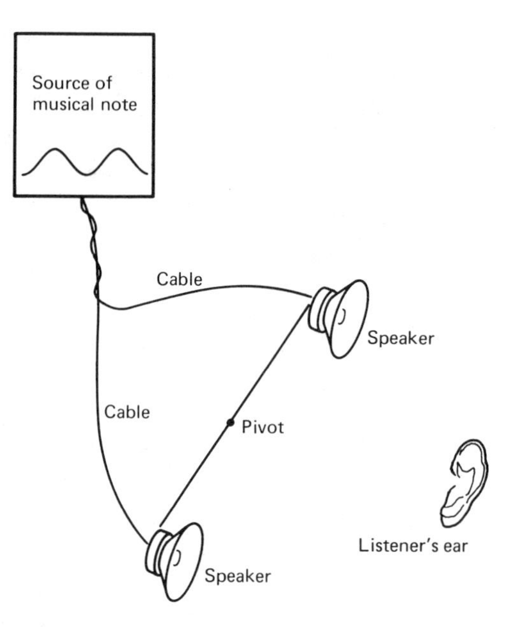

Figure 6-8

Suppose the wave train from one of the speakers alone arrives at our ear in a time pattern like that shown in Fig. 6-9(a). Suppose the wavetrain coming from the second speaker has a pattern like that shown in Fig. 6-9(b). The total wave pattern, when both speakers are on, is the *sum* of the two individual wave-patterns, shown in Fig. 6-9(c).

This is no wave pattern at all! The peaks of the wave train from the first speaker have just filled in the troughs of the wavetrain from the second speaker, and vice versa. There are no variations at all in the air pressure at the listener's ear. So no sound at all is heard.

The two wave trains that cancelled each other are said to arrive *out of phase*. One of them has a trough at just the point where the other one has a peak.

If we could delay one of them by one-half of the wave period, the two would be *in phase*. The peaks would come at the same time and the troughs would come together also. The combined wave would have double the peak, and double the trough, of the individual signals. Such a sound will be louder than the individual wave trains. When the waves arrive in phase they *reinforce* each other.

Why do certain positions in the class hear the weak cancelled sound, while others hear the reinforced sound? The cancelled sound is heard at just those positions where the wave from one speaker takes one-half period longer to reach the listener than the wave from the other speaker. These waves arrive one-half period out of phase and there is maximum cancellation.

The reinforced sound is heard by those students directly along the mid-line of the speaker system. For them the sound from both speakers takes the same time. The wave trains arrive in phase, and reinforce each other.

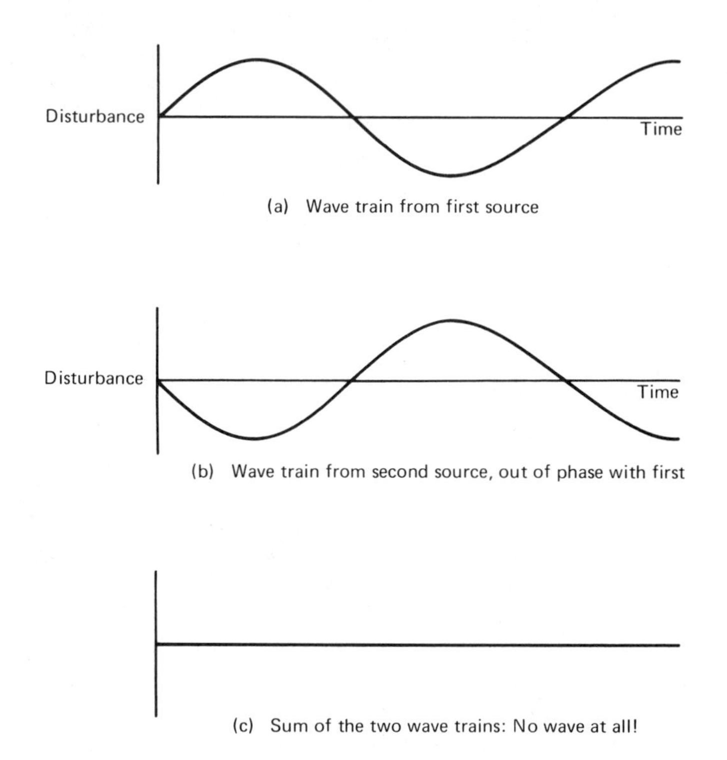

Figure 6-9

There is also reinforcement at positions such that the signal from one speaker takes a full period longer to arrive than the signal from the other speaker. In this case the peaks in the wave train from the farther speaker will overlap the peaks of the *next cycle* of the wave train from the closer speaker. Again there will be reinforcement.

In a similar way there will be cancellation when the difference in time is one and a half periods, and reinforcement when the difference is two full periods, and so on.

Figure 6-10 shows the pattern of surface waves from two point sources moving at the same frequency. The lines along which there is close to complete cancellation are easily visible.

Figure 6-10 Interference of two point sources in water waves (Photograph by Barrie Rokeach)

Bow waves and sonic boom

The speed of surface waves on water is not very fast, a few tens of meters per second. It is easy for a motorboat to travel as fast or faster than the waves it makes.

A boat moving just slightly slower than the waves it is making soon runs into a pile-up of waves in front of it. This accumulation of closely spaced waves forms a heavy *swell* in front of the boat that makes it hard for the boat to speed up any more.

This resistance to motion close to the wave speed is called the *wave barrier.* A very similar resistance is felt by airplanes moving at just under the speed of sound. That is known as the *sound barrier*.

A motorboat that is powerful enough can break through the wave barrier. Once it does, the resistance from this cause drops. The boat no longer has to fight the swells from its own waves.

The waves still do pile up, but not in front of the boat. Instead they come together to the side of, and behind, the boat's path.

You can draw a straight line that just touches all the waves that the moving boat sent out. Along this line the individual waves reinforce each other. Here the water disturbance is the greatest. At other places the waves tend to cancel each other, and there is little or no disturbance.

All the wave energy sent out by the boat, since it started going at its present velocity, is concentrated into this one wave front. A swimmer in the water

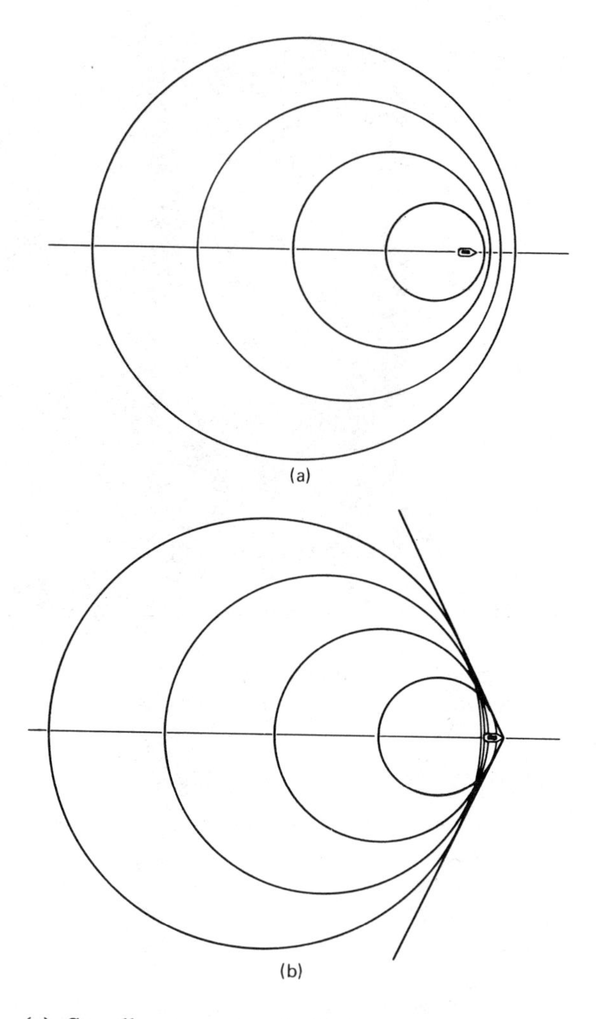

Figure 6-11 (*a*) Speedboat moving just slower than wave speed—waves pile up in front of boat, forming a "sonic barrier"; (*b*) Speedboat moving faster than wave speed—wave energy collects into bow wave

alongside the boat gets hit by the accumulated energy of these waves all at once. The *shock front* created in this way is known as a *bow wave*.

A very similar thing happens when an airplane travels through the air faster than the speed of sound. The shock wave that the plane makes is known as the *sonic boom*.

This sort of thing can happen with any kind of wave. Whenever an object that can produce waves passes through a material at a speed faster than the wave speed *in that material*, a shock wave can be produced.

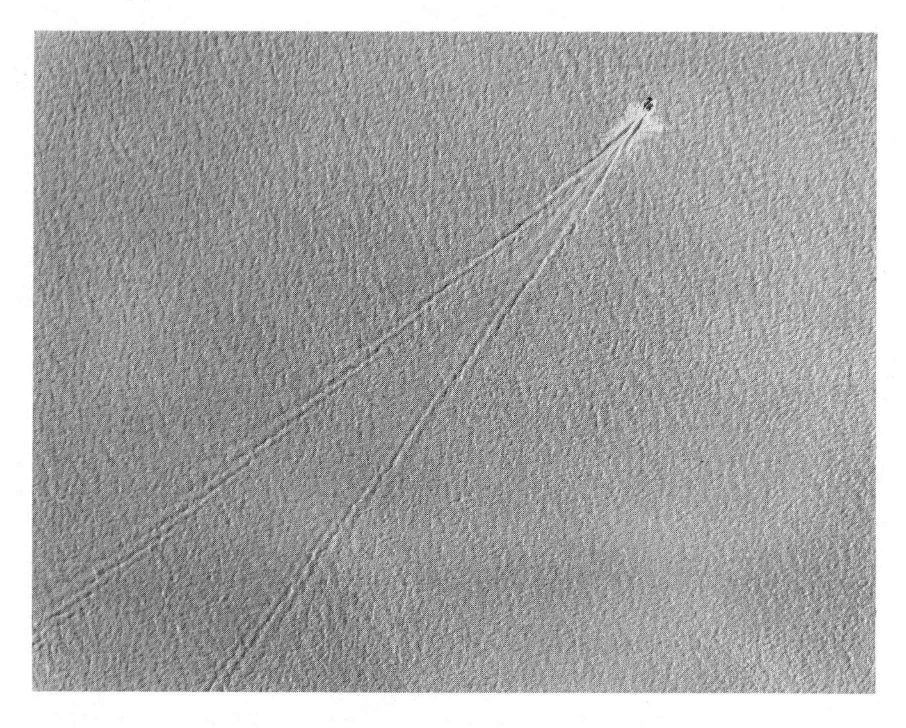

Figure 6-12 Bow waves from a speedboat (Photograph by Barrie Rokeach)

The speed of light waves in materials like water, or glass, is considerably less than the 300,000 kilometers per second that light travels through empty space. Even through a gaseous material, like air, the speed of light may be a meter or two per second less than it is in a vacuum.

The limit on the speeds of objects, required by Einstein's theory, is the speed of light *in vacuum*. It is thus possible for an electrically charged object to move through a material at a speed less than the relativistic limit, but faster than the speed of light waves in that material. From what we have learned in this section, we expect such a fast charged object to produce a shock wave of light. This does happen, and the light produced this way is called *Cerenkov light*.

Waves in physics

We have seen that wave motion shows up in many different ways in nature. We have talked about elastic waves in coils and strings. We have used surface waves in water to help us visualize some features of waves. We have studied sound waves. We have mentioned light waves. These are only a few of the many kinds of waves that can exist.

There are many aspects of wave motion that are common to all kinds of waves. Waves travel at speeds that are independent of the object that created

Figure 6-13 Cerenkov light from a water-cooled nuclear reactor. (Photo courtesy of the Dept. of Nuclear Engineering, University of California, Berkeley)

them. This fact leads to effects like the Doppler shift and to bow waves and sonic booms. Waves can reinforce each other or they can cancel each other. This comes about because waves are *disturbances* in a medium, rather than objects in their own right. The reinforcement or cancellation between waves is called *interference*. It is largely because of this feature of wave motion that they can sometimes cast sharp shadows and sometimes appear to bend around obstacles. The bending of waves around obstacles is called *diffraction*.

The widespread presence of waves in many branches of physics has on occasion inspired theories that propose that everything in nature can be explained as waves. Not the least of these theories is the 20th century idea of *quantum mechanics*, of which we shall learn more in a later chapter. We are not ready to claim that this will be the last word about the nature of matter, but we must be impressed with how universally the idea of waves seems to show up in nearly all the fields of science.

Glossary

bow wave—The shock front that forms on the surface of water when a speedboat is moving faster than the speed of the surface waves.

Cerenkov light—Light emitted in a shock cone when a charged object moves through some medium, like glass or water, at a speed faster than the speed of light in that medium (which may be quite a bit less than the speed of light in empty space, which no particle can exceed).

compressional wave—A wave in which the disturbance is along the direction of motion of the wave. Examples of compressional waves are (1) the "slinky" toy when a section is squeezed; (2) sound waves.

diffraction—The bending of waves around obstacles. Diffraction is most pronounced when the obstacle is much shorter than the wavelength.

disturbance—Any arrangement of the parts of a system that is different from its natural, resting condition.

Doppler effect—The shift in frequency and wavelength of a wave train when the source of the wave is moving toward (frequency gets higher, wavelength shorter) or away from (frequency gets lower, wavelength longer) the observer. The Doppler shift arises because the speed of the wave, once it leaves the source, does not depend on how the source was moving.

frequency—The number of times per second that a repetitive system, such as a train of waves, repeats its pattern.

hertz—The unit of frequency. One hertz corresponds to a repetition rate of 1 cycle per second.

interference—The cancellation (or reinforcement) between two or more coherent waves, arriving at a given place via different paths. If the crests of one wave arrive at the same time as the troughs of the other, the two waves can cancel each other. If the crests arrive together, the waves reinforce each other.

kilohertz—A unit of frequency corresponding to a repetition rate of 1000 cycles per second.

light—A transverse wave, detectable by our eyes, of electromagnetic nature.

longitudinal wave—Same as compressional wave.

megahertz—A unit of frequency corresponding to a repetition rate of 1 million cycles per second.

period—The time between repetitions of a system, such as a train of waves, that repeats itself at regular intervals.

phase wave—A wave (i.e., a moving disturbance) that carries no energy and information. In a phase wave there is no direct cause and effect between the disturbance at one place and the disturbance at the place next to it. An example of a phase wave is a flashlight beam from Earth being swept rapidly across the face of the moon.

physical wave—A wave that moves because of the direct effect that disturbances in one part of the medium have on the parts of the system next to it. Physical waves can carry energy and information. They move at a speed that is set by the properties of the medium.

pulse—A disturbance of short duration that does not necessarily repeat itself.

shadow—The region behind an obstacle that would not receive any wave energy if the waves travelled strictly in straight lines. When the wavelength is much shorter than the size of the obstacle, very little energy reaches the shadow region.

sonic boom—The cone-shaped shock wave that develops in air (or any other sound-carrying medium) when an object is moving through it faster than the speed of sound.

sound—A compressional wave, detectable by our ears, that arises from the organized motion of, and collisions between, large numbers of molecules in the medium that transmits it.

sound barrier—The difficulty encountered by an object, moving at just below the speed of sound, because the sound waves created by the object are accumulating directly in front of it.

speed of wave—The rate at which the patterns in a wave move from place to place. For a repetitious wave train the product of wavelength times frequency is always equal to the wave speed. The wave speed is determined by the properties of the medium, and does not usually depend on such factors as the motion of the source or the shape of the wave.

surface waves—Waves, generally transverse in nature, that appear on the surface of a liquid, such as water, when that surface is disturbed.

transverse wave—A wave in which the disturbance is at right angles to the direction of motion of the waves. Examples are: (1) the wave in the string of a guitar when it is plucked sideways; (2) surface waves on water.

wave—A disturbance that moves.

wave barrier—The difficulty encountered by a wave source moving at just below the wave speed, because the waves it has created are piling up in front of it.

wavelength—The distance in space between repetitions of a wave train.

wavetrain—A wave pattern that repeats itself at regular intervals.

Questions

1. On a crowded highway one car slows down, forcing cars behind it to slow down in turn. The lead car then speeds up again, allowing the following cars to speed up also. Show that this action causes a wave to travel through the traffic pattern. What determines the speed of this wave? Is it compressional or transverse?

2. Could a flashlight spot on the moon move faster than the speed of light? Would this violate Einstein's theory of relativity?

3. Give some examples of waves other than the ones mentioned in the text (sound, light, and surface waves). Are your examples *physical* waves? Do you know whether they are longitudinal or transverse?

4. Explain why several layers of sheetrock, or other solid, with air spaces between layers, make a good sound barrier.

5. What is the frequency of a wave train if its period is one millionth of a

second (1 microsecond)? What is the period if the frequency is 50 cycles per second (50 hertz)?

6. What is the wavelength in air of sound wave trains having the frequency and period of the examples in Question 5? The speed of sound in air is 350 meters per second. Will the wavelength in water, where the speed of sound is 1500 meters per second, be longer or shorter?

7. Sing a musical note. Feel the base of your throat with your fingers. Sing a high note; sing a low note. Whistle. Speak normally. Whisper. Shout. What changes in your voice-box (Larynx) enable you to make these sounds?

8. Suppose a car is honking its horn once per second while standing still. A second car, moving at 30 meters per second, drives past the first car. Will the second car notice a shift in frequency of the sound waves? When will the frequency received be higher than that of the source? Under what conditions will it be lower?

9. Will there be a Doppler shift for waves emitted *sideward*, or at an angle, from the source?

10. A man stands on a truck moving at 10 meters per second and throws baseballs forward at one-second intervals. He can throw the baseballs at a speed of 20 meters per second. How fast do the baseballs travel with respect to the ground? How frequently do the baseballs strike a fixed wall in front of the truck? How far apart are the baseballs in the air?

11. The bat is an animal that uses sound waves to help it avoid obstacles and find food. The sound waves the bat uses are so high-pitched that humans cannot hear them. What is the advantage to the bat of using such high-frequency sound waves?

12. Explain why there is cancellation between wave trains when the difference in source-to-listener travel times is $1\frac{1}{2}$, $2\frac{1}{2}$, $3\frac{1}{2}$, etc. periods of the wave. Why is there maximum reinforcement when the difference is a *whole number* of wave periods?

13. Some early thinkers thought of sound as a puff of air sent from speaker to listener, rather than a wave. Show that this model would fail to explain interference.

14. Draw pictures of the wavelets made by a boat moving (a) just slightly faster than the wave speed; and (b) very much faster than the wave speed. Show that the bow wave forms a much narrower "V" shape in case (b).

15. We often hear weathermen and others speak of a "heat wave." Can there be such a thing, with the word "wave" defined as in this chapter? Describe the nature of such a wave.

7

heat energy

Introduction

Isaac Newton and the scientists of the 18th century who followed him didn't know about the law of conservation of energy. They knew about kinetic energy and GPE, the sum of which they called the "vis viva."[1] But they didn't know that energy as a whole was *conserved.*

It is easy to understand why. Suppose we toss a pencil in the air. It has kinetic energy and it has gravitational potential energy. The pencil falls to the ground and stops. It has neither type of energy any more. There appears to be less energy in the system than there used to be.

We have been saying that the energy has been dissipated into sound and heat. But this is a fudge. The amount of heat generated in the fall of a pencil is too slight, and too widely diffused, to make it practical to measure. We do assert that if we were careful, and collected all the heat and sound energy produced, it would indeed equal the mechanical energy we lost. But we don't do this experiment all the time in everyday life.

From the modern point of view, heat and sound are forms of mechanical energy on the molecular level. So there is basically no difference from the forms of energy we have already been discussing. But molecules are also hard to observe.

We should not be surprised, then, to learn that, until the middle of the 19th century, heat was considered an entirely different substance. It was given names like "caloric" or "phlogiston." From hindsight we can see that it is a form of energy that we shall call *heat energy* from now on.

The 18th century scientists did not yet have a theory of conservation of energy, but there was a well-developed theory of conservation of heat energy. Such a theory, we know now, can be only partly valid. It works only for processes in which no other forms of energy are involved. But there are plenty of examples of such processes, so the theory of heat conservation had a long and useful life.

[1] Actually the term "vis viva" usually meant what we would now call *twice* the sum of the kinetic energy and the gravitational potential energy. If scientists had stuck to this definition, we would not have had the factor of one-half in our definition of kinetic energy.

In the past, before enough was understood to join it to other fields, each field of science named its own units. This is particularly true for the units of energy and its derivatives. Thus, the intensity of *sound* energy is expressed in *decibels*, that of *light* energy in *lumens*. *Nuclear energy* is measured in *MeV* ("millions of electron volts"), *radiation dose* in *rads*, and so forth. Heat energy was traditionally measured in *calories*.

Energy, in whatever form it takes, is still just energy. Any unit in which energy is measured ought to be convertible to *joules*. Whenever a new form of energy is being introduced the crucial questions that have to be answered are: How many joules does it take to make one unit of the new form of energy? Is this conversion factor always the same?

So the crucial experiment, to show that heat is a form of energy, is to measure how many joules there are in a calorie. As part of this chapter, we will conduct an experiment in which a known amount of mechanical energy is converted into heat energy. We hope to demonstrate that the ratio of the number of joules put in, to the number of calories of heat energy produced, is well defined. In fact, it should be the same ratio, 4.186, that scientists have been measuring for over a century.

Temperature

We are all familiar with the idea of temperature. Weather reports are always informing us of the temperature of the air. Baking recipes tell us the temperature at which to set the oven. When we feel sick, we measure the temperature of our body to help us tell how serious the illness is.

The general notion we have about temperature is that the warmer something feels, the higher the temperature is likely to be. We have a similar notion about heat. The more heat there is in some object, the hotter it feels. Many people (even some dictionaries!) do not make any clear distinction between the words "heat" and "temperature."

In physics there is a difference between heat energy and temperature. We can show what this difference is by thinking about 100 separate grams of water, all at the same temperature. If we put them all together, we have one hundred times as much energy as there is in a single gram. The additivity of energy makes this true. But they are all at the same temperature as the individual grams of water were.

Suppose we have a stove that supplies energy at a constant rate. It will take longer to boil a large pot of water on this stove than it will to boil a small cup of it. But the temperature at which water boils is the same in both cases.

All forms of energy, including heat energy, are what are called *extensive* quantities. They describe *how much* of the system we have. If you take two identical systems and put them together, the doubled system will have twice as much energy. It will have twice as much of every extensive quantity as the single system. Examples of extensive quantities are mass, energy, electric charge, and volume.

A quantity that describes the state of a system in such a way that it doesn't change when you double the system, is called an *intensive* quantity. Examples are temperature, position, pressure, voltage, and speed.

It is, of course, possible to double the temperature of a system. But this is not done by doubling the size of the system. To double the size of a glass of lukewarm water, you add a second glass of lukewarm water. To double its temperature, you would heat up the single glass. So temperature is an intensive quantity. You do not change the temperature when you add the second lukewarm glass.

If you put two objects in contact and wait until all changes stop, they will come to the *same* temperature. This is the most important fact about temperature. Using this fact, we are able to measure the temperature of anything.

A *thermometer* is a device whose size, shape, or some other feature changes in a recognizable and reproducible way when its temperature changes. A commonly used thermometer consists of a drop of liquid mercury inside a sealed glass tube. The higher the temperature, the more space the drop of mercury takes up. We can use these changes in size to measure the temperature of the mercury.

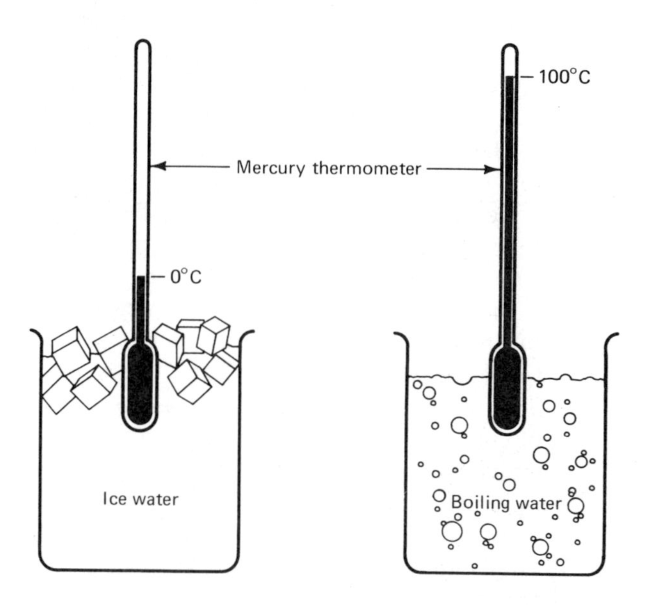

Figure 7-1 The fixed points on the Celsius scale

If we put one end of the thermometer in our mouth, it will soon come to the same temperature as our body. If we leave it in an open room, it will come to the temperature of the air. We can plunge the thermometer into ice water or into a boiling teapot. The mercury will contract or expand to reflect the temperature of whatever it is in contact with.

There are many ways to set the scale on a thermometer. The scale commonly used in Europe, and by most scientists, is called the *Celsius* (formerly called "Centigrade") scale. This scale uses the melting point of ice and the boiling point of water as its fixed points.

The thermometer is first plunged into melting ice, and the height of the mercury column is noted by scratching the outside of the glass. This point is called "zero on the Celsius scale," or 0°C. Then this thermometer is placed amid boiling water. Another scratch is made where the mercury reaches now. This point is called "one hundred on the Celsius scale," or 100°C.

The temperature at points in between these two can be defined by making 100 equally spaced scratches along the mercury column. Temperatures lower than 0°C, or higher than 100°C, can be defined by making scratches at the same spacing beyond the two fixed points.

The normal temperature of the human body is 37°C. We would consider the air temperature chilly below 10°C, and quite warm above 30°C. Mercury freezes at −40°C ("forty below zero on the Celsius scale"). Below that temperature we would have to switch to a different liquid in our thermometer.

We don't have to use liquids for our thermometers. We might use solid metal rods. The changes in length of such rods at normal temperatures are usually too small to measure accurately. However, a useful thermometer can be made of strips of two different metals bonded together. When the temperature rises, one of the metals will expand more than the other. The result will be that the bi-metallic strip will bend. The more temperature change, the more bending. Bi-metallic strips are found in many common devices, such as the oven thermometer or the thermostat in your house.

Heat energy

We are now ready to explore the connection between heat energy and temperature. It is clear that as we add heat energy to a substance, it will get hotter. The temperature will rise. But the amount of heat energy needed to make a given rise in temperature depends on several factors. It depends on the mass of the substance being heated, the material of which it is made, and the starting temperature. It depends on whether the substance is in its liquid, solid, or gaseous state. It may also depend on the pressure. The heat energy required to bring a substance to a given temperature is a quantity we shall have to measure.

We begin by defining 1 *calorie* as the amount of heat energy required to raise the temperature of 1 gram of water from 15°C to 16°C.

In principle we can then measure the amount of energy required to raise the temperature of 1 gram of water from any temperature to any other temperature.

For example, start with 1 gram of water at 0°C and another gram at 16°C. Place them in contact. Wait until the second gram has cooled to 15°C. In so doing it will have released 1 calorie of heat energy. Measure the temperature of the first gram at this moment. It will be nearly, but not exactly, 1°C. This then is the temperature rise produced by 1 calorie of heat energy in 1 gram of water at 0°C.

Or start with 1 gram of water at 40°C and a second gram at 15°C in contact with each other. Wait until the second gram has warmed up to 16°C. This takes 1 calorie of heat energy. See what the temperature of the first gram has dropped down to. It will be close to 39°C. So this is the temperature drop caused by the loss of 1 calorie of heat energy by 1 gram of water at 40°C.

We can proceed in this manner to patch together the whole curve of heat energy versus temperature. We find that between 0°C and 100°C it takes close to 1 calorie to raise the temperature of 1 gram of water each degree Celsius.

When the water is in the form of ice, the results are quite different. One calorie of heat energy will raise the temperature of 1 gram of ice by *two* degrees Celsius, e.g., from −10°C to −8°C. When the ice reaches 0°C, its

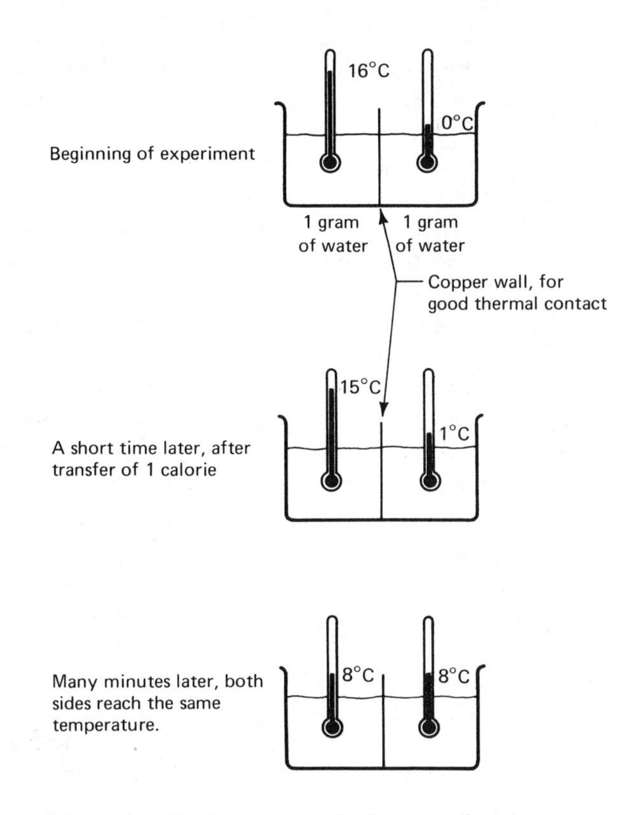

Figure 7-2 Measuring the heat energy in 1 gram of water at various temperatures

melting point, an even greater change takes place. We find that it takes 80 calories of heat energy (that is, 80 grams of water cooling from 16°C to 15°C) to melt 1 gram of ice, without any temperature change at all.

At 100°C, the boiling point of water, it takes 540 calories to boil 1 gram of water before the temperature can rise any higher.

These results can all be summarized in a graph like the one in Fig. 7-3. This graph shows the heat energy that must be added or taken away from a gram of water at 0°C to bring it to the temperature and state shown.

A similar graph can be drawn for 100 grams of water or for any other amount. Since heat energy is an extensive quantity, all the numbers on the vertical scale would be multiplied by 100. Since temperature is an intensive quantity, the numbers on the horizontal scale stay the same.

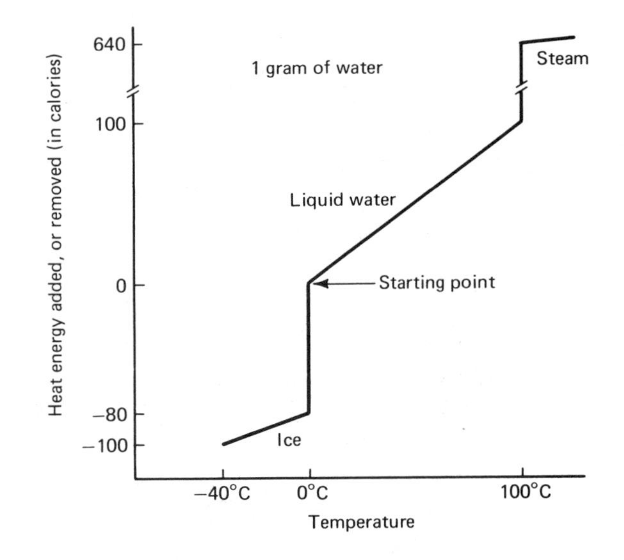

Figure 7-3 The heat energy in 1 gram of water, as a function of temperature

TEMPERATURE MIXING PROBLEMS

One of the features that make science so useful is the ability to predict "what will happen if . . . ." Perhaps nowhere is this predicting power illustrated in such a simple and straightforward way as in solving temperature mixing problems. "What will be the final temperature," such a problem asks, "if we mix so much water (or anything) at one temperature with a certain amount at another temperature?"

The law of conservation of energy, in this case heat energy, helps us to answer these questions easily.

Problem 1. Mix 50 grams of water at 0°C with 50 grams of water at 100°C. What will the final temperature be?

We express the law of conservation of energy in the form:

total energy before = total energy after

In this problem heat energy is the only form of energy involved.

Heat energy before = 50 grams @ 0°C + 50 grams @ 100°C
= 0 calories + 5000 calories
= 5000 calories

All energies are given as the amount needed to heat that many grams of water from 0°C to the given temperature.

At the end of the mixing process all 100 grams of water are at the same temperature; call it T.

Heat energy after = 100 grams at temperature T
= 100 times T calories.

This has to equal 5000 calories, the heat energy we started with. It will be that if T is 50°C, for 100 times 50 is just 5000. So we are led to the conclusion, perhaps not surprising, that if we mix equal amounts of water, half at 0°C and half at 100°C, the mixture will come to a final temperature of 50°C.

Problem 2. Mix 50 grams of water at 70°C with 10 grams of water at 10°C. This problem is not so easy to answer because we are mixing unequal amounts of water. We expect that the final temperature will be closer to 70°C than to 10°C.

Heat energy before = 50 grams @ 70°C + 10 grams @ 10°C
= 3500 calories + 100 calories
= 3600 calories
Heat energy after = 60 grams at temperature T
= (60 times T) calories.

If T is 60°C, then the heat energy after mixing comes to 3600 calories, the same as we started with. So 60°C is the final temperature.

Problem 3. Mix 50 grams of *ice* at 0°C with 50 grams of water at 100°C.

The heat energy ina gram of ice at 0°C is 80 calories *less* than that in liquid water at the same temperature. Since we are using liquid water at 0°C as our reference level of energy, the heat energy in the ice will be less than zero. It will be a negative number.

$$\text{Heat energy before} = 50 \text{ grams @ } (-80 \text{ calories per gram}) + 50 \text{ grams @ } 100°\text{C}$$
$$= -4000 \text{ calories} + 5000 \text{ calories}$$
$$= 1000 \text{ calories}$$

It takes 5000 calories to heat 50 grams of water from 0°C to 100°C. But you must *remove* 4000 calories to turn 50 grams of liquid water at 0°C into ice.

The total heat energy at the start of this problem is 1000 calories more than if the whole 100 grams had been liquid water at 0°C.

$$\text{Heat energy after} = 100 \text{ grams at temperature } T$$
$$= (100 \text{ times } T) \text{ calories}$$

If this is to equal 1000 calories, the heat energy we started with, T must be 10°C.

Notice that this temperature is much lower than the 50°C we found in problem #1. So it demonstrates that it does take quite a bit of heat energy to melt ice. We could turn this experiment around, and use it as a way to measure what the heat of melting of ice is.

Specific heat

Most materials don't require as much heat energy per gram to warm them to a given temperature as water does. We have seen that a gram of ice takes only half a calorie to warm up 1 degree Celsius. Consider 1 gram of copper metal, compared to 1 gram of water.

It takes only 9.2 calories to warm 1 gram of copper from 0°C to 100°C. This fact accounts for why, when nearly empty, a copper teakettle heats up so much faster than when it is full of water. It explains why hot metals can be cooled so rapidly with small amounts of cold water.

The *specific heat* (also called "heat capacity") of a substance is the amount of heat energy, in calories, required to raise the temperature of 1 gram of that substance 1 degree Celsius.

The specific heat of water is 1. The specific heat of ice is $\frac{1}{2}$. The specific heat of copper is 0.092.

The heat energy required to raise a given amount of some material a certain number of degrees Celsius is:

Heat energy required (in calories) =

Mass (in grams) times temperature rise (in degrees Celsius) times specific heat of the substance

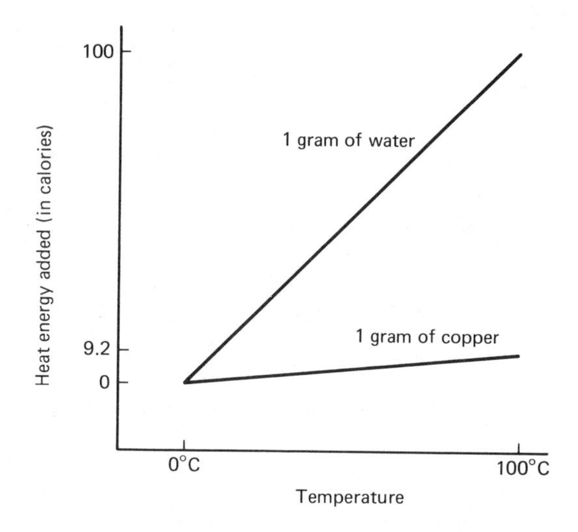

Figure 7-4 Comparing the heat energy in 1 gram of copper to that in 1 gram of water

The same formula can also be used to find the heat energy *released* when the material *cools* by a certain number of degrees Celsius.

Latent heat

Many changes that require input or release of heat energy take place without any change of temperature. The melting of ice is one example. The boiling of water is another. The heat energy required, per gram of material, to make such a change is called the *latent heat* of the process.

Here are a few examples of latent heats:

The latent *heat of fusion* (or "heat of melting") is the heat energy required to cause 1 gram of a solid substance to melt completely. The heat of fusion of ice is 80 calories per gram.

The *heat of evaporation* is the heat energy required to cause 1 gram of a liquid substance to change completely into its gaseous form. The heat of evaporation of water at 100°C is 540 calories per gram.

The *heat of sublimation* is the heat energy required to turn 1 gram of a solid directly into a gas. There are some solids that change directly into gas, without going through the liquid phase. One of the best known such sublimating materials is solid carbon dioxide, commonly known as "dry ice." At low enough pressure nearly any material can sublimate. A block of water ice placed in an evacuated container will emit some water vapor, for example.

The *heat of solution* is the heat energy released when 1 gram of a substance is dissolved in water.

Heat as a form of energy

Heat energy by itself is not always conserved. It is easy to create heat energy out of other forms. Rub a spoon vigorously with a piece of steel wool, or a screwdriver with sandpaper. Feel that the rubbed object gets warmer.

Benjamin Thompson, Count Rumford (1753–1814), who was born in America, but served the King of Bavaria in 1798, is usually credited with the observation that heat energy can be created out of mechanical motion. He noticed that boring out the insides of cannons made the metal very hot. This could in no way be explained by the then accepted caloric theory of heat.

Sir Humphry Davy (1778–1829) showed that he could melt ice by rubbing two pieces of it together. This demonstrated that there is no law of conservation of (what we call) heat energy alone. Somewhere new heat energy is being generated.

In 1847 Hermann von Helmholtz (1821–1894) published a paper proposing that heat is a form of energy, and that total energy is conserved. He was the first to advance such a law.

The crucial experiment demonstrating that heat is a form of energy was done by James Prescott Joule (1818–1889) in 1850. He showed that the same amount of heat energy was generated whenever a given amount of mechanical energy was dissipated. In essence he measured the number of joules in a calorie.

AN EXPERIMENT TO MEASURE HOW MANY JOULES IN A CALORIE

This experiment was carried out by the author in front of a class like yours. The numbers recorded here were the ones actually measured during one such performance. They give some feeling for the accuracy we can get in such a demonstration.

The idea of this experiment is to convert a known amount of mechanical energy—that is, a known number of joules—into heat energy. By seeing how many calories are produced, we can then figure how many joules there are in a calorie. The experiment is designed with two important features in mind:

(*a*) that we know exactly how many joules we put in;

(*b*) that as much as possible of this energy is converted into heat energy in the place we expect it.

The apparatus consists of a copper cylinder containing water. There is a thermometer to measure the water temperature. The cylinder is mounted on a shaft with a handle so it can be turned easily around its long axis.

A strip of copper braid is wrapped around the cylinder. A 5-kilogram mass hangs from the end of this braid. The friction between the cylinder and the braid should be sufficient to support the hanging weight. The upper end of the braid can be anchored lightly to a wall or table to keep it in position.

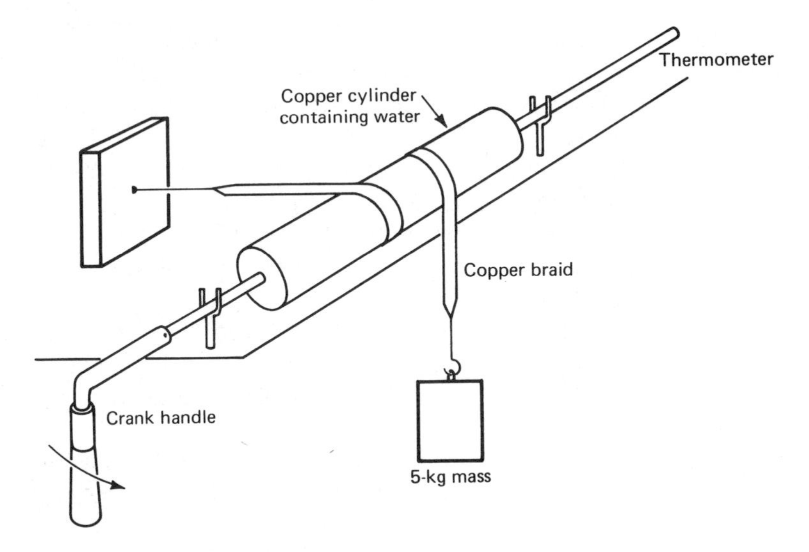

Figure 7-5 Experiment to find out how many joules are in 1 calorie

When you turn the crank handle through a small arc, the hanging weight rises a bit. Thus you have transferred a small amount of GPE to the system. The strip of braid soon slips, and the weight falls back to its original position. The GPE is converted into heat energy, mainly near the places where the braid rubs against the cylinder. This basic dissipation process is repeated many times as the crank turns.

How much energy is dissipated in one full turn of the crank? It is the GPE needed to lift the weight a height equal to the circumference of the cylinder.

We can measure the circumference by wrapping a piece of string around the cylinder. We make two marks on the string where it meets itself after going around once. Then we lay the string out straight and measure the distance between the marks with a meter stick. For the cylinder used in the author's class, the distance around the cylinder was 0.150 meter. In one turn of the crank, we raise the weight and then let it fall a cumulative distance of 0.150 meter.

The GPE required to raise a mass of 5 kilograms a height of 0.150 meter is:

GPE = (5 kg) times (g = 9.8 J/kg-m) times (0.150 m)

= 7.35 joules *each* turn

If we turn the crank 200 times we will dissipate 200 times 7.35, or 1470 joules.

Before turning the crank, we must weigh the strap, the cylinder, and the water in it, and take their temperatures. In the author's demonstration, the strap's mass was 20.95 grams, the cylinder had a mass of 122.25 grams, and the water in it measured 50.00 grams. All of these masses are accurate to within ± 0.05 gram, which is probably better than we need.

How many calories does it take to raise the temperature of this assembly by one degree Celsius? The water alone takes 50.00 calories. There are $122.25 + 20.95 = 143.20$ grams of copper. Since the specific heat of copper is 0.092, it takes 143.20 times $0.092 = 13.2$ calories to raise all the copper by one degree Celsius. We have rounded off the number (13.2) to the nearest tenth of a calorie.

There are other parts of the system (the thermometer, the air, the supports, etc.) that may also absorb some of the heat energy. We may guess at how much energy each of these elements absorbs, or we may neglect them entirely. In either case, there is a small uncertainty involved. It makes no sense to carry extra decimal digits when this additional uncertainty is probably more than one tenth of a calorie anyway.

We conclude that the water and copper require $50.0 + 13.2 = 63.2$ calories to rise by one degree Celsius.

We measured the temperature of the assembly before turning the crank to be 20.6°C. After 200 turns the temperature was 26.2°C, a rise of 5.6 degrees. From our calculation above, we know that it takes 5.6 times 63.2, or 354 calories, to raise the water and copper that many degrees. We have rounded off again, this time to the nearest calorie, because we feel that our accuracy is no better than that. The temperature change, for example, might easily have been as little as 5.5 degrees, or as much as 5.7 degrees, as well as we were able to read the thermometer.

We can then compare the 1470 joules that we say we dissipated in 200 turns of the crank with the 354 calories that we measured to be produced. Dividing 1470 by 354 gives a ratio of 4.15 joules per calorie.

We should not be dismayed that the value we found is not exactly equal to the "accepted" value of 4.186 joules per calorie. Rather, we should be pleased that, considering the uncertainties of the measurements, the result we got differs by less than 1% from the average of all the careful determinations by many good scientists.

If the circumference of the cylinder was really only 0.147 meter instead of 0.150, or the temperature rise was closer to 5.55 than to 5.6 degrees, we would have had a numerical result closer to the "true" value. Such is the nature of experimental physics. No result is ever absolutely precise. The best we can hope to get is agreement "within the experimental errors". In the present case, we certainly have done that.

The fact that once again we get substantially the same value as others have for the conversion of joules into calories demonstrates that heat is a form of energy. The conversion factor is always the same, no matter when or where or how the energy transfer is made.

The "accepted" value for this conversion factor, the average of many careful experiments done in different ways, is

one calorie = 4.186 joules

Heat energy and kinetic theory

The modern view of matter is that everything is composed of very large numbers of very tiny objects called *molecules*. These molecules are supposed to be moving about rapidly, colliding frequently, and presenting a rather chaotic picture when viewed on a small enough scale.

If the molecules are fairly far apart and able to move independently of each other, the material they make up behaves as a *gas*. If they are in close contact, but still able to slide easily past each other, the material is a *liquid*. If they are locked together by fairly firm attractions, the material is a *solid*. Even in the solid state, the molecules are each able to jiggle around, oscillating about their stable positions.

According to the kinetic theory of matter, the heat energy of an object is precisely the mechanical energy of all the molecules in it. The more energy the molecules have, the hotter the object will be. This mechanical energy is largely in the form of kinetic energy, especially for gases. In the case of liquids and solids, considerable amounts of elastic and electrical potential energy are also involved.

The latent heat needed to change the material from one state to another represents the energy needed to break the bonds of attraction between the molecules. So the latent heat is largely the collective potential energy of the molecules.

The temperature of a material, according to this theory, is related to the average kinetic energy of the molecules. The faster the molecules are moving on the average, the higher the temperature.

The evidence for the correctness of kinetic theory is all indirect. Nobody has ever seen a molecule of water. But a scientist can use kinetic theory to calculate the specific heat, or other thermal properties, of a substance. When these predictions come out right again and again, our belief in the validity of the model becomes ever stronger.

Absolute zero An interesting question arises out of the kinetic theory, as it relates to heat energy and temperature. Suppose we cool a substance until all the molecules are in their lowest possible energy state. The kinetic energy would be practically zero, and the potential energy would have its minimum value. This would be the lowest temperature possible. We could not cool the material any lower.

Another way to approach this idea is through the law of Gay-Lussac. This law, you will recall, states that a gas will expand in proportion to any rise in temperature. Conversely, a gas will contract if the temperature is

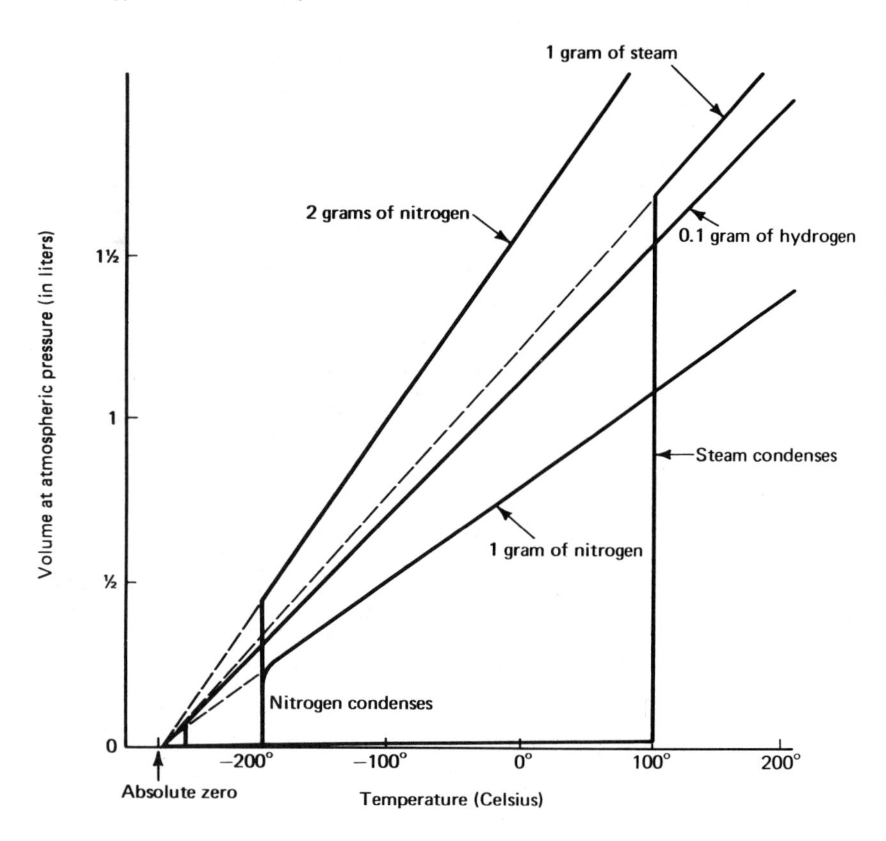

Figure 7-6 The volume of various amounts of different gases. All the lines extrapolate to zero volume at absolute zero temperature.

lowered. By examining the graph of the volume of a gas versus its temperature, we can extrapolate the line back to a temperature cold enough that the gas will have no volume at all!

All real gases will condense into a liquid at a higher temperature than the one at which the volume extrapolates to zero. But it is interesting that this temperature seems to be the same for all the gases that we know about, in whatever quantity and at whatever pressure the graph was measured.

The temperature at which the volume of all gases extrapolates to zero is 273.16° below zero on the Celsius scale. This is also the lower limit to the temperature that any scientists have been able to produce in the laboratory. This temperature is known as *absolute zero*.

No experiment has ever attained a temperature of exactly absolute zero. It is probably not possible. But it has been claimed that a temperature within

1 microdegree (one one-millionth of a degree Celsius) of absolute zero has been achieved. Since nearly all materials are solid at this temperature, it is not easy to make a reliable thermometer this cold.

There is a scale of temperature called the Absolute or Kelvin scale that uses Absolute Zero as its fixed point. The degrees on this scale are the same magnitude as on the Celsius scale. Thus the melting point of ice is +273.16 on the Absolute scale, and the boiling point of water is +373.16°K ("degrees on the Kelvin scale").

Glossary

absolute zero—The lowest possible temperature. At this temperature all molecules are in their lowest energy state, all gases will have condensed. No experiment has ever reached exactly absolute zero of temperature, but we can get very close to it.

calorie—The amount of heat energy required to raise 1 gram of water from 15° to 16° on the Celsius scale. Measurement shows that 1 calorie is equivalent to 4.186 joules.

Celsius scale (formerly called "Centigrade")—The scale of temperature on which water freezes at 0° and boils at 100°.

Davy, Sir Humphrey—English scientist (1778–1829) who showed, by melting two pieces of ice by rubbing them together, that heat energy by itself is not conserved.

evaporation, heat of—The heat energy required to change 1 gram of a substance from the liquid state into gas. The heat of evaporation of water is 540 calories per gram.

extensive quantity—A quantitative property of the state of a system that doubles when you double the size of the system. Examples of extensive quantities are mass, energy, volume, electric charge.

fusion, heat of (also called "heat of melting")—The heat energy required to change 1 gram of a substance from a solid to a liquid state. The heat of fusion of ice is 80 calories per gram.

heat energy—A form of energy that shows itself, for example, by raising the temperature of water and other substances. According to the kinetic theory, heat energy is the mechanical energy of a substance's molecules in random motion.

Helmholtz, Hermann von—German physicist (1821–1894) who first proposed the idea that heat is a form of energy, and that total energy is conserved.

intensive quantity—A quantitative property of a system that does not change its value when the system is doubled. Examples of intensive quantities are temperature, position, pressure, voltage, and speed.

Joule, James Prescott—English physicist (1818–1889) who first measured the number of joules in a calorie. The fact that this number is always the same, when measured carefully, shows that heat is a form of energy.

Kelvin scale (also called "absolute scale")—The scale of temperature in which absolute zero is at 0°, and the melting point of ice is 273.16°. The degrees on the Kelvin scale are the same size as degrees on the Celsius scale, but the starting point is shifted.

latent heat—The heat energy required (or released) to change the state of a system, such as to melt a solid or boil a liquid, without changing its temperature.

Rumford, Count, Benjamin Thompson—American-born (1753–1814) expatriate engineer and scientist who first showed that the old caloric theory of heat could not account for the unlimited amount of heat energy that can be produced while boring out a cannon.

solution, heat of—The heat energy released when 1 gram of a substance is dissolved in water or some other solvent.

specific heat (also called "heat capacity")—The heat energy required to raise 1 gram of a substance 1 degree Celsius. The specific heat of water is 1, by definition. The specific heat of copper is 0.092 calories per gram per Celsius degree.

sublimation, heat of—The heat energy required to change 1 gram of a substance directly from the solid state to the vapor.

temperature—An intensive property of a system that describes how hot or cold it is. Any two systems left in contact long enough will come to the same temperature. This fact enables us to measure the temperature of any system by putting it in contact with some thermometer. In the kinetic theory, temperature is identified with the average kinetic energy of the molecules.

thermometer—A device that changes its outward appearance in some way when its temperature changes, so that it can be used to measure temperature. Examples of thermometers are: an expanding column of mercury, a bimetallic strip, a thermocouple (whose voltage output changes with temperature).

Questions

1. Give five examples of extensive quantities and five examples of intensive quantities. Try to make them different from the examples given in the text.
2. Which of the following quantities are extensive and which intensive? Force, EPE, height, surface tension, weight, electric current, number of molecules.
3. Draw a graph of heat energy added versus temperature for 100 grams of ice, starting at −40°C.
4. Weigh a cupful of warm water. Remember to subtract the weight of the cup. Measure its temperature. Do the same for a cupful of cold water. Mix the two together and measure the final temperature. Is this result what you would have predicted, using the law of conservation of energy?
5. Describe a way to measure the heat of vaporization of water.

6. Describe a way to measure the specific heat of copper.

7. In the experiment to measure the conversion of joules to calories, we neglected the heat absorbed by the air, the thermometer, the supports, etc. If a maximum of 10 grams of material shared the heat energy with the water and copper parts, and if its specific heat averaged 0.1, how much difference would this make in the final result? Would the number of joules per calorie be larger or smaller than what we got?

8. Consider the following information: Things feel warm only when they are at higher temperatures than our bodies. They feel cool when their temperatures are lower than 37°C. Good heat carriers, such as metals, tend to feel hotter (or colder) at a given temperature than poor heat carriers, like wood. Would you say that our feelings of "hot" or "cold" are related to the *rate* of heat energy transfer to or from our bodies?

9. What would you say the kinetic theory predicts about the rate at which heat energy spreads through still air (no wind blowing)? How should this rate change with the temperature of the air? What would be the difference if we substituted for air a gas with much heavier or lighter molecules?

10. Can there be any temperature colder than absolute zero? What would the kinetic theory say?

11. A British Thermal Unit (Btu) is defined as the amount of heat energy needed to raise the temperature of 454 grams of water 5/9 of a Celsius degree. How many calories is 1 Btu? How many joules? Why do you suppose such peculiar amounts are used in this definition?

12. Suppose you have a cooking stove that supplies heat energy at a constant rate. Your object is to heat each of the following systems from 0°C to the normal boiling point of water. Which of these systems will reach that temperature first? Which will be last?
    (a) 100 grams of water in a 100-gram copper pot
    (b) 50 grams of water in a 500-gram copper kettle
    (c) 50 grams of ice in a 100-gram copper pot

13. What was proved by the experiment to measure how many joules there are in a calorie?

14. Hammer a nail into a block of wood. Feel the nail both before and after the hammering. Where did the heat energy come from?

15. It is observed that a human body will drop in temperature one Celsius degree each 200 minutes after death. Estimate how many calories it takes each day to keep a living 70-kilogram man (you may assume that most of that mass is water) from having his body go cold.

8

chemical energy

ATOMS AND PHOTOSYNTHESIS

Introduction and summary

There are some processes we are all familiar with that seem to release a great deal of energy. We can burn a stick of wood. Great quantities of heat and light energy are produced in the fire. We can eat some food. Our body finds from it the energy to do all kinds of work, and also to keep itself warm. An electrical battery can be taken off the shelf and connected to a circuit. It then releases the energy to turn on lights, or turn over an automobile engine, or produce sound from a radio.

All of these processes have some common features. Obviously, they all release energy. Where the energy came from is not so obvious. None of the forms of energy we have studied so far were present in large quantities before the reactions took place.

But the energy source is not inexhaustible. The fuel for the fire can be burned only once. The food cannot be eaten again. The battery must eventually be recharged or replaced. This suggests that there is a form of energy associated with the unburned fuel, the uneaten food, the charged state of the battery.

We call this new form of energy *chemical energy*.

There is a model that explains chemical energy in terms of the attractive forces between *atoms*. In this model the burning of a lump of coal consists of trillions of trillions of carbon atoms combining with oxygen molecules from the air to form a molecule of carbon dioxide. The formation of each such molecule releases a small amount of energy. The overall energy release from all these molecules results in the fire we see.

Chemical energy represents a very efficient way to *store* energy. The amount of energy stored per unit mass is much higher than in other forms of energy, such as heat or kinetic energy, that we have studied. The energy stored this way can be kept indefinitely, so long as the fuel is not allowed to combine with oxygen (or whatever). But the chemical energy can be released quite easily when we want it, as by starting a fire.

It is also possible to convert other forms of energy *into* chemical energy. We can recharge a storage battery, for example. The most important process to form chemical energy on Earth is the conversion of the energy of sunlight into sugars and starches by living plants. This process is called *photo-synthesis.*

The food we eat, the wood we burn, received their chemical energy from photosynthesis in plants that were recently alive. The energy we use from burning *fossil fuels*—coal, oil, and natural gas—was stored by plants that lived millions of years ago.

The amount of chemical energy stored on Earth in forms like these is very great, but it is not infinite. If we consume it faster than it is being replaced, we must sooner or later use it all.

Heat of combustion

Demonstration: A match is struck and fire is set to a piece of paper. Several forms of energy are released. Most noticeable are light and heat energy. Can you identify any others?

What was the source of the energy released in the burning of the paper?

The paper was at rest on the table, not under stress, before it was set afire. So there was no mechanical energy to be released.

A small amount of mechanical energy was supplied by the demonstrator in striking the match. But this was surely a small amount compared to the light and heat that we saw emitted. And we know very well that we could have burned ten times as much paper, and released that much more heat and light, from the striking of the same match.

The paper was at room temperature before we burned it. The ashes and gases left over after the burning are, if anything, warmer than the paper was before. So heat energy in the paper cannot be the source of the energy released.

Perhaps there is a new form of energy present in the unburned piece of paper. If there is, we have to calibrate this new form.

We can do the following experiment. Weigh a certain amount of paper to be burned. Put the paper inside a container, surrounded by a water bath. This is designed so that all the energy released in the burning is absorbed by the container walls and by the water. We should be careful to provide enough air inside the container so that the burning is complete.

We measure the temperature of the water before and after the burning. From the increase in the heat energy of the water, we can determine how much energy is released by burning.

It turns out that: (1) When the same weight of paper is burned, the same amount of energy is released each time; (2) when more paper is burned, more energy is released. The energy released goes up in proportion to the mass of paper burned.

The energy released per gram of material burned is called the *heat of combustion* of the material.

The heat of combustion of paper is about 4000 calories per gram. The exact value depends on the kind of paper. This is a lot of energy for such a small mass. Remember that it takes only 540 calories to boil 1 gram of water.

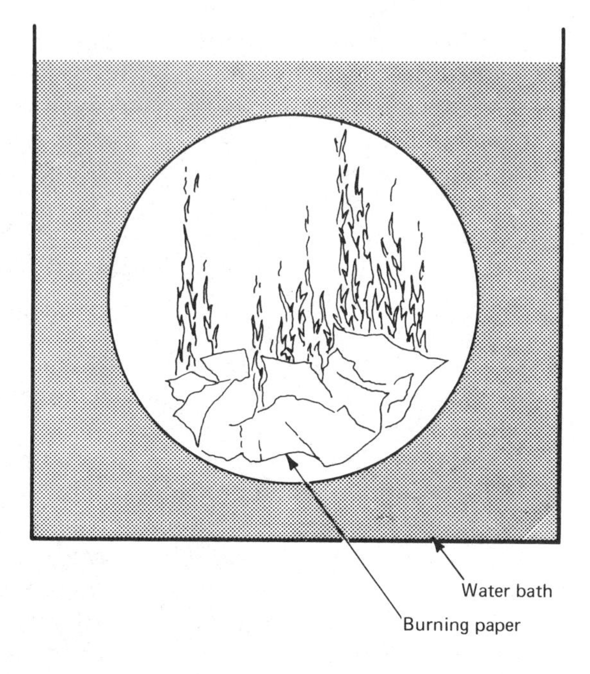

Figure 8-1

Chemical energy The evidence points to a new form of energy associated with the paper when it has not yet been burned. This energy is released in the process of burning it.

Any material that can be burned—wood, oil, coal, and so forth—possesses some of this form of energy. The materials left over after the burning—ashes, smoke, water vapor, carbon dioxide—do not have energy of this kind left in them. At least they do not have as much of it as the materials we started with. The original energy has been mostly converted into light and heat energy during the burning.

There is a simple model to explain burning that we were all taught in grade school. The material being burned combines with *oxygen* gas from the air. A new *compound* is formed, linking the fuel material to the oxygen. There is less energy in the compound than there was in the unburned fuel. The rest of the energy was released during the burning.

The final compound cannot be burned again. It already has all the oxygen in it that it can take. There is no more energy to be released.

There are reactions that, like burning, release energy, but do not involve oxygen. For example, the metal *zinc* can combine with the gas *chlorine*, emitting a bright flame as the compound *zinc chloride* is formed.

Burning is an uncontrolled release of energy. There are ways to make the same reactions proceed in a slow controlled manner. Instead of forming

> zinc chloride so explosively, by combining zinc and chlorine directly, we can let it form gradually at one terminal of a zinc-acid battery. The same total energy is released in the long run. In the battery, much of it is converted into electrical energy, rather than light and heat.

As already stated, we call this new form of energy *chemical energy.* A material is said to possess chemical energy if that energy can be released by combining the material with other substances, like oxygen, to form new compounds. We shall be more precise about defining chemical energy in the following sections.

Atoms and molecules

In earlier chapters, we mentioned the Kinetic Theory. According to this model, every material is composed of a very large number of very tiny *molecules*, moving and colliding, attracting and repelling each other, like passengers on a very crowded bus. Using this model, we are able to explain the pressure and the temperature behavior of any substance, particularly of gases.

There is a different kind of molecule for every kind of material. There are molecules of water, of salt, of sugar, of rubber, of chalk, and so forth. There may be millions of kinds of materials in nature. There must be an equally large number of kinds of molecules. There are thick handbooks, used by scientists, that are filled with their names.

A very great simplification can be made by supposing that there are smaller building blocks out of which the molecules are made. We suppose that there are such small objects, which we call *atoms*, but there are not many kinds of atoms. Following certain rules, however, we can put two or more atoms together to make compound molecules. Thus, there is virtually no limit to the number of ways atoms can be combined to make different kinds of molecules.

There are less than 100 different kinds of atoms found in nature. A few additional kinds can be created by man-made devices. The complete list of known kinds of atoms is given in Table 8-1 at the end of this chapter.

The mass of each kind of atom is given in *atomic mass units.* Six trillion trillion atomic mass units weigh 10 grams. An atom is very tiny indeed.

Only about thirty kinds of atoms are needed to make up most of the materials we see every day. There are different atoms for each of the common metals. There are atoms of iron, and copper, tin and aluminum. There are atoms of the precious metals like gold, silver, and platinum. There are atoms of metals that are nearly always found combined into molecules: sodium, potassium, magnesium, calcium.

The air and water around us, and the material of our bodies themselves, are largely made up of atoms of nitrogen, oxygen, hydrogen, and carbon. Most of the atoms in every living thing are of these four kinds. The materials that are made from living things, wood and oil, cotton and fur, are likewise composed largely of these four kinds of atoms.

A few nonmetallic atoms are fairly common. Silicon atoms are present in sand and rock. Sulfur, phosphorus, and chlorine play a small but important role in our bodies and those of other living creatures. The once-rare atom, germanium, on the border between being a metal or a nonmetal, is now widely distributed in transistors and other electronic divices.

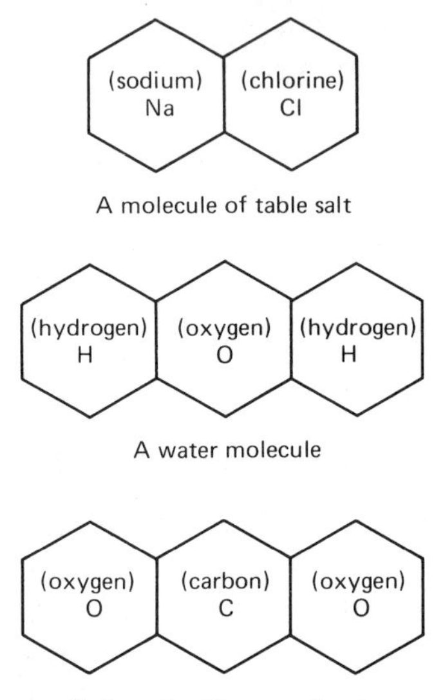

Figure 8-2

A molecule of table salt is made of 1 sodium atom and 1 chlorine atom. A water molecule has 2 hydrogen atoms and 1 atom of oxygen. The carbon dioxide gas that we exhale with every breath has molecules made of 1 carbon and 2 oxygen atoms.

In some cases a molecule can be a single atom. Molecules of the liquid metal mercury are single mercury atoms. Molecules of the light gas helium are single atoms of helium.

At the other end of the scale, many molecules found in living matter contain hundreds of atoms. One of the most complicated molecules whose complete structure is known, that of the hormone insulin in cows, has 777 atoms: 254 of carbon, 377 of hydrogen, 65 of nitrogen, 75 of oxygen, and 6 of sulfur. Molecules of DNA, the material that carries the genetic code in the cells of most living creatures, may have many thousands of atoms.

Nobody, however, has ever seen a single atom. Under some circumstances it may be possible to take a picture of atoms, not with visible light but with

X-rays or an electron microscope, but such a feat adds little to what we already know about atoms.

Our confidence in the atomic model does not rest on seeing atoms directly. The argument for this model is that it gives such a simple explanation of so many facts. The almost limitless variety of materials found in nature can be explained as combinations of a small number of kinds of atoms. The mass of each kind of molecule can be explained as the sum of the masses of the atoms it contains. The size and shape of the molecules can be explained in terms of the arrangement of the atoms within them.

The energy released in a chemical reaction, such as the burning of a piece of paper, can be explained in terms of the energy arising from the attractive forces between atoms. We shall explore this further in the next section.

The study of the ways in which atoms can be combined and arranged (and recombined and rearranged) to make different kinds of molecules is the science of *chemistry*. In a sense, this whole science is based on the atomic model for which there is very little direct evidence. The success of chemistry in making sense of a vast number of experiments, carried out over the past two centuries or more, argues for the validity of this model. By using the atomic model we are able to explain easily a great many facts that no other rival theory has ever approached.

Atoms and chemical energy The atomic model gives a simple explanation for the energy that is released in a reaction like the burning of a piece of coal.

We suppose that various atoms exert attractive forces on one another. These forces are basically electrical. They arise from the attractions and repulsions among the electrically charged parts within the atom. (We will not go into the details here.)

The fact that the forces are attractive means that there is less potential energy when the atoms are closer together than when they are farther apart. Energy must be supplied to pull the atoms apart. Energy can be released when the atoms come together.

We can sketch a curve of potential energy versus distance apart between two atoms or groups of atoms. To be specific, let one of them be an atom of carbon, such as is found in a lump of coal. Let the other group be a pair of oxygen atoms, a molecule of oxygen as it is found in the air.

When the two atomic groups are very far apart, they exert no forces on each other. It takes no energy to change their separation. So the left-hand part of the potential energy curve is a horizontal straight line. At large separations the atoms can get a little closer together, or a little farther apart, without any change in energy.

When the atoms get close enough to almost touch, there is an attractive force. This is represented by a dip in the potential energy curve. If the atoms

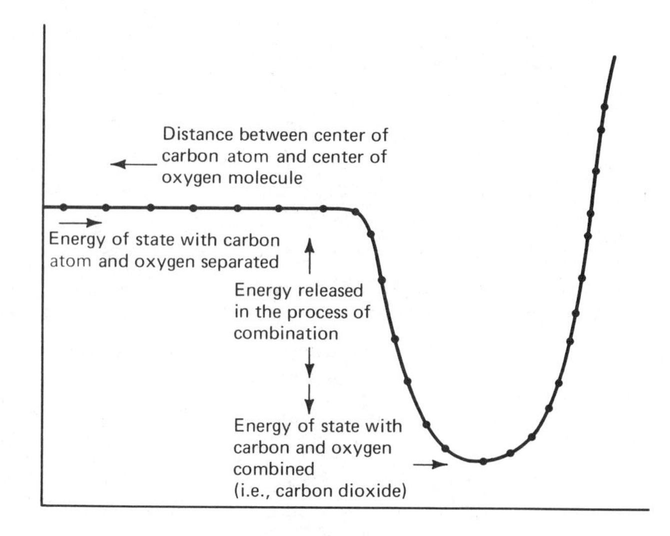

Figure 8-3 Potential energy due to the forces between atoms

are this close together, energy must be supplied to pull them apart. If they happen to pass this close to each other, after having been apart, their kinetic energy will increase to match the decrease in potential energy.

At even closer distances, the potential energy rises. The atoms have a finite size. Their centers cannot be brought closer than a minimum distance without squeezing the atoms. To do this would take more energy. So the force between atoms at close distances is a force of repulsion.

When a carbon atom and a oxygen molecule pass close to each other, they speed up a bit as the potential energy drops. Many times the two molecules will continue on past each other. The kinetic energy will return to its original value as the potential energy goes back to zero.

However, if some of the energy is dissipated during a collision, the carbon and oxygen molecules cannot separate again. The total energy of the two molecules would be less than what it takes to get very far apart from each other. The dissipation might take place by the release of energy in the form of light, or by collisions with another molecule.

Having lost some of their combined energy, the carbon and the oxygen molecules are stuck to each other. They now form a compound molecule, carbon dioxide.

The compound molecule can continue to dissipate energy, just as the mass on the spring did in Chapter 5. Eventually it reaches a resting condition with its energy near the minimum of the potential energy curve. The total energy released in the reaction is the difference between the initial potential energy, when the molecules were far apart, and the potential energy at the final resting point.

When a lump of coal burns, the process just described takes place trillions of trillions of times. Each time a molecule of carbon dioxide is formed, a certain amount of energy is relased. The number of molecules in a gram of carbon dioxide is a large but very definite number. Whenever a gram of carbon dioxide is formed, that many molecules have released their energy of combination. So the amount of heat and light released in the formation of 1 gram of carbon dioxide is always the same. If you burn twice as much coal, you form twice as many molecules of carbon dioxide, and so will release twice as much energy.

Definition of chemical energy We can now define the *formation energy* of a molecule. We start by defining the formation energy of a free single atom, of any kind, as zero. The formation energy of a molecule is the difference between the energy of the molecule in its resting condition, and the energy of the state in which all the atoms that make up the molecule are running free. Put another way, the negative of the formation energy of a molecule is the energy released if all the free atoms that make up the molecule were to come together at once to form the molecule.

The formation energy of any stable molecule is a negative number, less than zero. If the formation energy were positive, energy could be released by letting the molecule fly apart into the atoms that make it.

The *chemical energy* of any material is the sum of the formation energies of all the molecules that make it.

Thus, we see immediately that the chemical energy of a lump of coal, plus the oxygen gas that it can combine with, is higher than the chemical energy of the carbon dioxide formed by burning it. The difference is just equal to the amount of heat and light energy released in the burning process.

In a similar way the energy released in any chemical reaction is the difference between the chemical energy of the materials that go into the reaction minus the chemical energy of the products formed that remain after the reaction is complete.

The storage of energy

The storage of energy in a convenient form, to be used when needed, has always presented a problem. The main difficulty has been to keep the energy from being dissipated while it is being stored.

Kinetic energy is easily dissipated by frictional forces. A modest amount of kinetic energy can be stored for a short time in a device like a *flywheel.* A flywheel is a massive rotating wheel that energy can be supplied to or withdrawn from. A potter's wheel is an example. Unless extra special care is taken, the flywheel can dissipate its energy in a short time through friction at the bearings that support it, and with the air around it.

Gravitational potential energy can be stored by lifting weights. On a large scale, it can be done by pumping water above a high dam. Energy stored in this form can be kept for long periods and in large amounts. Converting it into other forms is sometimes a complicated and difficult job.

Elastic potential energy can be stored by winding up a spring. The ordinary wristwatch is an example of energy stored in this way. Very large springs are not common. The amount of elastic potential energy that can be stored in one place is not large.

Electrical energy can be stored in a battery or similar device. The "shelf life" of a typical dry cell battery is a few months or years. The amount of energy stored is modest. To store enough energy to run a small car for one day requires hundreds of kilograms of battery material.

Heat energy can be stored for a limited time in a heat reservoir, such as the boiler of a steam engine. The heat has a tendency to leak out through the walls of any such heat storage device. Good design can slow the rate of heat loss, but never completely eliminate it. A heat reservoir that can hold its heat energy for more than a few days is considered to be very well insulated.

Nobody has yet designed a way to store energy in the form of light or sound for more than a few seconds.

By contrast to the above methods of energy storage, chemical energy offers immense advantages. The chemical energy stored in a lump of coal can be (and has been) kept for millions of years, as long as the coal is not allowed to burn by combining with oxygen. The energy that can be stored in this way is unlimited. There is enough coal in the ground to supply all present human energy needs many decades. The energy can be easily converted, when it is needed, by simply burning the coal.

The reverse process, in which energy is converted from other forms *into* chemical energy, can also take place. One way to do this is to pass an electric current through a sample of ordinary water. The electrical energy driving the current can separate the hydrogen from the oxygen atoms in each molecule of water. Hydrogen gas collects at the negative electric terminal and oxygen gas at the positive end. The process of separating molecules into their elements by means of an electric current is called *electrolysis*.

The hydrogen separated by electrolysis can be stored, out of contact with oxygen, until the energy is needed. The chemical energy so stored can be released by setting fire to the hydrogen. It burns with a fierce blue flame. The energy released in the burning of hydrogen, 143,000 joules per gram of hydrogen burned, is by far the highest heat of combustion of any material known.

The chemical energy can also be released in a more controlled manner than by burning. Hydrogen and oxygen (or many other reactants) can be combined in a *fuel cell*, which is essentially an electrolysis system run backward, to convert the chemical energy directly into electrical energy.

HOW TO START A FIRE

You have certainly noticed that a lump of coal in contact with oxygen in the air does not immediately break forth into flames. You have to *set fire* to it before it will burn.

The explanation we offer is that most of the carbon atoms are hidden deep within the lump of coal and out of contact with the air. It is certainly easier to start a fire with powdered charcoal, or with thin paper, where far fewer carbon atoms are hidden in the interior.

Most metal atoms can combine with oxygen in a manner similar to carbon atoms. Usually we do not think of metal "burning," although the chemical reaction very much resembles burning.

Atoms on the surface of a metal like aluminum can combine with oxygen to form aluminum oxide molecules. These molecules remain stuck to the surface of the metal. When all the surface aluminum atoms have been *oxidized* in this manner, the reaction stops. The oxygen cannot get to the atoms below the surface.

Pure iron metal, as opposed to combinations (like steel) of iron and other materials, can combine completely with oxygen. The iron oxide that forms has a reddish color that we call *rust*. In the case of pure iron, the surface layer does not protect the interior, and the iron can rust all the way through. When the iron is finely divided, as in iron filings or "steel wool," the metal can be set afire and burns quite well.

The difference between slow oxidation, as in rusting metals, and burning has to do with how easily the atoms of the material to be burned can be shaken loose from their surroundings. Often it is a matter, as in the lump of coal, of breaking the atom off the surface of a solid.

In other cases, as in the burning of hydrogen gas or liquid petroleum, it is necessary to break the atoms loose from the molecules they are already in. They can then combine with oxygen to form molecules with lower total chemical energy than the molecules we started with.

According to the kinetic theory, the higher the temperature is, the more violently the molecules, and the atoms within them, are moving about. When this motion is energetic enough, the atoms can break loose from the molecule or solid they are in. So a certain temperature, depending on the material, must be reached before burning can begin. Once the reaction starts, enough energy is released in the burning itself to keep the reaction going.

The energy needed to shake an atom loose from its surroundings, so that it can participate in a reaction, is called the *activation* energy for that material. The temperature at which a sufficient number of atoms have the activation energy, so that the reaction can proceed, is called the *kindling* temperature for that process.

The energy in our bodies

Our bodies need energy for all the things we do. We need energy to walk, to talk, to do work, even to breathe. We need energy to keep our bodies warm. We use energy for such internal functions as pumping our blood around, fighting disease, and growing. The energy needed by our bodies for all these purposes can be more than 10 million joules per day.

We get this energy from the food we eat. For example, the combustion of 1 gram of sugar releases 15,000 joules of chemical energy. Nearly every item of food we can eat is able to release some energy by being "burned."

> The "Calories" usually referred to in diet discussions are really kilocalories, or 1000 of the calories defined in Chapter 7. Sometimes a capital C is used to distinguish the kilocalorie. The 15,000 joules released in combustion of a gram of sugar, is just less than 4000 calories, or 4 kilocalories. A nutritionist would say that the dietary value of the gram of sugar is about 4 Calories.
>
> To avoid confusion in this text, we will normally refer to chemical energy in terms of joules.

The body does not burn its food by direct combination with oxygen. The energy is released in a controlled manner. Usually the molecules go through the changes in several steps, releasing some of their energy at each step.

The final products of all these changes, mainly carbon dioxide and water, are the same as if the food had been directly set on fire. The total energy released, which is the difference in chemical energy between the initial and final states, is the same whether the process goes by direct burning or controlled changes.

When we eat food with high-energy content, much of it gets changed into sugar in our stomachs. In this form it can pass through the stomach walls and dissolve in the blood.

If the body does not need the energy right away, the sugar is changed into fat. A fat is a substance with high chemical energy that does not dissolve. The body can store large amounts of chemical energy in the form of fat. When the energy is needed, the fat can be changed back to sugar again.

A cell that needs energy to perform its work absorbs sugar from the blood. Inside the cell the energy from the sugar is transferred to molecules of the chemical adenosine tri-phosphate (ATP).

ATP is a compound that has a high level of chemical energy. When it is formed energy is held in temporary storage for use by the cell. When the energy is released, the ATP changes to molecules of lower energy.

In the process of transferring energy from the sugar to the ATP, oxygen is consumed. In effect, the sugar has been "burned" within the cell. But the burning releases little heat and no light. Most of the energy released in the sugar burning is converted directly into the chemical energy of ATP.

How the energy is finally used depends on the type of cell. In a muscle cell, the energy released by the ATP enables the molecules of the cell to curl up and shorten in length. The cell contracts and pulls with it the neighboring cells. We can say that the elastic potential energy of the cell increases, just like that of a coiled spring.

The energy of the coiled muscle cells can be converted directly into other forms of mechanical energy. We can throw a ball or lift a weight or push a bicycle pedal.

Thus we see how the body can use chemical energy in the form of food to produce mechanical energy. The chain from raw food to sugar, to fat, back to sugar, to ATP, and then to muscular energy is a remarkably efficient way to use the energy in the food we eat.

Energy is not the only useful thing we get from food. Food also contains building materials—proteins, vitamins, minerals—that we can use to make new living matter. Thus, food is necessary for us to grow and to maintain or replace existing cells. These processes usually require energy as well as building materials. So the energy content of our food plays a role even when the prime function is body-building.

Photosynthesis

We now explore the question of how the chemical energy got into the food in the first place.

When we eat a piece of steak, the chemical energy we receive comes mostly in the form of fatty tissue that the steer's body had been storing for its own use. The steer, of course, got the energy for its fat from the food it ate. The steer's diet is largely grass and other plant material. So the chemical energy in the meat we eat comes originally from plants.

We can also eat vegetables and get the chemical energy directly from the plants. So we see that all the energy in our food comes to us, directly or indirectly, from plants.

Plants need energy too. They need energy to grow and to point toward the sun. Some very small plants, such as one-celled algae, need energy to swim.

Plants can get energy from sunlight, as we shall see, but they need to store some of it. The energy so stored can be used at night and on cloudy days when there is not enough sunlight. The stored energy can also be used by sprouting seeds, in the days before the young plant is able to gather enough energy on its own from sunlight.

The chemical energy stored by plants goes mainly into the form of plant *starches* and sugars. It is fortunate for the entire animal kingdom that plants store a great deal more chemical energy than they will ever use. Animals cannot use sunlight directly. They derive their energy by eating plants or by eating the meat of other animals who have eaten plants.

The process by which plants convert the energy of sunlight into chemical energy is known as *photosynthesis*. In a sense photosynthesis is the opposite of burning. This process takes water and carbon dioxide, the molecules that are produced by the burning of fuel, and turns them into sugars and starches and other materials with high chemical energy. In the process, oxygen gas is released into the air.

When we breathe, we take in oxygen and exhale carbon dioxide. When we burn fuel the same overall changes take place. Animals—and plants, too, when they need to use energy—are constantly using the oxygen in the air and producing carbon dioxide. It is clear that all creatures owe a great debt to living plants.

The steps by which photosynthesis takes place, though complicated, have been worked out by biochemists. The process takes place in a part of plant cells called a *chloroplast*. It is mediated by the green substance chlorophyll, whose presence gives its color to leaves and grass.

At a crucial stage in the photosynthesis process, energy must be put in to allow a certain molecular change to take place. This energy is supplied by sunlight. The energy coming to Earth from the sun, in the form of visible light, is absorbed directly by the plant in the course of its photosynthesis reactions. The energy thus absorbed is changed into the chemical energy of plant sugars and starches.

The sources of our energy

The grand result of the whole cycle of photosynthesis and eating and breathing can be summarized simply. The molecules like oxygen and carbon dioxide and chlorophyll return at last to their original state. Though energy was stored for a time in some chemicals, it was not supplied nor consumed by them. Energy was absorbed by the plants from sunlight. And at a later stage it was released by the burning, direct or indirect, of food and fuel.

The fuel we burn for our energy needs comes generally from the parts of plants where chemical energy has been stored. Sometimes, as when we burn logs, we use energy stored by a tree that was recently alive. In other cases we use the energy stored by plants, and animals, that lived millions of years ago.

Fossil fuels, coal and petroleum and natural gas, derive from the remains of creatures that lived and died long before man appeared on Earth. Coal was formed from ancient trees and plants that fell and were buried. Oil and gas are derived from the fatty tissue of animals, including insects and sea creatures, that were similarly buried. Millions of years of being pressed down by the earth above them have shaped these tissues into their present form.

When we take these fossil fuels from the ground and burn them for our energy needs, we are consuming in a brief time what it has taken the natural cycles of the earth hundreds of millions of years to store up. When they are finally all used, they cannot be replaced on any human time scale.

This chapter should have made clear by now that nearly every source of energy we use regularly on Earth, the food we eat, the fuel we burn, depends on the energy delivered by sunlight. The photosynthesis in plants, trapping the sunlight and storing its energy in plant chemicals, now and in the past, is the major channel for most of the energy we use today.

There is some energy that comes from the sun but does not pass through the photosynthesis channel. Sunshine falling on lakes and oceans can evaporate some of the water and warm the water molecules. These warm molecules then rise in the air, converting the heat energy into gravitational potential energy. Eventually they may condense again into liquid water droplets that form clouds and fall again as rain. The gravitational energy released by rainwater running downhill can be used to turn wheels and can also be converted into electrical and other forms of energy.

The energy of the winds and of ocean waves derives also from the action of sunlight. If we use this energy to drive windmills or sailboats, we are again benefitting indirectly from the sun.

There are some sources of energy that do not derive from sunlight. There is the energy stored in the earth itself. The heat energy of the hot material deep within Earth, which drives geysers and volcanos, can be tapped to run *geothermal* power plants. The kinetic energy of Earth's rotation, which is responsible for the ocean tides, is another reservoir of energy that might be turned to our needs. Nuclear energy, which will be discussed in a later chapter, is also stored in some of the materials found on Earth. None of these non-solar sources of energy have been used extensively until very recently.

Humans have used many seemingly different sources of energy in our history. We have used our own muscles, and those of horses and oxen. We have burned wood. We have trimmed sails on our boats. We have used water-wheels and windmills. We have burned coal and oil to drive our engines. It is remarkable that all these sources commonly used by men have derived their energy from a common source—our sun.

Table 8-1 THE ATOMS OF THE KNOWN CHEMICAL ELEMENTS

	Name of atom	*Symbol*	*Mass**	
1	Hydrogen	H	1.0	in water, living matter, stars
2	Helium	He	4.0	a rare gas on earth, but abundant in the sun and other stars
3	Lithium	Li	6.9	in rocks and minerals
4	Beryllium	Be	9.0	a rare metal
5	Boron	B	10.8	in boric acid
6	Carbon	C	12.0	in living matter, coal, diamonds
7	Nitrogen	N	14.0	in air, living matter
8	Oxygen	O	16.0	in air, water, living matter
9	Fluorine	F	19.0	in "Freon," used as a refrigerant
10	Neon	Ne	20.2	neon lights
11	Sodium	Na	23.0	in rocks and minerals, common-salt
12	Magnesium	Mg	24.3	in rocks and minerals

*Masses of atoms are given in atomic mass units. Six trillion trillion atomic mass units have a mass equal to 10 grams.

Table 8-1 (Cont.)

	Name of atom	*Symbol*	*Mass**	
13	Aluminum	Al	27.0	a common metal; in clays
14	Silicon	Si	28.1	in sand, rocks, electronic devices
15	Phosphorus	P	31.0	in living matter, fertilizer
16	Sulphur	S	32.1	in sulfuric acid
17	Chlorine	Cl	35.5	in common salt, cleaning fluid
18	Argon	A	39.9	about 1% of the air
19	Potassium	K	39.1	in many salts, nerve cells
20	Calcium	Ca	40.1	in bones, chalk, limestone
21	Scandium	Sc	45.0	metals mixed into some forms of steel
22	Titanium	Ti	47.9	
23	Vanadium	V	50.9	
24	Chromium	Cr	52.0	
25	Manganese	Mn	54.9	
26	Iron	Fe	55.8	a common metal; in steel; in blood
27	Cobalt	Co	58.9	fairly common metals
28	Nickel	Ni	58.7	
29	Copper	Cu	63.5	
30	Zinc	Zn	65.4	
31	Gallium	Ga	69.7	used in electronic devices
32	Germanium	Ge	72.6	
33	Arsenic	As	74.9	
34	Selenium	Se	79.0	
35	Bromine	Br	79.9	used in photographic film
36	Krypton	Kr	83.8	a rare gas
37	Rubidium	Rb	85.5	fairly common metals, found in some minerals
38	Strontium	Sr	87.6	
39	Yttrium	Y	88.9	
40	Zirconium	Zr	91.2	
41	Niobium	Nb	92.9	
42	Molybdenum	Mo	95.9	
43	Technetium	Tc	98.9	man-made; not found in nature
44	Ruthenium	Ru	101.1	fairly rare metals
45	Rhodium	Rh	102.9	
46	Palladium	Pd	106.4	
47	Silver	Ag	107.9	a precious metal; used in coins and photographic film
48	Cadmium	Cd	112.4	fairly rare metals
49	Indium	In	114.8	
50	Tin	Sn	118.7	a common metal
51	Antimony	Sb	121.8	nonmetals
52	Tellurium	Te	127.6	
53	Iodine	I	126.9	
54	Xenon	Xe	131.3	a rare gas

Table 8-1 (Cont.)

	Name of atom	*Symbol*	*Mass**	
55	Cesium	Cs	132.9	fairly uncommon metals
56	Barium	Ba	137.3	
57	Lanthanum	La	138.9	
58	Cerium	Ce	140.1	
59	Praseodymium	Pr	140.9	
60	Neodymium	Nd	144.2	
61	Prometheum	Pm	(145)	man-made; not found in nature
62	Samarium	Sm	150.4	
63	Europium	Eu	152.0	rare metallic elements, very similar to each other; known collectively as "rare earths"
64	Gadolinium	Gd	157.2	
65	Terbium	Tb	158.9	
66	Dysprosium	Dy	162.5	
67	Holmium	Ho	164.9	
68	Erbium	Er	167.3	
69	Thulium	Tm	168.9	
70	Ytterbium	Yb	173.0	
71	Lutetium	Lu	175.0	
72	Hafnium	Hf	178.5	fairly uncommon metals
73	Tantalum	Ta	180.9	
74	Tungsten	W	183.9	used in cathode-ray tubes (as in TV or X rays)
75	Rhenium	Re	186.2	fairly uncommon metals
76	Osmium	Os	190.2	
77	Iridium	Ir	192.2	
78	Platinum	Pt	195.1	precious metals
79	Gold	Au	197.0	
80	Mercury	Hg	200.6	a liquid metal; used in thermometers
81	Thallium	Tl	204.4	an uncommon metal
82	Lead	Pb	207.2	a common metal
83	Bismuth	Bi	209.0	a fairly uncommon metal
84	Polonium	Po	(209)	
85	Astatine	At	(210)	
86	Radon	Rn	(222)	a gas
87	Francium	Fr	(223)	radioactive elements, found in association with uranium, or man-made
88	Radium	Ra	(226)	
89	Actinium	Ac	(227)	
90	Thorium	Th	(232.0)	a metal, possibly a future nuclear reactor fuel
91	Protactinium	Pa	231.0	man-made
92	Uranium	U	238.0	a metal, used as fuel in nuclear reactors
93	Neptunium	Np	237.0	man-made; not found in nature

Table 8-1 (Cont.)

	Name of atom	*Symbol*	*Mass**	
94	Plutonium	Pu	(244)	man-made; used in bombs; possible future use as reactor fuel
95	Americium	Am	(243)	man-made elements not found in nature, usually short-lived
96	Curium	Cu	(247)	
97	Berkelium	Bk	(247)	
98	Californium	Cf	(251)	
99	Einsteinium	Es	(254)	
100	Fermium	Fm	(257)	
101	Mendelevium	Md	(256)	
102	Nobelium	No	(254)	
103	Lawrencium	Lr		
104	Rutherfordium	Rf		
105	Hahnium	Ha		
106	(not yet named)			

Glossary

activation energy—The energy needed to shake an atom loose from its surroundings so that it can participate in a reaction.

adenosine tri-phosphate (ATP)—An energy-rich compound found in cells. ATP gets its energy from the oxidation of sugar brought to the cell by the blood. It releases its energy to enable the cell to do its work, for example, to enable a muscle cell to contract.

atomic mass unit (abbreviated amu)—The unit in which the masses of atoms and molecules are commonly measured. Thus, a carbon atom has a mass of 12 amu. It takes 600 trillion trillion atomic mass units to make 1 kilogram.

atoms—The extremely small particles that can be put together to build all the molecules of normal matter. Nobody can ever see an atom directly. They are so small that the waves of ordinary light will diffract around them. Our belief in the existence of atoms is based on the success of this model in explaining so many facts about, for example, chemistry or heat.

carbon dioxide—A gas found in small amounts in the atmosphere. It is produced whenever a carbon-containing material—such as coal, oil, wood, or sugar—reacts with oxygen to release energy. Carbon dioxide is produced in most burning. It is exhaled when we breathe.

chemical energy—The energy that a certain amount of material has by virtue of its state of chemical combination; the sum of the formation energies of all its molecules. The chemical energy of a material consisting entirely of atoms of the elements in their uncombined state can be taken as zero. If energy is released when the atoms combine, the energy of the resulting combination is less than zero, by just the amount of energy released. Although, by this convention, the chemical energy of any stable compound is a negative quantity, some are more negative than others. For example, a system of unburned wood plus oxygen gas has a higher chemical energy than the state of carbon dioxide, water vapour, and ash that remains after it has been burned.

chemistry—The branch of physical science that deals with the ways in which atoms can be combined and arranged to make different kinds of molecules.

chloroplast—The part of a plant cell in which photosynthesis takes place.

compound—A material composed of molecules containing atoms of more than one element.

electrolysis—The process by which water, or some other liquid, can be separated into its elements by passing an electric current through the liquid. Energy must be supplied to break up a stable compound such as water.

element—One of 106 now known substances whose molecules are made of a single kind of atom. No chemical process can reduce an element to any simpler substance. All other substances are made of combinations of atoms of the various elements.

fat—A compound, high in chemical energy, not soluble in water, which is used to store excess energy in the bodies of animals and humans.

flywheel—A massive rotating wheel in which kinetic energy can be stored for a short time.

fire—A chemical reaction in which elements and compounds high in chemical energy combine rapidly—usually with oxygen gas—to form compounds like carbon dioxide that are lower in chemical energy. The difference in chemical energy is released mainly in the form of heat and light.

formation energy, of a molecule—The difference between the energy of the molecule, in its resting condition, and the energy of the state in which all the atoms that make up the molecule are free. The formation energy of any stable molecule is a negative quantity; energy must be supplied to break it up. But some molecules have more negative formation energy than others.

fossil fuels—Coal, oil, natural gas; fuels derived from the fats and starches of animals and plants that died long ago. The energy-rich compounds in their bodies escaped consumption and have been preserved in the ground for millions of years. There may be enough fossil fuels on Earth to supply human energy needs for several generations. But they are not being replaced as fast as they are consumed, and so they must eventually run out.

fuel cell—A device in which a chemical reaction, similar to the burning of fuel, is allowed to proceed in a controlled way, releasing energy in an electrical form rather than as heat and light. In a hydrogen fuel cell, hydrogen and oxygen are kept apart, so they do not combine explosively, but only through the intermediary of a current passing through a water solution. This fuel cell resembles an electrolysis process run backward.

geothermal—Referring to energy derived from the internal heat energy of Earth. Geothermal energy sources include geysers and other places where hot material from within Earth comes near the surface. The origin of the heat energy of Earth is: (1) gravitational energy released when the rocks and dust that formed Earth first collapsed together; and (2) nuclear energy released when radioactive materials, like uranium, within Earth undergo decay.

heat of combustion—The energy released, per gram of fuel material, when a substance is burned. The heat of combustion of many carbon-containing materials—coal, wood, paper, sugar, oil—is about 15,000 joules per gram.

kindling temperature—The temperature needed to start a fire. At the kindling point the average mechanical energy of the molecules is high enough to break atoms loose from their surroundings so that they can combine with oxygen. The burning releases more energy that heats up the remaining molecules to keep the process going.

molecule—The smallest particle of material that can exist alone and still behave like that material. All normal matter is thought to consist of large numbers of very small molecules. Molecules themselves are made of one or more atoms bound together.

oxidize—To combine with oxygen, usually releasing energy. Oxidation can take place rapidly, as in a fire, or slowly, as in the rusting of iron.

photosynthesis—The process by which plants use the energy from sunlight to help them create energy-rich compounds, such as sugars, out of carbon dioxide and water. The net reaction is to reverse the oxidation of sugar—either by direct burning or by animal use of food energy. Carbon dioxide is removed from the air, oxygen is released back into it, and energy from the sun is transformed into the chemical energy of sugars and starches stored in the plant.

rust—The oxidation, or slow burning, of iron. Also, the reddish compound, iron oxide, that forms when iron rusts.

starch—An energy-rich compound containing carbon, hydrogen, and oxygen atoms, not soluble in water, which is used to store excess energy in plants.

sugar—An energy-rich compound containing carbon, hydrogen, and oxygen atoms, soluble in water, which is used to store and transport chemical energy in plants and animals.

Questions

1. What forms of energy appear when a piece of paper is burned?
2. Where does the energy come from when it is released in the burning of a piece of paper?
3. Why can't the paper be burned twice?
4. How much ice can be melted with the energy released in the burning of 1 gram of paper?
5. Suppose there were only 20 kinds of atoms, and that each molecule was made of exactly 2 atoms. How many different kinds of molecules could we make under these rules?
6. What makes us believe that there are such things as atoms and molecules?
7. How much energy can be stored in a kilogram of coal by lifting it to a height of 100 meters? How much energy is conveyed to the kilogram by setting it in motion at a speed of 100 meters per second? Compare this with the energy that can be released by burning 1000 grams of coal.
8. Why doesn't a piece of paper, in contact with air, break into flame spontaneously?
9. How can you keep a piece of iron from rusting?
10. Give three reasons why our bodies need energy. How are these energy needs supplied?
11. Give three reasons why plants need energy. How are these energy needs supplied? Under what conditions will plants use more oxygen than they produce?
12. Consider the process of a boy throwing a ball. Trace the energy used in this process from its origin in the sun until it is finally dissipated.
13. Consider the process of a battery-powered car being driven up a hill, coasting down, and then stopping. Suppose that the local source of electric power is derived from water flowing over a dam. Trace the energy used in this process from its origin in the sun until it is finally dissipated.
14. Name 10 different sources of energy that can be used to meet society's energy needs. What is the original source of the energy in each case?
15. Is there a limit to how much energy human society can consume per year? What sets this limit? (Hint: In the long run, all the reserve fuels—nuclear as well as fossil fuels—may be used up. After that we cannot consume energy any faster than it can be replaced.)

9

heat engines

THE SECOND LAW

Introduction and summary

Up to now, every time we have talked about changing between heat energy and mechanical energy, the transfer has always gone one way. Mechanical energy has been dissipated into heat energy. Is it possible to make the transfer go the other way? Can heat energy ever be converted into more useful forms?

The answer is yes, at least partially. There are devices that take heat energy as input and convert some of it into mechanical energy. Such a device is called a *heat engine*. An automobile motor is one type of heat engine. A steam turbine, such as is used by electrical power companies, is another. The fact that heat engines exist is proof that we really can extract useful forms of energy from heat sources.

But it is also true that the transfer from heat to mechanical energy is not as simple as the other way around. You can dissipate mechanical energy into heat by rubbing your hands together. Making a working heat engine takes knowledge and skill. The first steam engine was not invented (by a British engineer, James Watt) until the mid-18th century. Before then, there were no practical heat engines.

Is there a reason why energy flow is so much easier in one direction than the reverse? The law of conservation of energy gives no clue why this should be true. According to that law, transfer of energy in either direction is possible. That, we have seen, is correct. But a second law of heat science (or *thermodynamics*) is needed to explain why heat energy is harder to convert to mechanical energy than the other way around.

The Second Law of Thermodynamics can be expressed in several different ways. One statement of the law says, basically, that heat energy will not convert itself spontaneously and completely into mechanical energy. Another statement of the Second Law says that heat energy will not flow naturally from a cold source to a warmer one. These very different statements turn out to be equivalent. If either one were false, the other statement would have to be false also. There are other ways to express the Second Law. Most of them take the form that there are certain kinds of processes that simply do not occur by themselves in nature.

The great trend of nature, according to the Second Law, is all downhill.

Mechanical energy is dissipated into heat energy, and cannot be completely recovered. Hot and cold substances, when mixed together, come to the same lukewarm temperature, and don't unmix again. Air, water, and the environment are readily polluted, and can be cleaned up only at great expense. We can scramble an egg easily, but we don't know how to unscramble one. In the long run, this law predicts, everything comes to equilibrium at the same temperature, uniformly mixed, and spread out all over.

According to the kinetic theory we have argued before, on a molecular scale there is no difference between heat energy and mechanical energy. Heat energy *is* the mechanical energy associated with the random motion of the molecules. Where does the Second Law fit into such a theory? We can argue that the Second Law is really a law of *probabilities*. It is far more likely, since there are so many more combinations possible, that the molecules of a substance will move in random directions (which we would call "heat") than that large numbers of them should all move together (which we might consider as "mechanical" energy). When sound is produced, it is more likely to be "noise" than "music." On the molecular scale, the Second Law can be expressed in the idea that it is much more probable to produce disorder from order in molecular motions than the other way around.

C.P. Snow, the British novelist and scientist, in his essay, "The Two Cultures," has expressed the opinion that every educated person should be as familiar with the Second Law of Thermodynamics as with the plays of Shakespeare. It is the content of this chapter to acquaint the student with the meaning and implications of the Second Law.

Heat engines

Demonstration: A simple heat engine. We make use of a toy "dipping-bird" that can be purchased in novelty shops. Inside the "bird" is a liquid, such as methylene chloride, whose boiling point is just above room temperature. The "body" is mounted on a pivot, with most of the liquid in the "tail." The action is started by rubbing the tail with your hand, or by wetting the "beak" with cool water. This sets up a temperature difference between tail and beak.

The bird still sits stationary after rubbing or wetting. Then, quite suddenly, the beak tilts downward and drops into the "drinking cup." Immediately as it reaches the cup, the tail tilts back down again. The bird comes to its original position, where it remains for many seconds, until it is ready to repeat the cycle. This repeated action can continue for hours. If the tail is kept warm and the drinking cup full of cool water, the action can go on "forever."

How does it work? The liquid in the tail, which is kept warmer than the rest of the toy, keeps evaporating, turning into vapor. This vapor condenses in the beak, which is kept cooler. Some of the vapor collects in the tail, which is so warm that it does not condense there. The warm vapor in the tail builds up to a higher pressure than the cool vapor in the beak. This means that energy can be released if the vapor in the tail expands and forces some of the liquid up the neck of the bird. This is what happens.

When enough liquid has been transferred from tail to beak, the system gets unbalanced and tips over. Once it has tipped over, the excess liquid in the beak can run down the neck of the bird and back into the tail. The tail tips back down again, and the cycle starts over.

This toy is actually an example of a *heat engine*. It is a device for converting heat energy into more useful forms, such as mechanical energy. This toy could be used, for example, to wind up a clock or ring a bell. It has several features in common with all heat engines, which we should notice here.

(a) *Heat energy must be supplied to keep it going*. In this case we must provide the latent heat of vaporization to the fluid. This comes from the heat

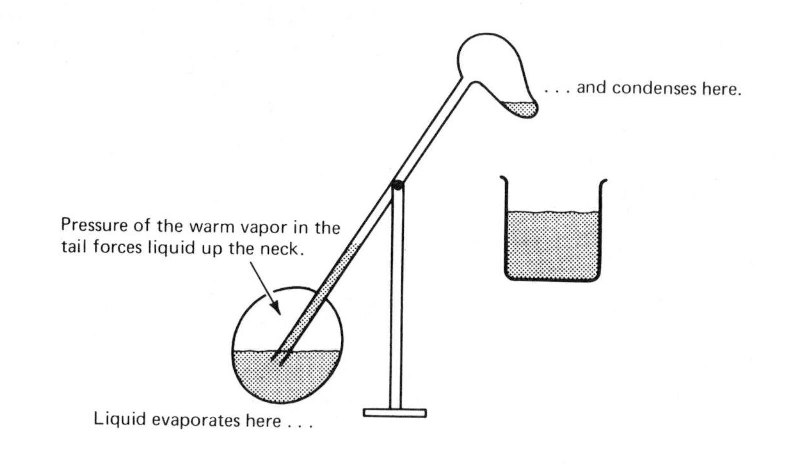

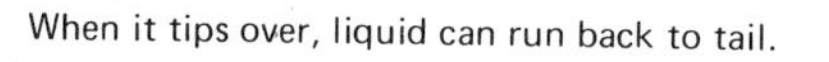

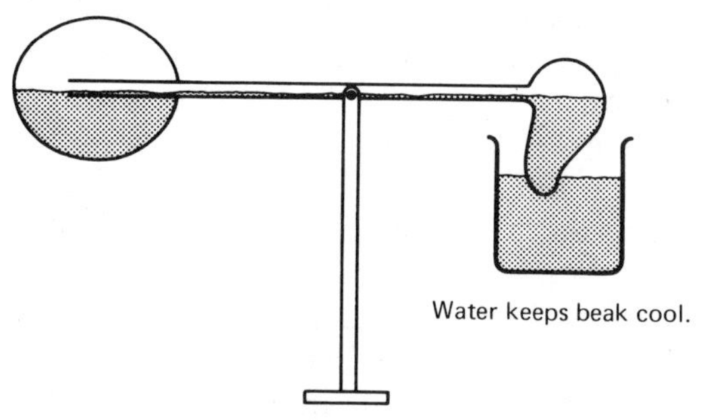

Figure 9-1 How the "Dipping-Bird" toy works

of our hand, or sunlight, or simply the warm room air. The toy will not work in a very cool place.

(b) *Mechanical energy appears, where none was present before.* Gravitational potential energy appears first, when the liquid is transferred from the lower tail to the higher beak. Then this GPE is converted into kinetic energy, as the beak moves downward and the tail up.

(c) *A temperature difference must be maintained.* The beak must be cooler than the tail, so that more vapor will condense in the beak. The action of the toy will stop if we dip the tail in the same cool water that the beak touches.

(d) *Much of the heat energy is not converted into mechanical energy, but is simply transferred from the warmer to the cooler region.* When the vapor condenses in the beak, the latent heat of vaporization is released. This goes largely to warm up the water that sticks to the outside of the beak. (Usually the beak is covered with cotton or felt to soak up some of the water from the cup.) More cool water must be supplied from the drinking cup each cycle to absorb the heat energy thus transferred.

(e) *The working substance inside the heat engine returns to its original condition after each cycle.* The liquid goes back to the tail ready to be warmed up and evaporate again. This is a necessary feature if we are to have repeated action. We could certainly build a "one shot" engine that would release its energy on the first cycle and then not work again. But a useful heat engine should be capable of converting as much heat energy as we can supply. This can happen only if the working substance returns to its original state after each cycle. The source of the energy is not the working substance itself, but an external heat supply.

Characteristics of heat engines

A heat engine, as we have said above, is a device for converting heat energy, at least partially, into useful mechanical or electrical energy.

A heat engine generally makes use of a *working substance*, a material whose size or shape changes when heat energy is added or taken away. The working substance returns, after each cycle of the heat engine, to its original state so it can be used again to convert more heat energy. The energy of the working substance is the same at the end of the cycle as it was at the beginning. *In any complete cycle of a heat engine, no net energy is supplied or absorbed by the working substance.*

Gases are typically used as the working substance of a heat engine. The volume of a gas can change dramatically upon rather modest changes in temperature or pressure. Even more effective is a liquid that can vaporize into a gas, changing its volume by a factor of a thousand. The steam engine and our dipping-bird are examples. A liquid, such as gasoline, that can change into gases by a chemical reaction, can be both the working substance and the source of energy at the same time.[1] Heat engines can also be made to use solid substances that expand or change shape when heated.

During part of the cycle of any heat engine, heat energy is supplied to the working substance to make it expand. This heat energy can come from a part of the system that is kept at high temperature, such as the *boiler* of a steam engine.

At another point in the cycle, heat energy must be extracted from the working substance, so that it can return to its original condition to be used again. This can be done by bringing the working substance in contact with a cool part of the system, such as the *exhaust* of a steam engine.

These are general features of heat engines. There is a source of heat energy, a *warm reservoir*, at a high temperature. And there is a *cool reservoir* to absorb the excess heat energy at the end of the cycle. Every heat engine known depends on there being a temperature difference between the two reservoirs. It is obvious that a heat engine needs a source of heat energy. But it is equally true that a cool reservoir is a needed part of the system. The cool reservoir can be, for example, water drawn from a nearby lake or river. Or it can be simply the cool air surrounding the engine.

We can summarize these features of a heat engine by the diagram in Fig. 9-2. During each cycle, an amount of heat energy, H_1, is extracted from the

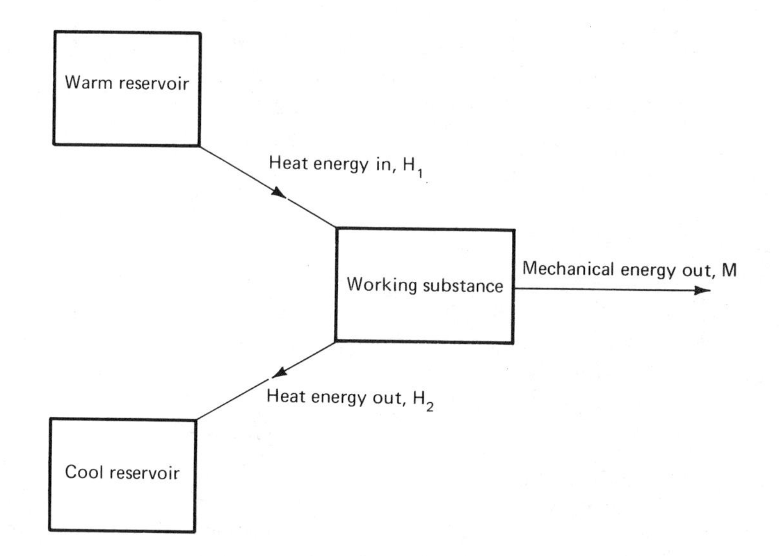

Figure 9-2 Energy flow in a typical heat engine

[1] In an *internal combustion* engine, such as the motor of an automobile, the working substance does not return to its original state. The analysis is more complicated than described in this chapter.

warm reservoir. At the end of each cycle an amount of heat energy, H_2, is deposited in the cool reservoir. Call M the amount of mechanical energy produced during the cycle.

The law of conservation of energy requires that

$$H_1 = H_2 + M$$

All the heat energy put into the system must be accounted for. Part of it goes into useful mechanical energy, such as lifting a weight or turning a wheel. The rest goes into the heat energy of the reservoir that cools the exhaust part of the engine.

M is always less than H_1, because H_2 is not zero. So a heat engine does not convert *all* of the heat energy supplied into useful mechanical energy. We can define the *efficiency* of a heat engine as M divided by H_1. The efficiency is a number between 0 and 1. It can also be expressed as a percent.

Often the efficiency of heat engines is not very high. The efficiency of an automobile engine is less than 5%. The engineers at your local power company would be quite pleased if the efficiency of the steam turbines, which produce electrical energy from the heat energy of burning fuel, were as high as 40%.

Consequences of the inefficiency of heat engines

The fact that a heat engine is not 100% efficient has significant consequences, some of immediate practical importance, and others of a philosophical nature:

1. The necessity of having a cooler reservoir available places a limitation on the location of industrial plant sites. Adequate cooling is a necessity for operating a heat engine of any type.

2. *Thermal pollution.* The exhaust of heat energy always goes along with any form of heat engine. This will be true even if we someday discover a virtually inexhaustible, clean source of energy. The warming of rivers and lakes may affect the animals and plants living in them. The heating of the air and the water may eventually cause unwanted changes in the climate.

Even the mechanical energy produced by a heat engine will usually be dissipated into heat energy before long. The brakes of your automobile heat up when you stop the car. Television sets, electric light bulbs, power tools, dishwashers—all convert electrical energy into heat. The fate of nearly all mechanical and electrical energy in our experience is to be turned into heat energy.

The very fact that we use energy creates thermal pollution. Even the useful mechanical and electrical energy is soon dissipated. But since the heat engine that produced the useful energy probably had low efficiency, an even larger amount of heat energy was tranferred to the environment right at the power plant. The only way to avoid thermal pollution is to use less energy.

3. *Impossibility of perpetual motion.* Since ancient times, mechanics and philosophers have been fascinated by the idea of building a machine that would persist forever in undiminished motion. Such a machine would make a fascinating toy. It might even, so it was thought, have useful functions such as driving a mill wheel, or at least a clock, that would never run down.

We know now, from the law of conservation of energy, that we cannot extract useful mechanical energy from any system without slowing it down in some way. So it is hopeless to expect to extract useful energy from a perpetual motion machine.

Even a perpetual motion toy, from which no energy is taken, will not work. Every real system in motion has some friction, some form of dissipation. With even the slightest amount of dissipation, some mechanical energy will be converted into heat energy. When this has gone on long enough, all the kinetic energy of any system will be dissipated, and the motion will stop.

With our understanding that heat is a form of energy, we can think of a perpetual motion machine of the *second kind.* In such a machine, the heat energy generated by friction is used to run a heat engine. In this way we hope to recover the mechanical energy lost through dissipation.

Since no heat engine is perfectly efficient, even this scheme will fail. At least part of the heat energy is transferred to cool reservoirs, and is not recovered as mechanical energy. Energy that has once been converted to heat energy cannot be completely recovered.

4. The eventual "*heat death*" of the universe. Heat energy is constantly being generated by friction and other forms of dissipation. Even the best heat engines cannot recover all of it. Eventually, some people believe, we must come to the state where all the available energy is transformed into heat energy. In such a state, all material will be evenly mixed up and distributed. Everything will be at the same temperature. Aside from the effects of chance fluctuation, nothing useful can happen further.

THE CARNOT HEAT ENGINE

The Carnot engine is an idealized engine designed in such a way that its efficiency can be calculated exactly. Real heat engines will not have efficiency as high as a Carnot engine operating between the same temperature limits.

The features that are special about the Carnot engine are:

1. The working substance is an "ideal gas." This is a gas whose molecules have negligible size, and do not attract or repel each other except in the brief moments when they collide. For such a gas, the kinetic theory can make exact predictions for the relations between pressure, temperature, and volume.

Most gases at low pressure and high temperature behave very much like ideal gases.

2. The warm and the cool reservoirs are so massive that they can provide, or absorb, all the heat energy required by the engine without any significant change in the temperature of the reservoirs.

3. Each step of the engine cycle is carried out so slowly and carefully that the working substance has time to adjust completely to the changed conditions. Thus, no energy is wasted by needlessly stirring up the working gas.

The engine consists of a fixed amount of ideal gas confined in a metal cylinder in which a sliding *piston* is free to move up and down. The piston fits

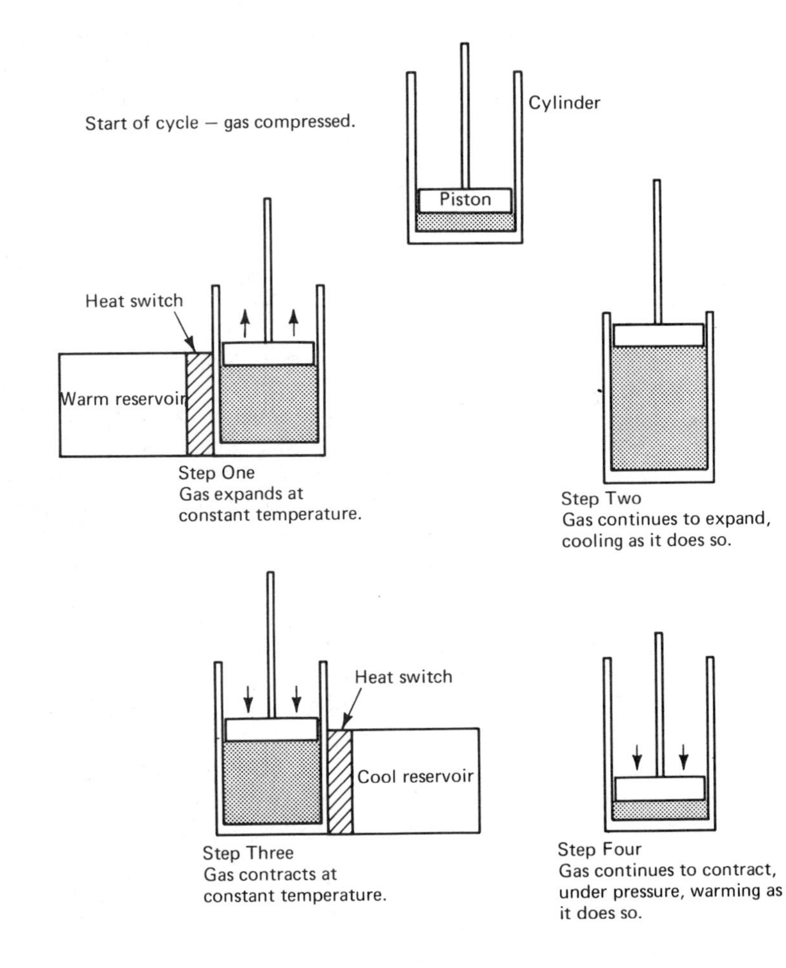

Figure 9-3 A Carnot cycle heat engine

snugly in the cylinder, so no gas can escape or enter around its edges. Various weights can be placed on the piston to increase or decrease the pressure on the gas.

At different parts of the cycle the cylinder can be put in contact with either of the two reservoirs. This is done by connecting a *heat switch* that allows heat energy, but no material, to flow between the working gas and one of the heat reservoirs. A strip of a good conducting metal, like copper or silver, in tight contact with the cylinder, makes a good heat switch.

We can start the cycle at any point. Let us begin at the point where the working gas is compressed to its smallest volume at the highest pressure of the cycle.

In the first step of the cycle, the heat switch is closed between the working gas and the warm reservoir. The gas is made to expand at the constant temperature of this reservoir. Mechanical energy is extracted as the weights on the piston are lifted. This energy has to come from somewhere. It does not come from the gas itself. Since its temperature is kept constant, the kinetic energy of its molecules stays the same. So the energy for raising the weights must come out of the heat energy of the warm reservoir.

The area shaded in cross-hatch in the pressure-volume graph of Fig. 9-4 represents the energy, H_1, extracted from the warm reservoir during this part of the cycle.

In the second step of the cycle, the cylinder is isolated from both reservoirs. It is allowed to expand further, as the weights are gradually removed. More mechanical energy is extracted. The source of this energy can be only the internal energy of the working gas itself. So the gas cools down. The kinetic energy of its molecules decreases. This step of the cycle continues until the gas reaches the temperature of the cool reservoir.

In step three, the cylinder is placed in contact with the cool reservoir. Weights are gradually added to the piston to make the gas contract at constant temperature. Mechanical energy is released as the weights fall. This energy must go somewhere. In the Carnot cycle we are careful not to let it go into kinetic energy of the piston. Such energy would be soon dissipated by frictional forces, and lead to waste.

The energy released does not go into the working gas. The gas is kept at constant temperature during this phase, which means the kinetic energy of its molecules does not change.

So the mechanical energy released during step 3 is all transferred into heat energy of the cool reservoir. The amount of energy thus released in this step is represented by the area shaded in gray in the pressure-volume graph.

In step 4, weights are added to the piston to compress the gas back to its starting conditions. This step is carried out with the cylinder isolated from both the heat reservoirs. The energy released by the falling weights goes into speeding up the internal motions of the gas molecules. So the temperature of the gas rises. The separation point between steps 3 and 4 is set in such a way that the gas comes back to exactly its starting temperature, pressure, and volume. The cycle can then begin again.

Because the Carnot cycle is carried out under such idealized conditions, it

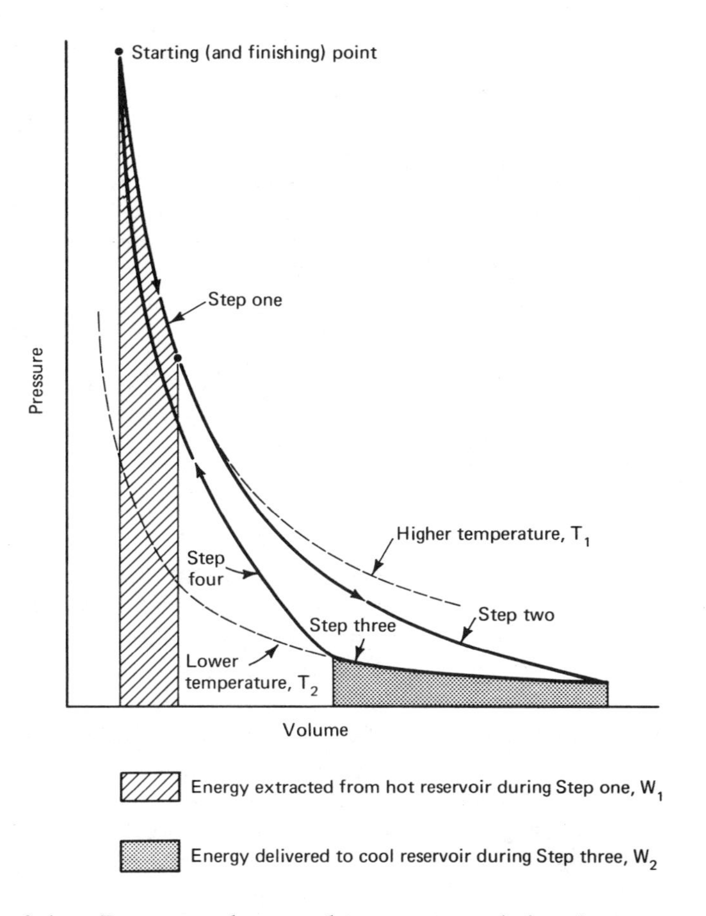

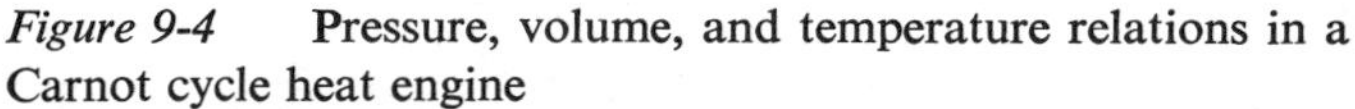

Figure 9-4 Pressure, volume, and temperature relations in a Carnot cycle heat engine

is possible to calculate exactly how well the Carnot engine works. It turns out that the ratio of energy transferred from the warm reservoir, H_1, to the energy transferred to the cool reservor, H_2, is equal to the ratio of the *absolute* temperatures of the two reservoirs.

If the two heat reservoirs are at the *same* temperature, H_1 and H_2 will be equal. There will then be no net mechanical energy delivered by the engine during a complete cycle.

If the temperature difference is small, H_2 is almost as large as H_1. Only a small fraction of the heat energy supplied from the warm reservoir comes out as mechanical energy. Most of it becomes heat energy added to the cool reservoir. In such a case the efficiency of the engine is low.

The way to get high efficiency from this kind of heat engine is to operate the warm reservoir at the highest possible temperature, and to keep the temperature difference between the two reservoirs as large as possible.

For example, suppose the warm reservoir is kept at 600° *absolute* (327°C, far above the boiling point of water), and the cool reservoir is at 300° absolute (about normal room temperature). In a Carnot engine operated between these two temperatures, H_2 is one-half of H_1. The efficiency of such an engine is 50%. Only half of the heat energy delivered by the warm reservoir can come out as useful mechanical energy.

A Carnot engine cannot operate with 100% efficiency if the cool reservoir is at any temperature above absolute zero. Absolute zero has never been achieved, and probably cannot be. All real gases condense into liquids at temperatures above absolute zero. So it is not likely that any Carnot engine can be made to operate at 100% efficiency.

As we shall see in the next box, no real engine, operating between two reservoirs at given temperatures, can be more efficient than an idealized Carnot engine served by the same reservoirs.

This state of affairs is called the "heat death" of the universe. Calculations of the time scale for the "heat death" show that it is so far in the future that many other predictable catastrophes are likely to have ended all life on Earth long before the "heat death" comes about.

Refrigerators

A heat engine can be run in reverse. Mechanical energy can be applied to force heat energy to flow from a cooler to a warmer reservoir. An engine used in this fashion is called a refrigerator.

We can draw a diagram to show the energy flow in a refrigerator by reversing all the arrows in the diagram for the heat engine. In your home refrigerator, working under steady conditions, the "cool reservoir" is the meat, milk, vegetables, and air inside the refrigerator. These materials get heat energy that leaks into the refrigerator from the room outside, either through the insulation, which is never perfect, or when the door is opened.

The "warm reservoir" is the air of the room outside. It gets into contact with the inside parts, through the vanes at the back or top of your refrigerator. Feel the vanes on your own home refrigerator. Do they get warm?

The "working substance" is Freon or some other refrigerating substance, a liquid/gas with a low boiling point. This material is pumped around inside the coils of the refrigerator. The mechanical energy for the refrigerator motor is supplied by plugging it into the local electrical outlet.

It is clear that H_1, the heat energy supplied to the warm reservoir in a refrigerator, is larger than M, the mechanical energy supplied. In typical conditions of your home refrigerator it can be 10 times as large.

We will call the ratio of M divided by H_1 the *counterefficiency* of the refrigerator. The smaller the counterefficiency, the less mechanical energy is needed to pump a given amount of heat.

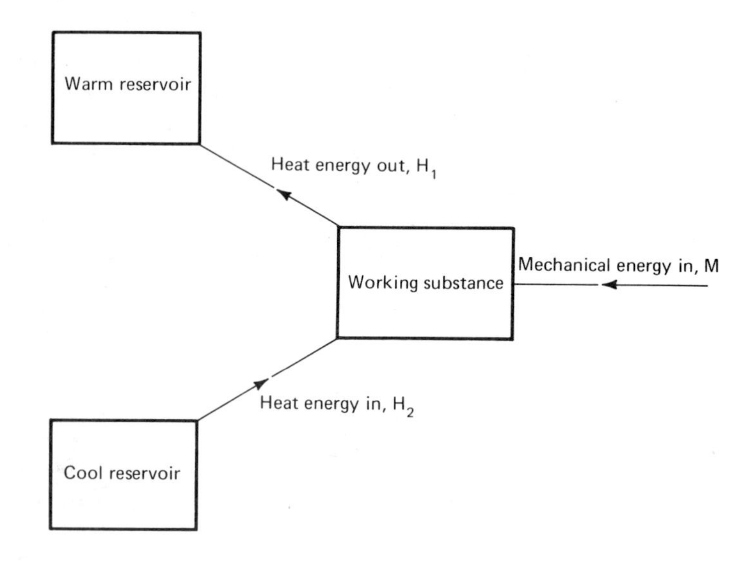

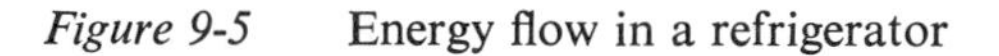
Figure 9-5 Energy flow in a refrigerator

A very poor heat engine makes a very good refrigerator when run in reverse. Clearly, refrigerators work best when they operate at low temperatures and between reservoirs that have a small temperature difference.

When only electrical energy is available, a very efficient device for heating your home is a *heat pump*. A heat pump is a refrigerator whose "cool reservoir" is the outdoors and whose "warm reservoir" is the inside of your house. It is as if you mounted a refrigerator in your window with the door open and facing outward. Your attempt to cool down the world outside would result in the transfer of heat energy to the vanes of the refrigerator, which are inside the house.

Suppose you have electrical energy available to heat your house. You can simply dissipate it in an electrical heater. The heat energy thus produced will equal the electrical energy dissipated. Or you can use the electrical energy to run a heat pump. The heat energy transferred to your house then will, depending on the counterefficiency of the heat pump, be several times the electrical energy used.

The Second Law

In Chapter 7 we calculated that, if we mixed 50 grams of water at 0°C with 50 grams of water at 100°C, the mixture would settle at a temperature of 50°C. It would not contradict the law of conservation of energy for a 100-g beaker of water at 50°C to separate itself into 50 g at 0°C and 50 g at 100°C. Nor would such a process violate any other principle of mechanics.

But we know very well that such things do not happen spontaneously in nature. There is a natural direction in which processes take place. Heat flows from warm bodies to cold bodies, and not vice versa. We can think of many other examples of processes that have this irreversible property.

A new principle of physics is needed to account for the irreversibility of such processes. It is called the *Second Law of Thermodynamics.* (The first law of thermodynamics is the law of conservation of energy, with heat energy recognized as one form of energy.)

This new principle can be stated in many ways. We will use the following statement, first proposed by Rudolph Clausius (1822–1888):

> "*There is no process which takes place in which the final net result is that heat energy has moved from a cooler to a warmer body and nothing else has changed.*"

In other words, there is no such thing as an ideal refrigerator, one that works with no input of mechanical or electrical energy.

This statement of the second law means that no such process is possible, no matter how indirectly it is accomplished, nor how complicated an apparatus we use. This gives the statement of the Second Law very powerful consequences. Using it we can show that a large number of other processes also cannot take place. For if they did, it would be possible to use them to build an ideal refrigerator.

To illustrate the power of the Second Law, we show, in the accompanying box, that one cannot build a perfectly efficient heat engine, one in which no heat energy needs to be transferred to a cool reservoir. This fact can be summarized in an alternate statement of the Second Law, one proposed by Lord Kelvin (1824–1907).

> "*There is no process which takes place in which the final net result is that a single warm object has lost heat energy, an equal amount of mechanical energy has been generated, and no other change has taken place.*"

The "no other change" provision means in particular that the working substance in any heat engine has gone back to its starting condition.

Applications of the Second Law The second law, no matter how stated, leads to several important results. For one thing, the Kelvin statement of the Second Law tells us that, even in principle, there is no such thing as a 100% efficient heat engine. This leads to all the consequences that were outlined in an earlier section of this chapter. Thermal pollution must always accompany the operation of any heat engine. Perpetual motion machines do not work. And so forth.

We can use the Second Law to prove that there has to be a difference in temperature to make a heat engine run. (Can you supply the proof?) This fact places a severe restriction on the availability of heat energy to do useful work.

For example, there are billions of joules of heat energy that would be released if the water in the ocean could all cool down by just 1 degree Celsius. But this energy is not available to run machines. There is no cooler reservoir into which the exhaust energy could be conveniently deposited.[2]

If heat energy keeps flowing from warmer to cooler reservoirs, eventually all the warmer bodies will cool down, and the cooler ones will warm up. To keep our engines running we must rely on the continual supply of heat energy *at high temperatures*. It is this feature that makes a fuel that burns at high temperature so valuable.

PROOF THAT THE CLAUSIUS STATEMENT OF THE SECOND LAW LEADS TO THE KELVIN STATEMENT

We shall prove in this section that if the Clausius statement of the Second Law (that there is no ideal refrigerator) is correct, then the Kelvin statement (that there is no perfectly efficient heat engine) must be true.

Our method of proof will be to assume the opposite. We will suppose the Kelvin statement is false, that there is a way to make a perfectly efficient heat engine. Then we will show that, if this were the case, we could use it as part of an ideal refrigerator. But the Clausius statement says that there is no way, however indirect, to build such a refrigerator. Therefore, we will see that if the Kelvin statement were false, the Clausius statement would also be false. So, by a well-known step in logic, we will conclude that if the Clausius statement is true, the Kelvin statement must also be true.

The construction of the ideal refrigerator, given that there is a perfectly efficient heat engine, is simple. Suppose there is such a heat engine that can deliver, say, 100 joules of mechanical energy in 1 cycle by extracting an equal amount of heat energy at, say, 100°C.

Let us also suppose we know how to make an *ordinary* refrigerator that pumps heat energy from a reservoir at, say, 0°C to another reservoir at 100°C. Let the counterefficiency of this refrigerator be such that, if 100 joules of mechanical energy are put into it, about 200 joules of heat energy will be transferred from the cool reservoir, thus delivering a total of 300 joules to the warm reservoir. The exact numbers are not important here, only the fact that *some* heat energy is transferred from the cool reservoir.

The way to make the ideal refrigerator is to connect the output of the perfectly efficient heat engine to the input of the ordinary refrigerator. The warm reservoir of both devices can be the same body at 100°C.

[2] Proposals have been made to make use of the fact that not all the ocean water is at the same temperature. A heat engine could be run whose warm reservoir is the water at the surface of the ocean. The cool reservoir would be the water, only a few degrees lower in temperature, at a depth of 100 meters or so. Such a heat engine would have very low efficiency. But it would tap a great source of energy that is now going to waste.

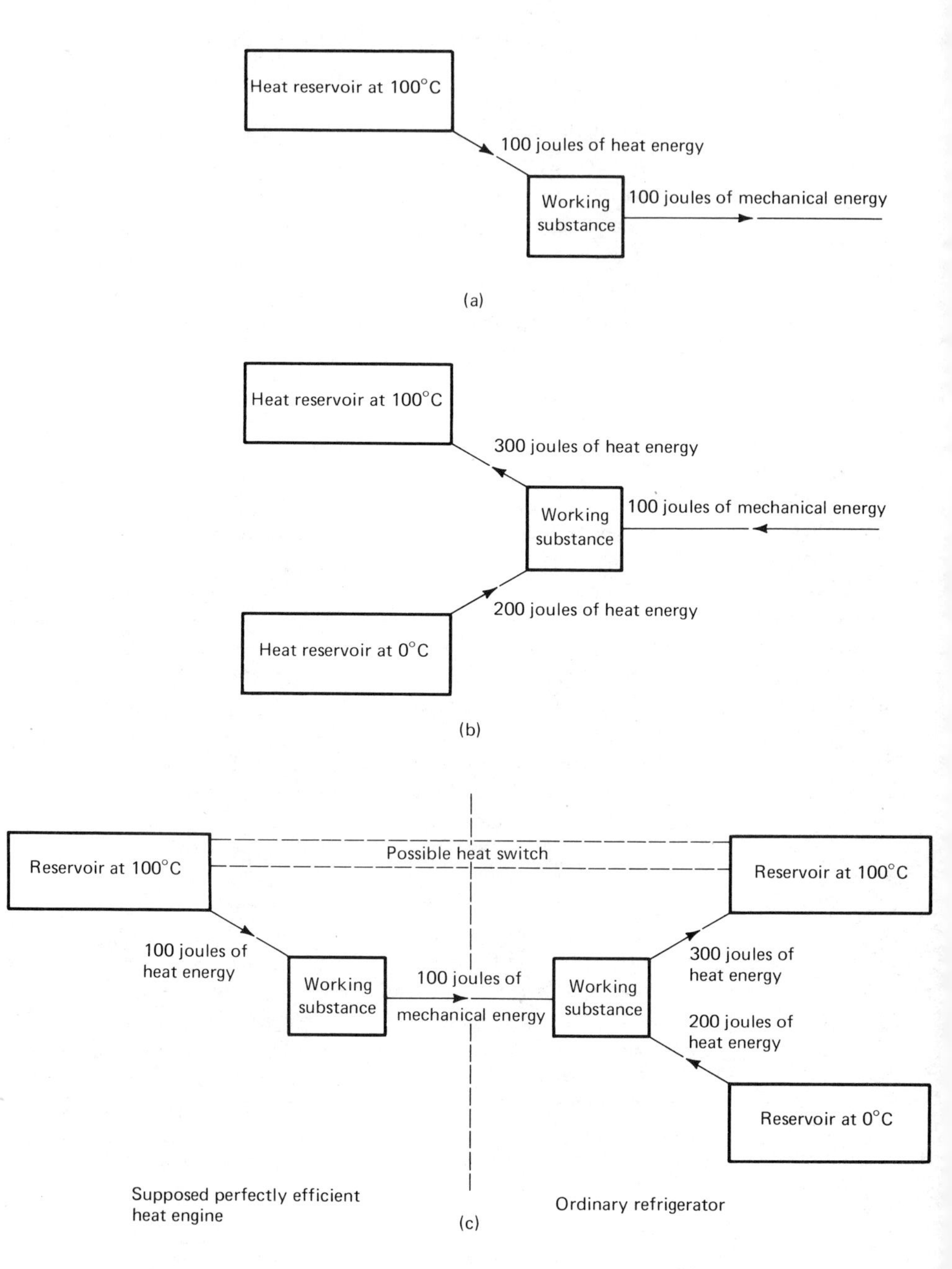

Figure 9-6 (a) The supposed perfectly efficient heat engine (b) The ordinary refrigerator (c) The ideal refrigerator that could be made by connecting the supposed perfectly efficient heat engine to the ordinary refrigerator

In 1 cycle, 100 joules of heat energy will be extracted from the warm reservoir. This energy, converted into mechanical form, will be used to drive 1 cycle of the ordinary refrigerator. The net result will be that 200 joules of heat energy will be transferred from the cool reservoir, and a net 200 joules of heat energy will wind up in the warm reservoir. There will be no net input of mechanical energy.

A device that could do this would be what we have called an ideal refrigerator. Its very existence would, of course, violate the Clausius statement of the Second Law.

We conclude that, if the Clausius statement is true, the Kelvin statement cannot be false. There are no perfectly efficient heat engines.

It is also possible to prove the converse of this. If the Kelvin statement is true, the Clausius statement cannot be false. We would prove this by showing that if we could make an ideal refrigerator, we could use it to help build a perfectly efficient heat engine. The details are left as an exercise for the reader.

Thus, the two statements of the Second Law are *equivalent*. If either is false, so is the other. Therefore, if either is true, so is the other.

There are other ways of stating the Second Law of Thermodynamics. All of them can be shown to be equivalent to the Clausius and Kelvin statements.

We can prove something even stronger than the foregoing statements. Suppose there is known to be an ordinary refrigerator, operating between two given reservoirs, with a certain counterefficiency. We can prove that there is no heat engine, operating between the same two temperatures, whose efficiency is higher than the counterefficiency of this refrigerator.

Suppose the counterefficiency of the refrigerator is 25%. This means that it can take 25 joules of mechanical energy, add to it 75 joules of heat energy from the cool reservoir, and deliver 100 joules to the warm reservoir.

Suppose someone else claims to have built a heat engine, using the same two reservoirs, that is 40% efficient. Such an engine could take 125 joules of heat energy from the warm reservoir, deliver 50 joules of mechanical energy, and deposit the remaining 75 joules into the cool reservoir.

Now connect the two devices together. First, 125 joules are taken from the warm reservoir; of this, 75 joules go to the cool reservoir, and the other 50 joules come out as mechanical energy. Use 25 joules of this mechanical energy to drive 1 cycle of the refrigerator. The 75 joules of heat energy are thus transferred back to the warm reservoir.

This leaves 25 joules of mechanical energy, converted from heat energy in the warm reservoir. There is no net transfer of energy to the cool reservoir. So the combined device is a perfectly efficient heat engine. This contradicts the Kelvin statement of the Second Law.

We conclude that there is no such device. There is no heat engine whose efficiency is higher than the lowest counterefficiency of any refrigerator we can build, operating between the same temperatures.

The Carnot cycle plays an important role in setting limits here. A Carnot cycle refrigerator can be built, in principle, to operate between two given temperature reservoirs. Its counterefficiency can be calculated. It is simply the ratio of the temperature difference between the reservoirs, divided by the absolute temperature of the warmer reservoir.

According to the argument in this section, no heat engine operating between those same temperatures can have a higher efficiency than this ratio that applies to the Carnot cycle.

Sunlight, which is an abundant source of energy, usually comes to us at rather modest temperatures. This is why it is so hard to use sunlight efficiently to run machines directly.

There are other consequences of the Second Law of thermodynamics that do not seem to involve heat energy at all. For example, if a hole opens in the side of an evacuated container, the outside air rushes in to fill the space. On the other hand, it does not happen that the air spontaneously moves out of one side of a box, leaving a vacuum in that half of it, while the other half has twice the normal amount of air. You are asked (in Question 12) to show whether or not, if such an effect took place, you could use it to build a heat engine that would violate the Second Law. Therefore, this effect does not take place.

The Second Law can even be used to get quantitative results. In Question 13 of this chapter, we show that there is a definite relation between the specific heat, the thermal expansion, and the tension in a metal, that is required by the Second Law.

The molecular view of the Second Law

We have been careful in this chapter always to say that certain processes "*do not* take place." There is a reason for our choice of words.

It is not forbidden by any law of physics that the air molecules in a room could at some instant all be moving toward the nearest wall. If this ever happened, the next few moments would see the center of the room become empty of air molecules. The unfortunate people in that spot would be left gasping for breath.

We know from experience that this sort of thing does not happen. Such an occurrence would be a violation of the Second Law, as we have seen. It does not happen, not because it is *impossible*, but because it is *extremely unlikely*.

From the molecular point of view, the Second Law of thermodynamics is the statement that events that are extremely unlikely do not in fact take place.

Consider a box that is so well evacuated that there are but 2 molecules in it. Let one be an oxygen molecule, say, and the other a nitrogen molecule. Each molecule spends about half its time in each half of the box. If we took snapshots of the box we would find that 25% of the time the oxygen molecule would be in the left half of the box and the nitrogen molecule in the right half. Another 25% of the time the oxygen would be in the right and the nitrogen in the left. We would find both molecules in the left, with the right half completely empty 25% of the time. The remaining 25% of the time would see both molecules in the right, with the left half of the box vacant.

In a system that has only two molecules, local vacuums occur rather often. The Second Law is violated regularly.

If there are 4 molecules in the box, the Second Law is violated less frequently. There is 1 chance in 16 that the right half of the box will be evacuated, and an equal probability of finding the left half empty.

When there are as many as 100 molecules, the violations of the Second Law become more rare. Two-thirds of the time there will be between 45 and 55 molecules in each half of the box. The probability that one side of the box will be completely vacant is less than 1 in 1,000,000,000,000,000,000,-000,000,000,000!

If there are a million molecules in the box, even small deviations from an

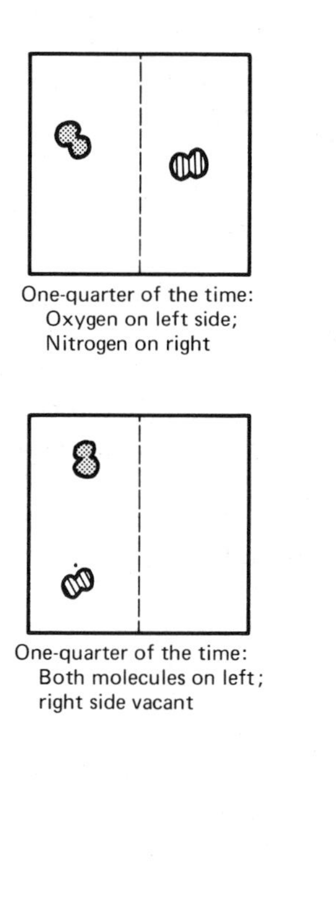

One-quarter of the time:
Oxygen on left side;
Nitrogen on right

One-quarter of the time:
Both molecules on left;
right side vacant

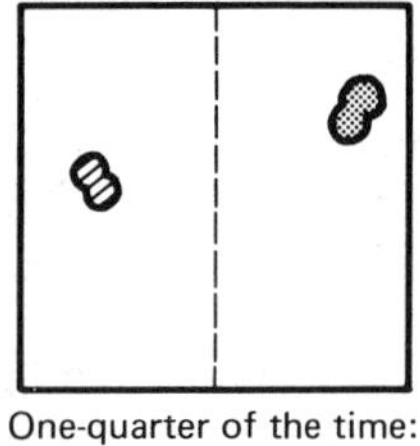

One-quarter of the time:
Oxygen on right side;
Nitrogen on left

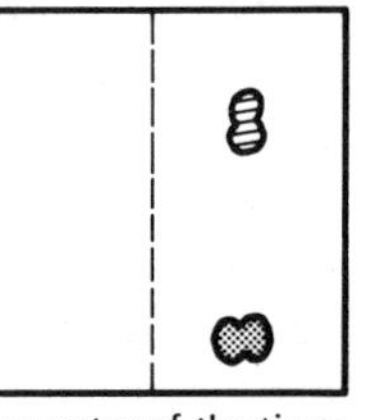

One-quarter of the time:
Both molecules on right;
left side vacant

Figure 9-7 Two molecules in a box

equal distribution of molecules are unlikely. We will find that 99.6% of the time there are between 498,500 and 501,500 in each half of the box. A state where there was as much as 1% difference between the number of molecules in each half would occur so seldom that we would call it a miracle if it did.

Imagine, then, how unlikely it is to get even a small nonuniformity in a normal gas, which has 16 billion billion molecules in every cubic centimenter. We will almost never see a measurable difference in the number of molecules in any visible fraction of the box.

The Second Law rests, then, on the very large number of molecules involved in any real process. The law of large numbers applies: The more molecules involved, the less likely we are to see any deviations from the expected average behavior. On the molecular scale, the Second Law is not really a "law" at all, but simply a statement of probabilities.

Another way of looking at the Second Law is to say that it predicts the transition from structure to randomness, from order into chaos.

A highly ordered condition is one in which there are relatively few combinations of molecular states that will give the same overall appearance. The molecules in a crystal at low temperature, all lined up in regular rows and columns, is one example. All the molecules in a box gathered into one corner is another.

A more chaotic condition is one in which there are many molecular combinations that can give the same appearance. The crystal can melt and let the molecules move about in random fashion. The gas molecules can spread out to fill the entire room. By definition, the chaotic condition has a higher probability than the more ordered condition.

The Second Law, then, predicts that order will give way to chaos. Molecules will become more uniformly mixed, heat energy will flow in such a direction to even out all the temperatures. Crystals will break and melt, and not easily form again.

It is possible to force local increases in the state of order. We can make a refrigerator that pumps heat energy from a cold body to a warmer one. We can make a vacuum pump that forces the air out of a container. But we do so only at a price. We must convert mechanical energy, which is usually highly structured, into the randomness of heat energy. And the amount of order we lose in this way is always greater than the amount of order we create with the refrigerator or vacuum pump.

We can make a mathematical definition of the amount of order (sometimes called the "information") or the amount of disorder (called the "entropy") of a system. The Second Law says that the changes in these quantities go always in the same direction. The total information decreases; the total entropy increases.

This, then, is the context of the Second Law of Thermodynamics. We can transform more random forms of energy, like heat energy, into more useful forms, like mechanical energy. But we do so always at a price, such as dumping some heat energy into cool reservoirs. And the price we pay, in terms of turning order into chaos, is always at least as much as the benefit we gain.

Glossary

Carnot engine—An idealized heat engine, useful in showing how such engines can work. Its importance lies in setting a limit to the efficiency that any real engine can have, operating between the same two temperatures.

Clausius, Rudolph (1822–1888)—German physicist, pioneer in the study of thermodynamics, responsible for the Clausius statement of the Second Law.

Clausius' Statement of the Second Law—"There is no process which takes place in which the final net result is that heat energy has moved from a cooler to a warmer body and nothing else has changed."

cool reservoir—A system into which heat energy can be dumped, large enough so that its temperature does not rise very much when it receives moderate amounts of heat energy.

counterefficiency—In a refrigerator, the ratio of the mechanical energy required to operate it for a cycle, divided by the heat energy transferred to the warm reservoir. A good refrigerator has a *low* counterefficiency.

efficiency—In a heat engine, the ratio of the mechanical energy produced in 1 cycle of the engine, divided by the heat energy transferred from the warm reservoir. The efficiency is a fraction between 0 and 1 (or a percentage between 0% and 100%). In a heat engine it is desirable to have the efficiency as *high* as possible.

entropy—A measure of the degree of disorder in a system. In any real process, according to the Second Law, the total entropy of all parts of the system always increases.

exhaust—The cool reservoir of a heat engine; the part of the system where the unused heat energy is dumped.

heat death—The eventuality, long in the future, when all systems come to equilibrium at the same temperature, all materials are uniformly mixed together, and nothing more of interest happens. Though the heat death of the universe may be long in coming, every real process moves us another step in that direction.

heat engine—A device to transform heat energy, at least partially, into mechanical or electrical energy, or some other useful form.

heat pump—A device to make best use of mechanical or electrical energy in heating a room. Rather than simply dissipating the available energy (as an electric heater does) a heat pump uses it to pump heat energy from the cool outdoor environment into the room being heated. In operation a heat pump works much like a refrigerator, but the "warm" and "cool" reservoirs are identified differently.

heat switch—A device that, when inserted in place, permits heat energy—but no material—to flow across it. A heat-conducting metal plate can make a good heat switch.

information—A measure of the degree of order in a system, the opposite of entropy. In any real process, according to the Second Law, the total information present in all parts of any system always decreases.

Kelvin (William Thomson, Baron Kelvin) (1824–1907)—British physicist and mathematician, who made discoveries in thermodynamics and in electromagnetism, and gave his name to the absolute scale of temperature.

Kelvin's statement of the Second Law—"There is no process which takes place in which the final net result is that a single warm object has lost heat energy, an equal amount of mechanical energy has been generated, and no other change has taken place." In other words, there are no 100% efficient heat engines.

perpetual motion machine—A device that continues to move forever, without the input of any energy. The first and second laws of thermodynamics, together with the unavoidability of some dissipation, guarantee that no perpetual motion machine can exist.

refrigerator—A device to pump heat energy from a cooler to a warmer reservoir, using mechanical energy as input. In outline, a refrigerator operates like a heat engine with all the energy flows reversed.

reservoir—A system into and from which heat energy can be transferred, and which is large enough so that its temperature does not change very much under moderate transfers of heat energy.

Second Law of Thermodynamics—A general statement that any process involving mixing, or energy flow, will take place naturally in one direction only; that most natural processes are irreversible. The exact content of the Second Law can be made in many equivalent statements. If any one of them is true, so are all the others:
Examples:

1. Clausius' statement (which see).

2. Kelvin's statement (which see).

3. Many other statements, such as: The air in the room will not spontaneously collect in one spot.

4. The entropy (degree of chaos) of any complete system never decreases; the information (degree of order) never spontaneously increases.

thermal pollution—The transfer of heat energy to the environment, which accompanies the operation of every heat engine.

thermodynamics—The study of processes that involve the flow of heat energy.

warm reservoir—A system from which heat energy can be extracted, large enough so that its temperature does not drop very much when moderate amounts of heat energy are removed.

working substance—A material that changes size (or other physical feature) under input of heat energy, so that it can be used as part of a heat engine, to convert heat energy partly into mechanical energy. In a complete cycle, the working substance returns to its initial state. Thus, it does not overall provide nor absorb any of the energy used.

Questions

1. Does the internal combustion engine, whose working substance does not return to its original state, offer a way to circumvent the Second Law?

2. Some of the small artificial satellites that have been put into orbit around the earth are expected to stay up for more than a million years. Can these satellites be considered perpetual motion machines?

3. Suppose you are trying to heat your house with only electrical energy available (not having gas, oil, or coal to burn). You can connect a heating coil, which dissipates all the energy supplied, directly to the electrical outlet. Or you can use a heat pump. Compare the effectiveness of the two heating methods. Why is the argument not valid if you have fuel to burn, rather than electrical energy, available?

4. Name five processes that always take place in one direction and not spontaneously in the opposite direction.

5. Prove that if the Kelvin statement of the Second Law (that there are no perfectly efficient heat engines) is true, the Clausius statement (that there are no ideal refrigerators) cannot be false.

6. Use the Second Law to prove that it is impossible to achieve absolute zero of temperature. (Think of a Carnot engine with a cool reservoir at absolute zero.)

7. On the basis of the Second Law, show that two objects free to exchange heat energy with each other, must come to equilibrium at the *same* temperature.

8. Show that the efficiency of a Carnot heat engine is equal to the ratio of the *difference* in temperature between its two reservoirs, divided by absolute temperature of the warmer reservoir, i.e., that

$$\frac{M}{H_1} = \frac{T_1 - T_2}{T_1}$$

9. Prove that there has to be a difference in temperature to make a heat engine run. (Hint: What is the counterefficiency of a refrigerator with zero temperature difference?)

10. What is the source of the energy that drives the heat engines (mentioned in the footnote on page 213) that operate on the difference in water temperature between the surface and the deeper part of the ocean?

11. The law of conservation of energy says that energy never disappears. We always have as much total energy at the end of any process as we ever started with. Why, then, is there so much concern about the "consumption of energy?" How can there be an "energy crisis" when energy is always conserved?

12. Suppose that the air in one half of a container could be induced to move on its own completely into the other half, leaving the first half evacuated. Show that this effect could be used to build a heat engine that violated the Second Law.

13. Most metals expand when heated and shrink when cooled. Consider a weight hanging from the ceiling by a cable made of such a metal. You might suppose that the cable could heat up and grow longer. The internal heat energy required might be supplied from the GPE released as the weight attached to it goes down.

Show that the Second Law doesn't allow this. The heat energy needed to expand the wire must always be more than the energy transferred in stretching it. (Hint: Use a cool reservoir to bring the system back to its original state. Show that the net result is an ideal refrigerator.)

In this example we have shown that the Second Law can lead to detailed relations between such properties as the specific heat, the thermal expansion rate, and the tension in a metal cable.

14. Suppose we have a box with a small number of molecules in it. We watch it carefully until all the molecules are in the left half, then quickly pull down a partition and trap all the molecules in that half. We then use the difference in pressure between the two halves of the box to make a piston move, and thus withdraw mechanical energy. Would this arrangement violate the Second Law? Do you think we could make widespread applications of this type of device?

15. A student tries to cool off on a hot day by sitting in front of the open refrigerator door. He forgets that the vanes of the refrigerator are also in the room with him. The doors and windows are closed. When the air in the room becomes thoroughly mixed, will it be warmer or cooler than when he first began?

section three

THE ELECTROMAGNETIC SPECTRUM

10

electromagnetic waves

FROM RADIO TO INFRARED

Introduction

In Chapter 3 we mentioned that it was possible to send signals through space at the speed, *c*, of 300,000 kilometers per second. We said that these signals were electromagnetic in nature. And we said that visible light was one example of this type of signal.

In Chapter 5 we defined *voltage*. The voltage difference between any two points is the energy it takes to move a coulomb of electric charge from one point to the other. The energy that exists in this form is known as *electrostatic potential energy*.

In Chapter 6 we studied the nature of *waves*. We defined a wave as a shape or form or disturbance that moves with time. We learned that waves can cancel or reinforce each other ("*interference*"), and that they can bend around barriers ("*diffract*"). We saw that a wave that repeated its pattern with a regular *frequency* in time will also repeat its pattern in space. The separation between places where the pattern repeats is called the *wavelength*.

In this chapter we put these ideas together to form the concept of an *electromagnetic wave*. An electromagnetic wave is a pattern of voltages in space that moves with time.

Our basic theory of electricity and magnetism, due to James Clerk Maxwell, tells how to create electromagnetic waves. They are made any time any electric charge is accelerated. The theory also tells us how these waves behave once they are formed. They propagate through empty space at a speed of 300,000 kilometers per second. They travel through many materials at slower speeds, depending on the material.

Electromagnetic waves can be detected because of the energy they carry. There is a pattern of voltages in space associated with each wave. We can place electrically charged objects in the path of such a wave, free to move in response to voltage differences that they meet. When the wave reaches the detector charges, they will begin to move. Thus, they will extract electromagnetic energy from the wave and turn it into their own kinetic energy. When we observe the detector charges in motion, we know that an electromagnetic wave has reached them.

The type of detector we use takes many forms. It may be electrons in the radio aerial of an automobile. Or it may be the electric charges within the atoms in the retina of your eye. The detector depends on the form of electromagnetic waves we are detecting, and these waves take many forms.

Basically, however, all electromagnetic waves are the same. They are created by electric charges in acceleration. They travel at the speed c in empty space. They are detected by their effect on other electric charges. They differ from each other mainly in how rapidly the wave pattern repeats itself. In other words, they differ in their frequency and in their wavelength.

There is a very wide spectrum of electromagnetic waves. They range in wavelength from very low-frequency radio waves, whose pattern repeats itself at intervals of thousands of miles, down to gamma rays whose wavelength is smaller than the nucleus of an atom.

The wavelength is very important as to how we regard these waves. An efficient system for making waves, or for detecting them, should be somewhat the same size as the wavelength of the waves being made or detected. A radio antenna is much too large to generate visible light waves. The atoms in your eye are much too small to detect radio waves.

In this chapter we begin the discussion of electromagnetic waves with the longest wavelengths. In following chapters we shall work our way down to the shorter wavelengths of the spectrum.

Measurement of voltage

Voltage has been defined as the transfer of energy per unit of electric charge. An obvious way to measure voltage is to allow some electric charges to move under its influence. We can then see how much kinetic energy the charges acquire.

A common way to measure the voltage difference between two points—say, between the two terminals of a battery—is to connect a rather poor conductor between those points. We use a *poor* conductor because we don't want to draw too much energy out of the system. We are trying to measure the energy. We don't want to change the thing we are measuring while we are doing the measurement.

Call this poor conductor "wire A." Since there is a voltage difference between the two ends of wire A, the electrons in the wire will begin to move. They will make many collisions with the fixed atoms in the wire. These collisions keep the electrons from moving too fast; that is why the wire is said to be a "poor" conductor.

The moving electrons make up an electric *current* in the wire. The greater the voltage difference, the faster the electrons move; hence, the more current there is.

We detect the presence of current in wire A by making use of *magnetic* forces. We know that two wires carrying electric current exert magnetic forces on each other. We provide a second wire, wire B, already carrying current, close to wire A.

There will not be any *electrostatic* forces between the two wires because

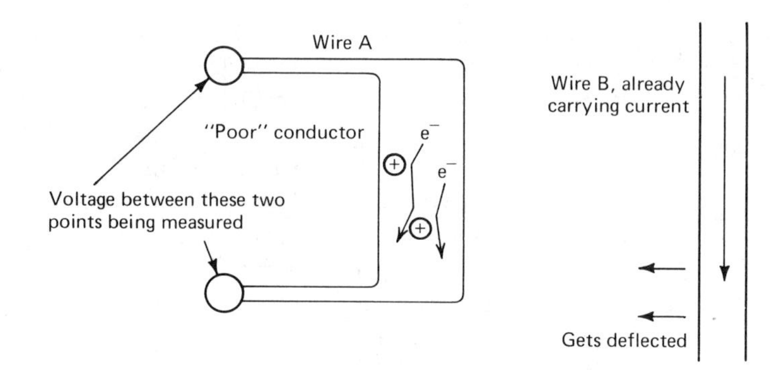

Figure 10-1

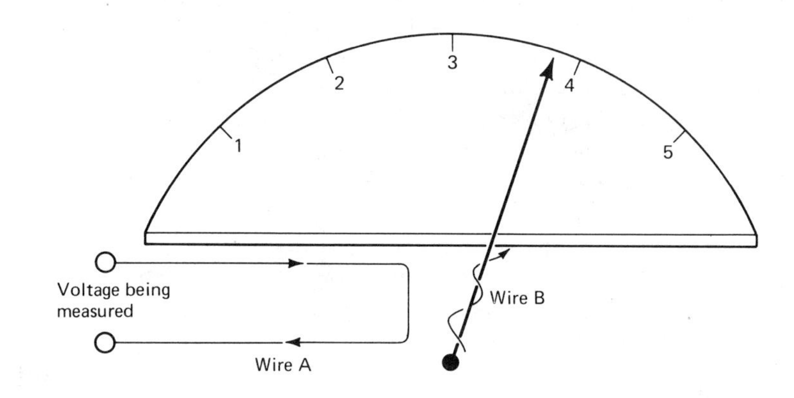

Figure 10-2 A simple voltmeter

each wire as a whole is electrically neutral. The positive charges of the atomic nuclei balance the negative charges of the electrons. But only the electrons, not the heavy atoms, are in motion, so each wire does have a nonzero electric current.

When the current begins to flow in wire A, the two wires mutually deflect each other. The amount of deflection in wire B depends on how much current there is in wire A. This, in turn, depends on the voltage difference between the two ends of wire A. So the amount of deflection of wire B is then a measure of the voltage we set out to determine.

Very often wire B is wound around the needle of an electric meter. When you read "the dial," what you are measuring is the deflection of wire B.

Wire B need not be a solid wire at all. It can be a beam of electrons travelling across an evacuated cathode-ray tube. Such a beam of electrons in motion also constitutes a current.

We see the spot on the face of the tube to indicate the place where the electron beam strikes the face. When the beam is deflected, by passing a current through wire A, the spot is observed to move.

Demonstration: Cathode-Ray Oscilloscope, with Spot Visible on Face: The input wire pair ("Wire A") is touched to the opposite terminals of a 45-volt battery. The spot moves downward. The input wire pair is then connected to the terminals of a 90-volt battery. The spot moves twice as far from its original position. The input wires are interchanged, so the current in wire A flows in the opposite direction. The spot is now deflected upward.

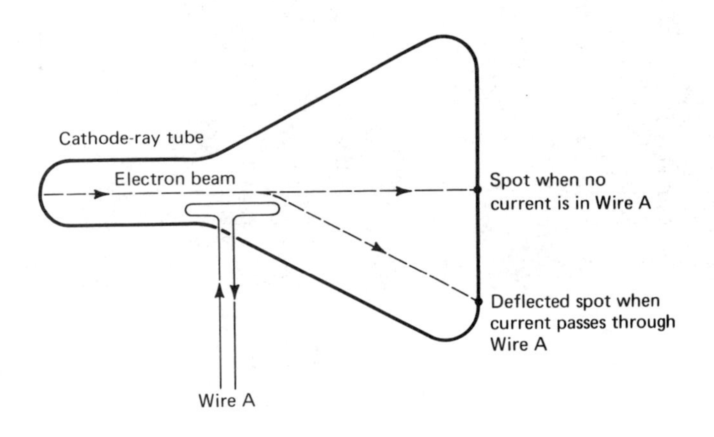

Figure 10-3 The oscilloscope—a cathode-ray tube used as a voltmeter

Thus, the deflection of a spot on the oscilloscope screen can be used to measure the voltage difference between two points.

The input wires are now connected to an oscillator, so that the voltage is changing with time. The spot on the screen moves up and down.

So we have a means of measuring voltages that are changing with time.

The oscillator frequency is gradually increased. The spot on the screens bobs up and down ever more rapidly. Finally, the spot blurs into a vertical line on the screen.

The horizontal sweep of the oscilloscope is turned on. This sweeps the beam from left to right, in proportion to the time elapsed. So we have a display on the screen of the voltage versus the time. At low frequencies we can see the spot trace out the wavy curve.

As the frequency increases, the loops of the curve crowd closer together and the motion of the spot itself fades into a continuous blur.

The generation of electromagnetic waves

A voltage that is changing rapidly with time is not yet a wave. Remember that in Chapter 6 a wave was defined as a "form, shape, or disturbance that moves." An *electromagnetic wave* is a pattern of voltages that travel across space.

The key to producing the electromagnetic waves lies in the oscillator box. This instrument produced the changing voltage that we saw on the oscilloscope. It did this by shuttling electric charges back and forth within its circuits. Because of this shaking of the electric charges, electromagnetic waves are being sent out.

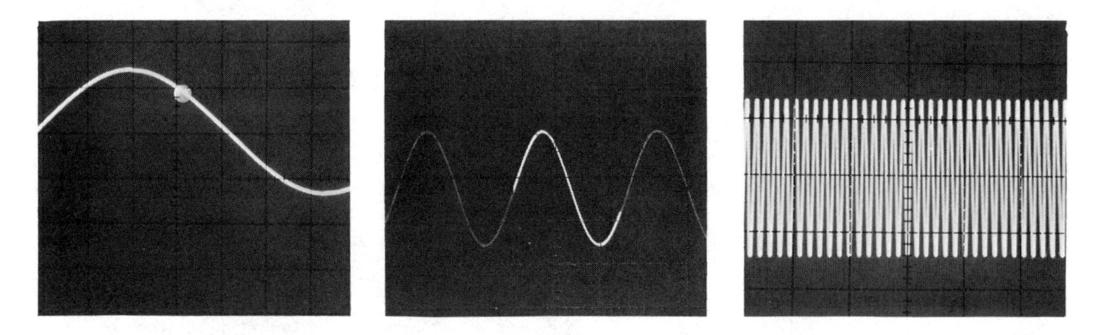

Figure 10-4 Oscilloscope traces of voltage signals of increasing frequency. Note how the moving spot becomes increasingly blurred. (Photographs by Barrie Rokeach)

HOW TO MAKE WAVES

An electrically charged object that has been in the same place for a while gives rise to a well-known pattern of electrostatic forces. It will repel other objects that have similar charges. It attracts those with opposite charges. The farther away the other object is, the weaker the attraction or repulsion will be.

Suppose that suddenly the charged object is jerked out of its resting state and sent rapidly in motion in some direction. The pattern of forces it exerts will now become different. It is closer to some objects and will repel, or attract, them more strongly. It is farther from others and will have a weaker force on them. In other cases, the *direction* of the force will change, since the attraction or repulsion is directed to its new position. Since the charge is now in motion, making a current, there will also be magnetic forces.

It is a principle of Einstein's theory of relativity that no signal, no information, can travel faster than speed c, 300,000 kilometers per second. This principle is built into all our modern theories, especially that of electricity and magnetism.

People far enough away from the charge, so far that no signal can have reached them yet, will not know that the charge has begun to move. The pattern of electrical forces they will observe is just as if the charge were still at rest in its original position.

Observers who are closer will have had time to get the message. They will see a pattern of forces that correspond to the new position and motion of the charge.

There is a boundary region between the people who can detect the new conditions, and those whom the information has not yet reached. This boundary is a rapidly expanding sphere, centered on the position of the central charge. The sphere is expanding at speed c.

In the vicinity of the boundary there is a sudden shift in the pattern of electric and magnetic forces. This sudden shift, this moving pattern, is the essence of the electromagnetic wave. It signals the arrival of the information that the charge has changed its state of motion.

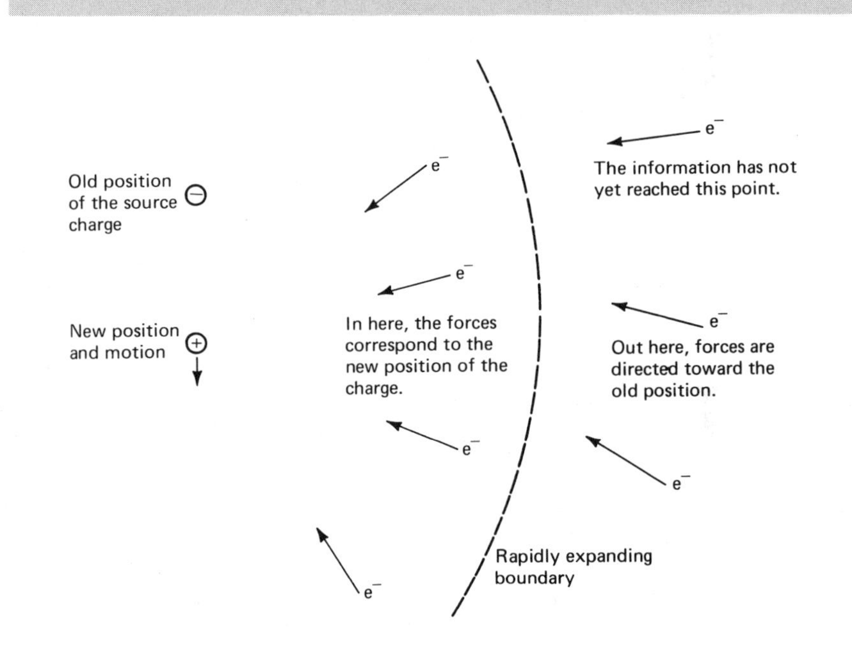

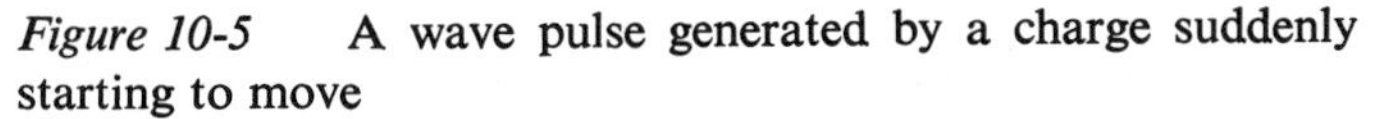
Figure 10-5 A wave pulse generated by a charge suddenly starting to move

As is explained in the box on the adjacent page, *electromagnetic waves are produced whenever electrically charged objects are accelerated.*

If the source charge is given a single jerk, and then left at rest again, the wave that results is just a single moving *pulse*. However, if the source is repeatedly accelerated back and forth, the resulting pattern is a *train of*

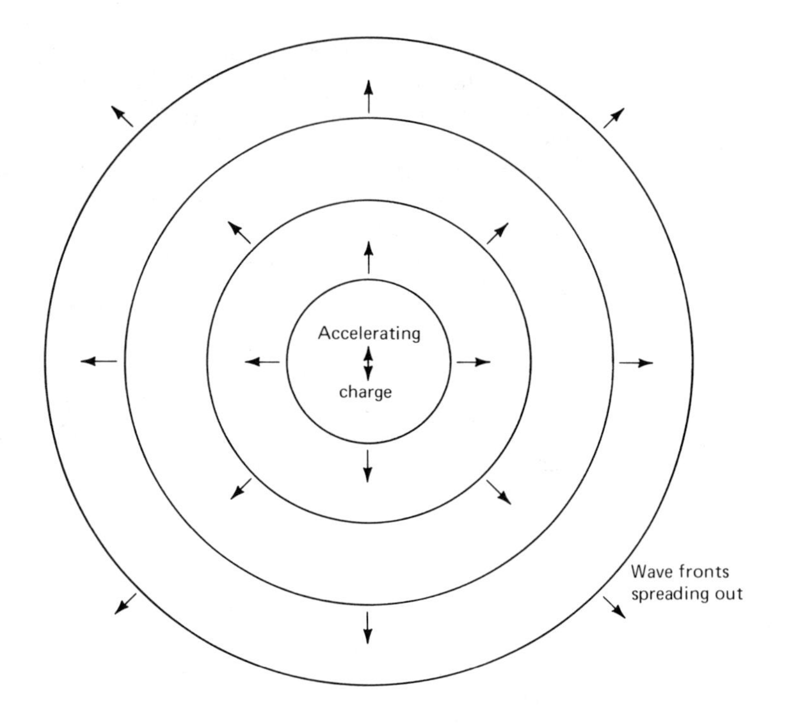

Figure 10-6 A wave train generated by a charge oscillating up and down

waves. Each time the source charge is accelerated, another *wave front* forms and propagates outward at speed c. Eventually we produce a series of wave fronts all spreading outward from the dancing charges at the center.

Frequency and wavelength

If we accelerate the source charge once per second, each wave front will move a distance of 300,000 kilometers away from the source before the next wave front is formed. We say that the *wavelength*, the distance between similar wave fronts, is 300,000 kilometers.

If the source is accelerated back and forth 60 times a second, each wave front can travel only 1/60 of 300,000, or 5000 kilometers, before the next wave front forms. Sixty cycles per second, or 60 *hertz*, as it is called, is the standard repetition rate of household electric current in the United States. Five thousand kilometers is roughly the distance form New York to Los Angeles. Signals in this range of repetition rate are known as very low frequency (VLF).

The reader will recall, from Chapter 6, the relation between wavelength and frequency. This is that the product of the wavelength of a wave times the frequency of its repetition is equal to the speed with which the wave propagates. Electromagnetic waves travel at speed c in empty space, so for these waves, wavelength times frequency equals c. The more rapid the frequency of repetition, the closer together the wave fronts; hence, the shorter the wavelength will be.

In discussing the diffraction of waves, we learned that a wave can bend around an obstacle comparable in size to, or smaller than, its wavelength. Very low-frequency waves have wavelengths comparable to the size of Earth.

Such a wave can reach all parts of Earth's surface by simply bending itself around all obstacles. It is easy to see why some military people have been interested in VLF for emergency long-distance communications. Shorter wavelength radio waves can travel around the earth only by bouncing off the electrically charged layer at the top of the atmosphere, which may not be dependable at all times.

There are disadvantages to using VLF for communications. The rate of information that can be sent is low, roughly only 60 dots or dashes per second. An efficient apparatus for generating waves has to be comparable in size to the wavelength. In the case of VLF waves the generating equipment would have to be hundreds of kilometers in length.

Radio waves and radar

By increasing the repetition rate we can generate radio waves of shorter wavelength. This is an advantage because electromagnetic waves are easier to generate and to detect, the more closely the size of the antenna matches the wavelength of the radiation.

Table 10-1 gives typical wavelengths and frequency for some commonly used types of electromagnetic waves. One hertz means 1 cycle per second. Broadcast band radio waves have no difficulty bending around hills, whereas FM radio and television stations are generally limited to the line of sight. On the other hand, long wavelength signals have difficulty penetrating holes that are much narrower than the wavelength.

If waves can bend around objects smaller than their wavelength, this also means that they are not satisfactory for examining the details of such objects. If we wish a radar instrument to detect an object like an airplane or a car,

Table 10-1 SOME MAN-MADE ELECTROMAGNETIC WAVES

Type of wave	*Typical frequency*	*Typical wavelength*
Very low frequency	60 hertz	5000 kilometers
Broadcast band radio	1,000,000 hertz (or 1000 kilohertz)	300 meters
FM radio and TV signals	100,000,000 hertz (or 100 megahertz)	3 meters
Microwaves, radar	10 billion hertz (or 10 gigahertz)	3 centimeters

much less give us information about its size and shape, we must use wavelengths that are appreciably shorter than the size of the object under examination. This is why such short wavelengths are needed for radar applications.

A radar signal works by sending out a burst of high-frequency, short-wavelength (a few centimeters) electromagnetic waves. These waves will bounce off objects that are (1) made of material that is at least partly an electrical conductor, and (2) much larger in size than the wavelength of the radar signal. By timing how long it takes for the reflected signal to bounce back, we can calculate how far away the detected object is. If the object is moving, there is also a Doppler shift (see Chapter 6). The frequency of the reflected signal will be different from the transmitted signal. This information can be used to determine the speed of the detected object.

AM AND FM RADIO

A radio signal that repeats the same pattern forever does not carry much information. At most it tells the receiver that the station sending the signal is on the air. In order to convey more information than that, we have to make variations in the signal that can be detected and understood by the person, receiving them. Such variations in the basic repeating signal are known as *modulations*.

The simplest kind of modulation is simply to turn the signal on and off for various lengths of time. A sequence of shorter or longer wave trains ("dots and dashes") can be transmitted, following Morse code or some other well-known system. This was the way electromagnetic signals were first used to send "wireless telegraph" messages over oceans and other long distances in the early 20th century. This method of transmitting information over waves is known as *pulse modulation.*

It is not necessary to turn the signal completely off to transmit information. We can easily vary the strength of the signal gradually up and down, over many cycles of the basic wave pattern. This method is known as *amplitude modulation* (AM).

One way to use AM is to take the pattern of *sound* waves representing an human voice and to modulate the strength of a high-frequency radio wave up and down in step with the ups and downs of the sound wave pattern.

The sound waves that humans can hear have patterns that change relatively slowly with time. The highest-pitched whistle you can hear fluctuates no more often than 20,000 times per second. On the other hand the electromagnetic wave that is used as a *carrier* may have a repetition rate of 1 million cycles per second. When such a radio wave is modulated with a voice pattern, there will be at least 50 repetitions of the carrier wave between even the fastest changes in the modulation signal.

When an amplitude modulated radio signal is detected, the problem of decoding it is more complicated than in the pulse modulated case. All of this decoding is done by electrical circuitry.

The antenna responds to the electromagnetic wave by letting its electrons

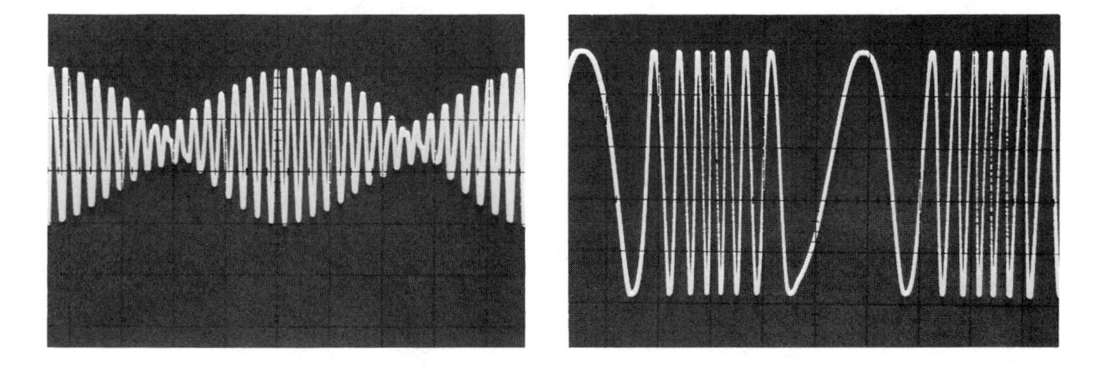

Figure 10-7 (a) An amplitude modulated (AM) signal (b) a frequency modulated (FM) signal (Photograph by Barrie Rokeach)

dance up and down in synchronism with the wave pattern. This produces a time-varying voltage within the circuits that mimics the wave pattern. A *tuned circuit* is used to select only those wave patterns whose carrier frequency is close to the one we want. In this way we "tune out" all the other stations that may be on the air at the same time. Next, a *rectifier and filter* circuit "smooths out" the high-frequency carrier wave so that only the slow modulation signal remains. An *amplifier* circuit then increases this signal in size, so that it is strong enough to drive the *speaker*. The speaker vibrates back and forth in step with the detected modulation signal. This sets up sound waves in the air much like the waves that were first used to shape the modulation signal. Finally, our ear detects the sound waves, and our brain does the final decoding.

Amplitude modulated electromagnetic waves can sometimes run into problems. The strength of the signal can vary for other reasons than that they were modulated by the sending station. Electrostatic charges in the atmosphere, for example, can have undesired effects on radio waves. This leads to distortion and "static" at the receiver.

One way to get around this problem is to use frequency modulated (FM) radio transmission. In this system the carrier wave does not have a single frequency. Its frequency varies depending on the strength of the modulating signal. If the modulating signal is strong, the carrier will have a slightly higher frequency than normal. If the modulating signal is low for a moment, the carrier will have a slightly lower frequency.

It takes a more complicated detecting circuit to decode an FM signal. But the final decoded modulation signal does not depend much on how strong the carrier signal is. So it is likely to be more free of atmospheric distortion.

A radio station that transmits voice signals does not use just a single frequency, but rather a *band* of frequencies. This should be obvious in the case of an FM station, but it is also true for AM stations. The band of frequencies

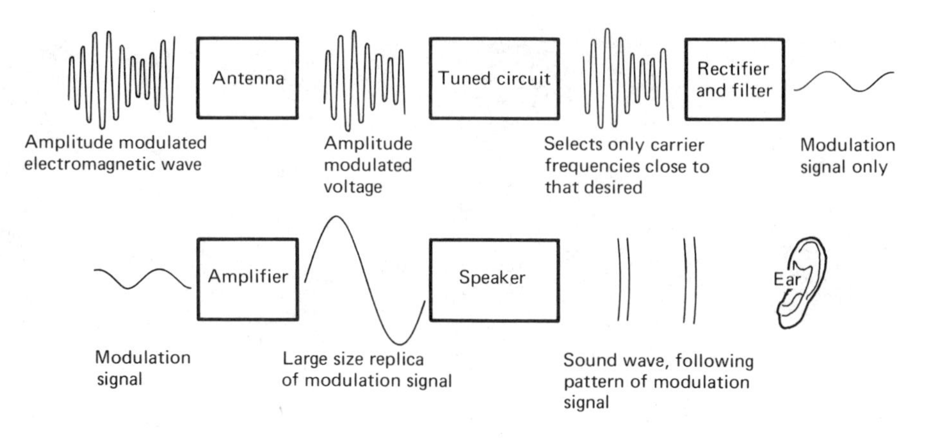

Figure 10-8 Inside an AM radio receiver

must be at least wide enough to accomodate the human voice, which ranges up to 20,000 hertz. To avoid interference between neighboring stations, they must have central frequencies at least 50,000 hertz apart.

The AM radio band, which runs from about 500,000 hertz to about 1,500,000 hertz, can accomodate at most about 20 stations separated by 50,000 hertz each. The same number of FM stations can operate between 100 megahertz and 101 megahertz, which is only a small fraction of the total FM band.

The detection of electromagnetic waves

In principle, electromagnetic waves can be detected by an apparatus not very different from that which generated them. It is necessary to put electric charges that are free to move in the path of the wave. These are most easily to be found in the interior of a piece of electrically conducting material, such as a strip of metal wire. This is called an *antenna*.

In response to the shifting pattern of voltages in the electromagnetic wave, the electrons in the antenna will dance up and down. This motion of the charges can be detected in many ways. The oscilloscope demonstrated earlier in this chapter is one such way. Another is in a radio receiver, as described in the accompanying box.

If the wavelength of the wave being detected is much shorter than the antenna, the pattern of voltages in the wave will reverse itself many times along the length of the detector. This means that some electrons will move in one direction, while others in a different part of the antenna will be moving the opposite way. Thus, they will cancel out each others' effect, and the net signal received will be small. It is clear that a detector cannot be longer than the wavelength of the wave you are measuring.

On the other hand, if the antenna is too short, it will not have enough electrons in it to yield a detectable signal. So the ideal antenna should be as long as possible up to about one-fourth of the wavelength of the waves being detected.

The electromagnetic waves discussed so far have all been generated by man-made apparatus. They are detected by human-scale apparatus. Thus, we can measure the voltage between points in the air by putting up an antenna, connecting it to an oscilloscope, and noting the deflection of the electron beam.

As we proceed to shorter wavelengths and higher frequencies this becomes more difficult to do. The limit of modern techniques to measure voltages directly is reached at wavelengths shorter than about one-tenth of a millimeter. Beyond that we must use indirect means to measure the energy involved, and we must make use of natural systems (such as atoms) to generate the radiation.

Infrared radiation

Infrared radiation occupies the region of the electromagnetic spectrum from a wavelength of about 0.01 centimeter down to about 0.0001 centimeter. At the upper limit we are dealing with sizes so small that the unaided human eye can barely resolve them. At the lower limit we are talking about sizes almost too small to be distinguished under the best optical microscope.

Infrared radiation cannot be seen. Neither can it be detected by ordinary receivers with antennas. It occupies a region of the spectrum between microwaves and visible light. It has some features in common with both.

Like visible light, infrared radiation is emitted by the sun. However, much of it does not reach the ground because it is absorbed by the air above us. Like microwaves, infrared radiation can be absorbed by many materials, the energy it carries being converted into heat. If it is intense enough, we can detect infrared radiation by the sensation of heat it produces in our skin.

Our skin is not a reliable detector of infrared, for there are many ways that heat can be carried to us. The one feature that distinguishes infrared from the other carriers of heat is that *only electromagnetic waves can travel across empty space.*

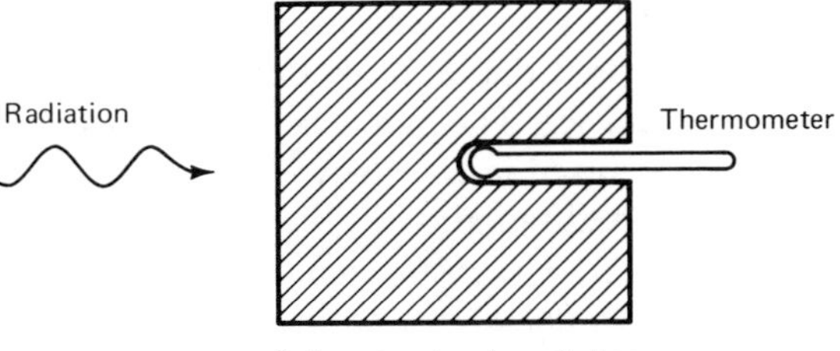

Figure 10-9 A bolometer

The bolometer

The instrument commonly used to detect infrared radiation is called a *bolometer. The principle of this instrument is so general it can be used to detect radiation of almost any sort.*

A bolometer is a piece of material that absorbs all the radiation that strikes it. Hence, a bolometer that works for visible light will appear black in color.

As a consequence of the absorption of the electromagnetic energy, the heat energy of the bolometer will increase. Hence, its temperature will rise. Knowing its specific heat, and knowing the conversion factor from calories to joules, we can calculate how much electromagnetic energy was absorbed. So our measurement of infrared radiation depends on our previous knowledge of heat as a form of energy.

The sources of infrared radiation

The most common source of infrared radiation is warm objects. The kinetic theory says that heat energy is really the mechanical energy of molecules. Remember that atoms and molecules are known to have electrically charged parts. If they are constantly moving back and forth, colliding, and otherwise accelerating, then by Maxwell's laws they will be radiating electromagnetically.

The higher the temperature, the faster the motion. Hence, we expect that as the temperature increases:

1. The total energy radiated will increase.
2. The average frequency will increase; hence, the average wavelength will get shorter.

Demonstration: Radiation from a Lamp Bulb. The thin wire in a lamp bulb is heated by dissipating electric current through it. When the temperature is moderate we can see no change in the wire. But the outside of the bulb does get warmer, indicating that some heat energy is being transmitted. The temperature is increased and the wire is seen to glow a dull red. The temperature is increased further and the light emitted is brighter and yellower. As the temperature continues to increase the lamp becomes a brilliant white color.

The radiation emitted by a warm body consists of many wavelengths at once. There is a *spectrum* of radiation emitted. The spectrum can be represented by the curve shown in Fig. 10-10.

The light that appears dull red has a lot of infrared radiation present that we can't see. The light that appears white has blue and red and all the other colors mixed in together.

Infrared radiation is about as nondescript as any form of electromagnetic radiation. The instrument used to detect it, a bolometer, would be equally useful to detect any form of radiation. The natural way to produce it, by radiation from warm bodies, could be used to produce any part of the spectrum, if the temperature were set properly.

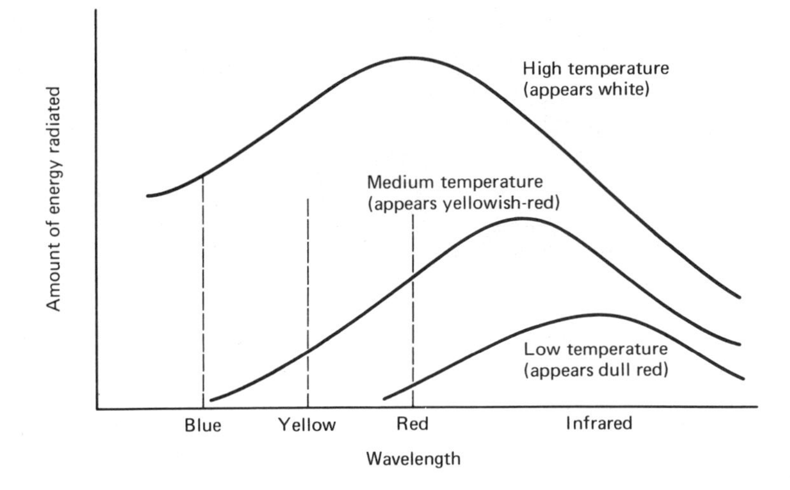

Figure 10-10 Typical radiation spectra from warm objects

And yet, infrared radiation is the most plentiful form of electromagnetic waves on Earth. A visitor from outer space who could see this radiation might be impressed by how brilliantly Earth and its inhabitants shine in the infrared. Perhaps our senses have been adapted to ignore this form of radiation, because it is so abundant, just as we ignore the presence of the air in which we move about, just as a fish must ignore the presence of the water.

Infrared waves, at the long wavelength end of that spectrum, may be the most abundant radiation in the universe. Since the mid-1960s we have been aware of the presence of a low-level radiation that seems to pervade all space. This universal radiation peaks in intensity at a wavelength of about 1 millimeter. It has a spectrum and intensity that correspond to the radiation emitted by a black body at a temperature of 2.7°K, that is, 2.7 degrees above absolute zero. Speculation about what this universal black body radiation could mean must wait for a later chapter.

INFRARED RADIATION FROM A MAN'S HAND

These pictures were taken with a detector that is sensitive to infrared radiation. The radiation is emitted by a man's warm hand. The warmer parts of the hand appear brighter in the photo.

The hand in the picture at the left appears uniformly bright. It shows the normal distribution of temperatures in a hand. He then smokes a cigarette for a few minutes. The biological effect is to reduce the circulation of his blood. So the extremities of his body are reduced in temperature. The picture at the right, taken under these conditions, shows his palm still bright but his finger-

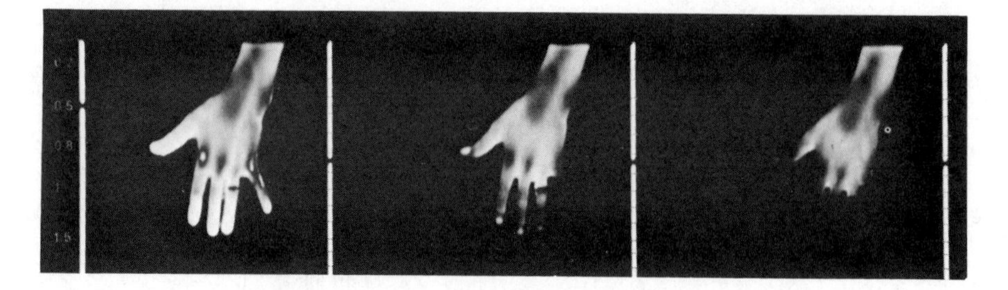

Figure 10-11 (a) Normal hand. (b) After smoking for three minutes. (c) After smoking six minutes. (Thermograph, courtesy AGA Corporation, Secaucus, NJ)

tips no longer visible. The amount of infrared radiation is quite sensitive to small changes in temperature.

Glossary

Some of the terms discussed in this chapter, having to do with wave motion, (such as frequency, wavelength, etc.) are in the glossary for Chapter 6.

amplifier—A device that makes a large copy of any signal that is fed to its input side. Most amplifiers work on electrical signals, but there are some mechanical amplifiers too. The energy of the amplified signal is taken from the power supply that every amplifier must have.

amplitude modulation (AM)—A common way of attaching a signal to a high-frequency carrier wave. The carrier wave is made stronger or weaker at various times, in proportion to the signal that is being sent.

antenna—A length of conducting material that can be used either to send out or to receive electromagnetic waves. Antennas work better the longer they are, until they reach one-quarter of the wavelength of the waves for which the antenna is used.

bolometer—A device for detecting any kind of radiation, by absorbing it completely and converting the energy into heat.

broadcast band—The range of frequencies from 500,000 to 1,500,000 hertz (wavelengths between 600 and 200 meters) of electromagnetic waves, used in this country by AM radio stations.

carrier wave—A wave of high frequency that can be made to carry information by having gradual changes ("modulations") made in its strength or shape.

cathode-ray tube—An evacuated glass-walled device, inside which a beam of electrons is directed onto a screen. The cathode ray tube can be used as a sensitive detector of electrical signals, as a television screen, and as an X-ray generator, among other uses.

electromagnetic wave—A pattern of voltages in space that moves with time. Radio waves and visible light are examples of electromagnetic waves.

filter—A device to remove certain frequencies from a signal, and to pass others. For example, the electronic filter in a radio circuit will block rapidly varying signals, such as the carrier wave, and pass slowly changing voltages, such as the modulation signal.

frequency modulation (FM)—An advanced method of attaching signal a to a carrier wave. The frequency of the carrier waves is changed slightly, in proportion to the signal that is being sent. Frequency modulated signals will not be distorted by effects, like static electricity in the air, that change the strength of the carrier wave.

infrared radiation—The range of electromagnetic waves whose wavelengths are too short to be generated by man-made antennas (less than a few tenths of a millimeter), and too long to be seen by our eyes (more than 0.7 microns).

microwaves—The range of electromagnetic radiation whose wavelength is shorter than about 1 meter (the limit of technology about 1940) and longer than a few tenths of a millimeter (the limit of our ability now). Microwaves are used in radar and communications. Their ability to penetrate dry tissues, but to be absorbed readily by water, makes them useful as a cooking device.

modulation—The gradual changes that can be made in the shape or strength of a high-frequency carrier wave, in order to make the wave carry information.

pulse modulation—The simplest method of attaching a signal to a carrier wave. The wave is turned on and off for various lengths of time. A signal in Morse code can be sent in this manner. A more powerful method is "pulse-code modulation," in which the size of the signal (such as a voice or a TV scan) is sampled many times, the measurements are converted into whole numbers, and the code for these numbers is the information sent out over the waves.

radar—("*ra*dio *d*etecting *a*nd *r*anging") A system that uses microwaves of a few centimeters wavelength to detect the presence of large metal objects, such as airplanes, ships, and cars. A pulse of microwaves is emitted by the radar source. When the pulse is reflected back from the metal object, the time of flight tells how far away the object is. The Doppler shift tells how fast it is moving.

radio waves—Electromagnetic waves having a wavelength longer than about 1 meter and up to hundreds of meters. They are much used in communication.

rectifier—An electronic device that passes current in one direction, but not in the other. One use of a rectifier, together with an electronic filter, is to produce an output equal to the slowly varying strength of a modulated signal, while throwing away the rapid variations of the carrier wave.

speaker—The part of a radio receiver or phonograph in which the electronic signals are converted into mechanical motion, setting up sound waves in the nearby air that the listener can hear.

spectrum—The way the energy of a source of waves is distributed among the different wavelengths. For example, we can say that "the spectrum of a hot tungsten wire has a peak in the infrared region." The word is also used to mean the whole range of possible wavelengths, as when we talk about the "electromagnetic spectrum."

tuned circuit—The part of a radio receiver, or other electronic device, that will accept only one narrow band of carrier wave frequencies (that is, one particular radio station) and reject all others.

Very Low Frequency—The range of electromagnetic waves with wavelengths on the order of hundreds of kilometers. Such waves can penetrate the deep ocean and bend around the earth by diffraction. They have a low rate of carrying information, however, and require enormous antennas.

voltmeter—A device for measuring the voltage difference between two points.

Questions

1. Explain why two wires carrying current do not exert any *electrostatic* forces on each other, although they do exert *magnetic* forces.
2. What is the difference between an electrical *current* and an electromagnetic *wave*?
3. Use the principle of relativity to show that a charged object moving at *constant* velocity does not emit waves. Hint: How does the pattern of its electrostatic attractions and repulsions appear to observers who are moving at the same velocity as the charged object?
4. Why would AM radio signals have difficulty penetrating a long tunnel? Explain why an antenna strung along the inside of the tunnel improves reception. Hint: The antenna is both a receiver and a transmitter.
5. Look at an AM radio and FM radio. What is the highest and the lowest carrier frequency on each dial? To what wavelength do these correspond?
6. What is the typical size for a television antenna? The first television channel in the United States operates at about 60 megahertz. What wavelength does this correspond to?
7. What is the typical size of an UHF (Ultra High Frequency, channels 14 and above) television antenna? A typical frequency for an UHF station is 200 megahertz. What wavelength does this correspond to?

8. What are some ways that heat can be transmitted other than by infrared radiation?

9. Infrared radiation was sometimes known as Radiant Heat. Why do you suppose this name was applied? Explain the difference between heat energy and infrared radiation.

10. What would ever make us believe that infrared radiation is a form of electromagnetic waves? What experiments might be done to tell whether this belief is correct or not?

11. It is generally true that electromagnetic waves do not penetrate through materials that are good electrical conductors. Why should this be so? What happens to the energy of an electromagnetic wave that is trying to penetrate a conductor?

12. A wire-mesh cage is nearly as good in shielding out low-frequency electromagnetic signals as a box of solid metal. Explain why. But also explain why the argument does not work for visible light, which is also an electromagnetic signal. (Can you see through a wire mesh cage? Will a radio work inside a wire mesh cage?)

13. Why should a frequency-modulated signal be less sensitive to "static" in the air than an amplitude-modulated signal?

14. Suppose your eyes could see infrared radiation. What do you imagine the room you are in would look like? What would be the brightest objects? Would you be able to see through walls?

15. The speed of light in glass is only about 200,000 kilometers per second. What will happen to the wavelength of a given type of electromagnetic wave when it begins to travel through glass? Hint: The frequency should not change, since that is determined by the rate of oscillation of the source charges.

11

visible light and ultraviolet

WAVES AND PARTICLES

Introduction and summary

The bulk of our information about the world around us comes to us through our eyes. What we see and what we read make up the most important sources for what we learn and what we know. No human is so severely handicapped in perception as one who has lost the sight of both eyes.

The human eye is a very efficient detector of a very narrow band in the electromagnetic spectrum, the band that we know as *visible light*. This band extends only from a wavelength of 0.00007 centimeters down to a wavelength of 0.00004 centimeters, less than a factor of two. Even the AM radio band, whose longest wavelengths are more than three times as long as its shortest, seems richer. Yet, just as the radio band includes many stations, so the visible light band comprises all the colors of the visible spectrum. Red, yellow, orange, green, blue, and violet light, and all the shades between, are seen to differ mainly in the wavelength and frequency of the waves that carry the energy of the light.

Isaac Newton did many experiments with light, about the year 1700. He understood that what appears as white light is a combination of all the colors of the rainbow. He recognized that objects having a definite color, have it because they reflect that color better than the others.

But Newton did not know that the difference between colors was their wavelength. He did not think of light as waves at all. Newton thought of a light beam as a stream of particles, like a spray of pellets from a shotgun.

Other scientists in Newton's time had suggested that light might consist of waves. But so great was Newton's authority that the particle theory of light persisted for a century after his death.

In the early 19th century other experiments showed that light does propagate like waves. Effects like *diffraction*, the bending of waves around obstacles, and *interference*, the cancellation (or reinforcement) between overlapping wave trains, can be explained only on the basis of a wave theory of light. By the middle of the 19th century, Maxwell's theory had claimed to show that visible light was an electromagnetic wave. It travelled at the same speed, and had other properties in common with radio and microwaves.

Then in the 20th century, the pendulum of the theory of light swung back toward the particle idea. Einstein showed that experiments like the *photoelectric effect*, in which electrons are knocked out of atoms by light beams, could best be explained if light were composed of little packets of energy called *quanta*.

The wave versus particle controversy continued for a quarter century. The resolution of the problem came with recognition that light moves like waves, but interacts with matter like particles. Moreover, the same statement is true for electrons and other material objects that had always been thought of as particles. The theory that embraces this idea is known as *quantum mechanics*.

Quantum mechanics has been very successful in explaining nearly every experiment that can be done with atoms. But its success has come at the expense of a loss in our ability to picture the atomic world. An object that moves like a wave cannot be located more closely than within 1 wavelength. The picture we have now of electrons in atoms is a sort of smear, rather than that of a point object. In many cases quantum mechanics cannot predict the exact outcome of an experiment, but only the probabilities of various possible outcomes.

The debate over the interpretation of quantum mechanics is still a lively one. Scientists have little doubt that the theory works. But philosophers are not agreed what it means. Does the fact that the theory sometimes gives only probabilities mean that it is really incomplete? Or does this very incompleteness leave some room for free will in a science that for three centuries had seemed so deterministic? These are open questions, to which we encourage every student of physics to contribute his own thoughts.

Colors

It was Isaac Newton who first showed that white light is a mixture of the various colors. He sat in a dark room into which a ray of sunshine came through a small hole in a curtain. He put a wedge of glass in the path of the beam. The light coming out of the glass wedge had all the colors of the rainbow.

Newton concluded that what we perceive as white light is really a mixture of all the natural colors.

He then put a screen with a hole in it behind the wedge. He could move this screen so that only one color at a time passed through it. He then sat behind this screen and held up various objects to the beam of separated colors.

A tree leaf held up to the red light beam looked red. An apple held up to the green light beam looked green. Held up to the blue light beam, both the leaf and the apple looked blue. But the leaf was brightest when exposed to the green light beam. And the apple shone best when placed in the red beam.

Newton concluded that an apple looks red to us because it *reflects* red light better than any other color. It also reflects green and blue and other colors, but not as well. If no red light is present, we see the apple dimly in

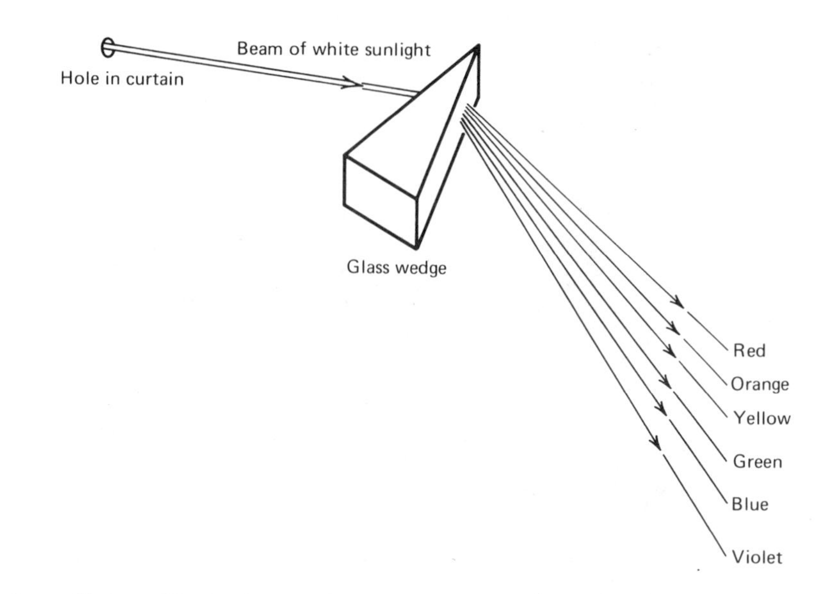

Figure 11-1 Newton's experiment to separate the colors in sunlight

the other colors. But if red light strikes the apple, it gets reflected so strongly that it outshines all the other colors, and we get the impression that the apple *is* red.

In the same way the tree leaf reflects green light more strongly than the other colors. We say that the leaf is green because, if green light is present, it gets reflected from the leaf so much more strongly than the other colors that we notice only the green.

WHY THE SKY IS BLUE

Throw up a cloud of chalk dust in the path of a beam of light. The track of the beam becomes outlined for us in the light that is *scattered* by the small bits of chalk dust.

Without the chalk dust, we could not see the light beam. It is not directed at our eyes. Each dust particle takes the small amount of light that strikes it, and scatters it in all directions. Some of it reaches our eyes. We say that we can "see" the light beam. What we are really seeing is the light that is scattered out of the beam.

If enough dust is placed in the path of the beam, nearly all of the light in it can get scattered. The beam itself is much weakened from passing through the dust cloud. Eventually there is no beam at all, only scattered light.

Water droplets in the air, such as in a cloud or a fog, can also scatter light. When we look at the sun through a thin cloud, we can often see a dimmed image of the sun's disk. The water droplets in the cloud have scattered much of the original beam. When the cloud is thick enough, we do not see any of the direct beam, but only the scattered light that seems to come from all over the cloud.

Chalk dust and fog droplets appear white when they are in a beam of white light. This is because they are equally good at scattering all colors of visible light. The scattered light is visible at almost any angle we look at it. This is because scattered light is sent nearly equally in all directions.

The molecules in the air can also scatter light. The amount of light scattered by each molecule is tiny. You cannot see even the light scattered by the enormous number of air molecules in a beam of light that passes across the room. But the combined effect of many molecules in a beam path several kilometers long begins to be visible.

The air molecules, unlike the chalk dust and the fog, are much smaller than the wavelength of visible light. This means that the molecules do not make good antennas for absorbing and rebroadcasting the light waves (which is basically how light-scattering works). The molecules work better as antennas, the more closely they are matched to the wavelength of the light. This means that they will scatter shorter wavelengths more easily than longer wavelengths. So the air molecules are much more effective at scattering shorter wavelength blue light than at scattering longer wavelength red light.

If we look at the sky overhead when the sun is in another part of the sky, the light we see is sunlight that has been scattered by the molecules high up in the atmosphere. Since these molecules are so much more effective at scattering blue light than longer wavelengths, the sky appears blue overhead.

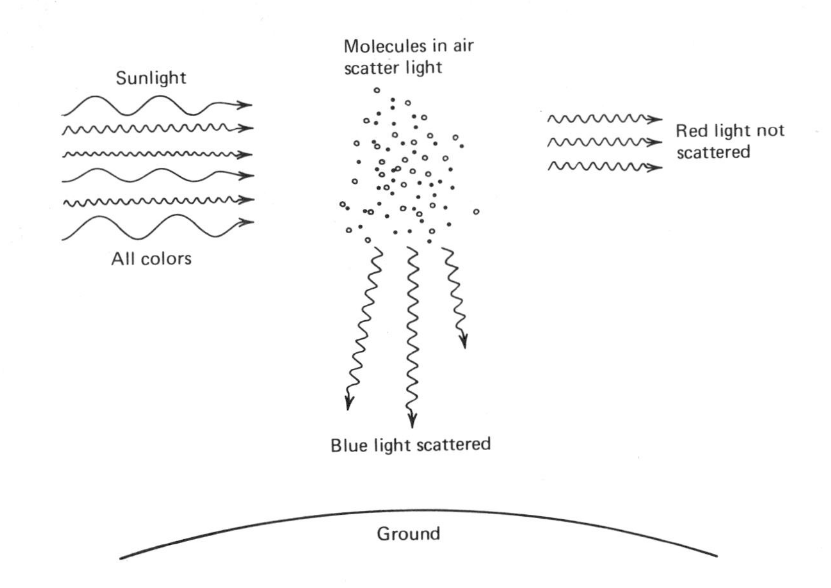

Figure 11-2

Why does the sky not appear violet? Violet light has even a shorter wavelength than blue, and should be scattered even more effectively. The answer in part is that some of the violet light gets *absorbed* by the air, its energy changed into heat and other forms. Also, our eyes are more sensitive to blue light than to violet. An equal mixture of blue and violet light would appear blue to our senses.

On the other hand, when we look toward the sunset, at sunlight that has come through many kilometers of air to reach our eyes, the light appears reddish. The blue light in the sunlight has been largely scattered out of the beam. What remains is the longer wavelength red light. So the sun appears red at sunset.

On the moon, or in outer space, there is no air to scatter the sunlight. The sun is an intense white—so bright that we dare not look directly at it—at all times of the lunar "day." The sky everywhere else is pitch black. Stars are visible in the daytime, since there is no scattered sunlight to drown them out. Any place that is in shadow is also absolutely black. Without an atmosphere to scatter the light, the contrasts from black to white are sharp everywhere.

Some molecules are better at scattering light than others. A type of molecule called *ozone*, which has three oxygen atoms rather than the usual two, can be found in the upper atmosphere. Being longer than ordinary air molecules, ozone molecules make better antennas for scattering visible light. Most of the effects that we have attributed to "air" molecules in this section are largely due to ozone. The ozone layer is also responsible for preventing most of the sun's ultraviolet radiation from reaching us on Earth.

Some "colors" that we are familiar with do not appear by themselves in the rainbow. We have seen that white is an equal mixture of all the pure colors. Black is the sensation we get when there is no visible light present. Various shades of brown are produced by mixing red, orange, yellow, and green light in different proportions. Purples and magentas are obtained by mixing red light with blue and violet.

Inks and paints work by absorbing all the light except the color they are supposed to be. Red paint absorbs most of the violet, blue, green, and yellow light that strikes it. It reflects red light well, and possibly some orange light less well.

Yellow paint absorbs violet, blue, and red light. It reflects yellow light strongly, and some of the neighboring colors, like green or orange, less strongly.

When we mix red and yellow paint, the yellow paint absorbs the red light, and the red paint absorbs the yellow light. The mixture will reflect light in the colors that both paints reflect in common; in this case, orange. So when we mix red and yellow paint, the mixture appears orange to our eyes.

The wavelength of visible light

The centimeter is an awkward unit for discussing wavelengths shorter than infrared. Let us make use of a more convenient one. We define an angstrom unit as follows:

1 angstrom unit = 0.000 000 01 centimeter

One hundred million angstrom units make 1 centimeter. The angstrom unit (abbreviated Å) is convenient in atomic physics because the diameter of a typical atom is about 1 Å.

Visible light occupies a very narrow region of the electromagnetic spectrum. There is less than a factor of two between the longest and the shortest wavelength of visible light.

Table 11-1 gives the ranges of wavelength usually associated with each color of the visible spectrum.

Table 11-1 THE WAVELENGTHS OF VISIBLE LIGHT

Color	*Range of wavelengths*
red	6470 to 7000 Å
orange	5850 to 6470 Å
yellow	5750 to 5850 Å
green	4912 to 5750 Å
blue	4240 to 4912 Å
violet	4000 to 4240 Å

Light as an electromagnetic wave

We have maintained since Chapter 3 that light consists of electromagnetic waves. But we do not pretend that we can really measure the voltage that varies rapidly as the wave passes us. The evidence for the electromagnetic wave nature of light is circumstantial.

What makes us think that light is *electromagnetic* in nature? For one thing, it travels at speed c, predicted for all electromagnetic waves and actually measured for radio and microwaves, which we know to be electromagnetic. For another, we know that light is produced in such obviously electrical discharges as lightning flashes and electric sparks. Light does not penetrate materials that are good electrical conductors, such as metals.

Light can be made to interact with various electrical devices. Photoelectric cells and photoconductors can signal the fact that light is striking them by beginning to conduct electric currents. Solar batteries put out useful voltages when struck by sunlight.

Light certainly carries energy and information. We can measure the energy carried by a light beam. We can use a bolometer to absorb all the energy in the beam, and see how much heat energy it turns into. As for information carried by light, the input we receive about the world through our eyes outweighs all our other senses by a wide margin.

What makes us think that light consists of *waves*? Here the evidence is all indirect. We do not actually observe the waveforms of light. But many of the features of waves discussed in Chapter 6 can be shown to apply to light.

If we look closely enough, we can detect light bending around obstacles, showing *diffraction*. We usually think that light travels in straight lines. The practice of surveying, for instance, depends on this fact. But this only appears to be true because the wavelength of visible light is so short. When the obstacles have sharp edges, with features smaller than the wavelength of the light, the light can bend around them. This diffraction is, of course, a characteristic of wave propagation.

Two or more light beams can cancel each other in the process known as *interference*. We can set up a situation in which each beam by itself produces a bright spot on a screen, but the two together produce no spot at all. Interference between *waves* can be understood. The two waves arrive at the spot in such a way that the crests of one wave train fill in the troughs of the other. The sum of the two wave trains is then an unchanging level, which is no wave at all. On the other hand, if we think of a light beam as a stream of particles of some sort, it is difficult to understand how interference can take place.

There are other features of light that help confirm our belief that light consists of electromagnetic waves. We can observe *refraction*, the sudden change in the direction of a light beam when it passes from air to a material like glass or water. The wave picture explains this effect on the basis that light travels slower in the heavy medium than in air. In Newton's particle picture, refraction was explained by having the particles travel *faster* in glass than in air. In fact, measurements show that light does travel slower in the heavy medium.

We can detect *polarization* of light. This is the difference in the waveform when the voltage pattern varies from top to bottom, or when it varies instead from left to right, as we observe it looking along the beam.

We can measure the *Doppler effect*, the change in wavelength and frequency when the source of light is moving toward or away from us.

Taken altogether, the arguments for the wave nature of light seem to be very strong.

Demonstration: Diffraction and Interference of Light. A laser is used to show the wave-like properties of light. The laser is not essential to the experiment. Any source of single-color light will do. But the laser makes a source bright enough for the whole class to see.

First, the laser light is passed through a pair of slits that are wide and spaced widely apart. We see mainly the image of the two slits. Even so, there are indications that the light doesn't follow exactly straight lines. There are dark and light fringes at the edge of the slit images, showing that light is

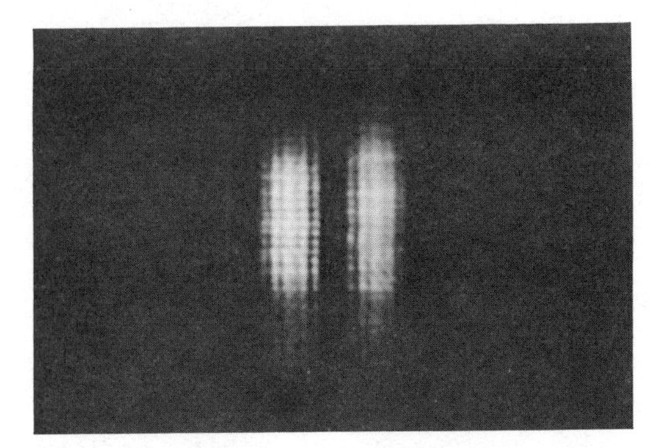

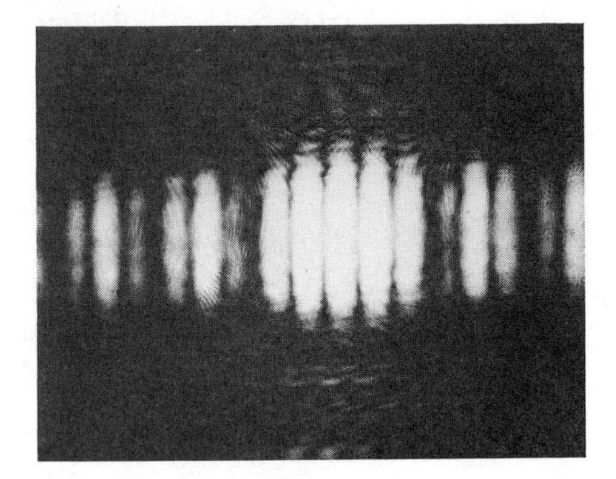

Figure 11-3 (*a*) Image of two wide slits well-spaced apart. Note diffraction fringes outside the region of the direct slit images. (*b*) Pattern observed from two narrow slits well-spaced apart. The diffraction pattern from each slit dominates over the direct image. (*c*) Pattern observed from two narrow slits close together. The waves emerging from the two slits now overlap, and interference is observed. The interference pattern of many equally spaced fringes bears little resemblance to the slit images. (Photographs by Barrie Rokeach)

getting into the region of the geometrical shadow. The first bright fringe occurs at such a position that light travelling from the far end of the slit has to travel two wavelengths farther than light coming from the near edge of the slit.

The fringes show the effect known as *diffraction*, of waves bending around obstacles. Diffraction experiments can be used to measure the wavelength of the waves.

As the slits are made narrower and closer together, the direct image, the one we would expect from drawing straight lines from the source through

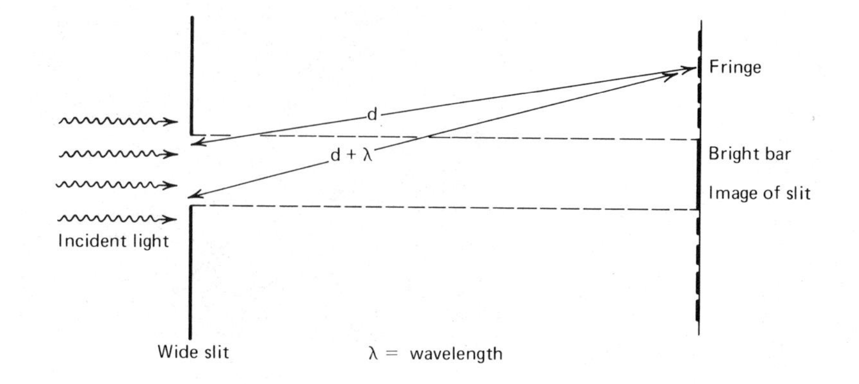

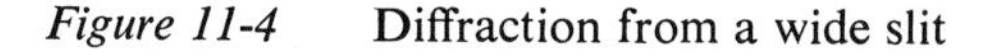
Figure 11-4 Diffraction from a wide slit

the slits to the screen, becomes less important, and the fringes become more pronounced. The wave properties of light are emphasized.

When the waves from the two slits begin to overlap, the effect of *interference* takes place. We may think of the waves as radiating from each slit as a series of concentric rings, alternating troughs and peaks. When the peak of one wave train happens to coincide with the peak of another, they reinforce each other, and we see a bright spot. When the peak of one wave hits the

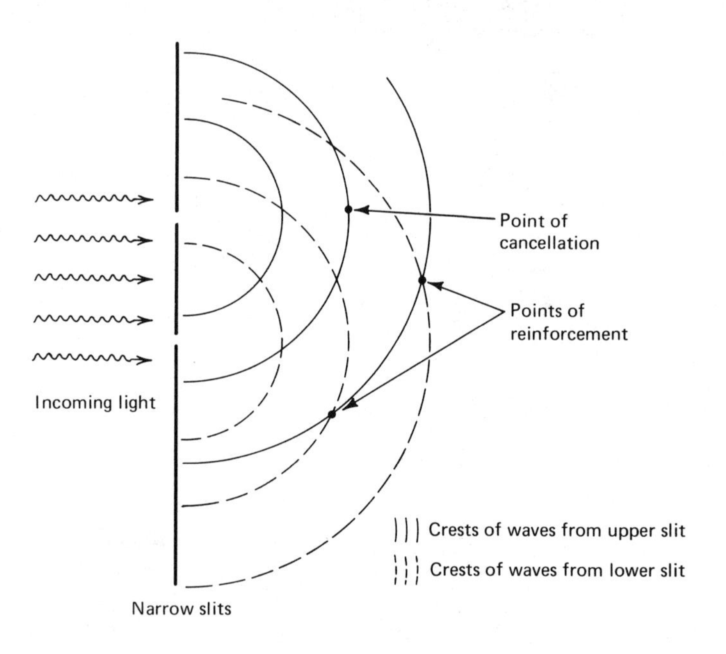

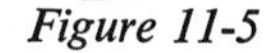
Figure 11-5 Interference between waves from two slits

trough of another they cancel each other, and we have a dark point in the pattern. So the overall pattern is one of alternating light and dark fringes.

Hold a pointer up to the screen in a position where there is a dark fringe in the overall interference pattern. Cover one slit. Notice whether some light now hits the pointer. Open the first slit and cover the other one. Notice that some light again reaches the pointer. But when both slits are uncovered the pointer is dark. The sum of two sources of light, each of which sends some light to the pointer when acting by itself, is thus shown to give a dark spot when the slits act together.

Interference and diffraction are considered to be characteristic properties of waves. So experiments like these are generally thought to be proof of the wave nature of light.

Ultraviolet

The range of wavelengths starting at the lower limit of the visible, below 4000 Å, is known as the *ultraviolet*. Ultraviolet light is very difficult to work with because it is so readily absorbed by so many materials. Glass will not pass most of the ultraviolet spectrum. Below about 3000 Å, even air will not transmit ultraviolet.

The reason ultraviolet is absorbed by so many materials is the existence of the *photoelectric effect*. Briefly, the photoelectric effect takes place when electromagnetic energy is absorbed by an atomic system in such a way that electrons are ejected from the atoms. We may study this effect by using a *photoelectric cell*.

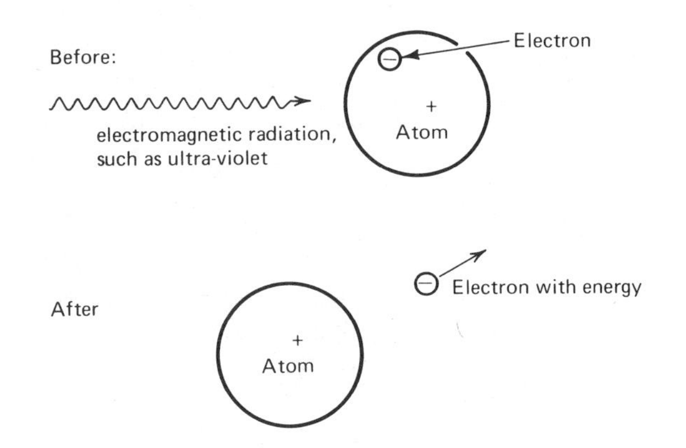

Figure 11-6 The photoelectric effect on an atom

A typical photoelectric cell has two conducting plates, called *electrodes*, in an evacuated bulb. We shine light on one of the electrodes, called the *photocathode*. Electrons escape from it. Some of them reach the other electrode, called the *anode*.

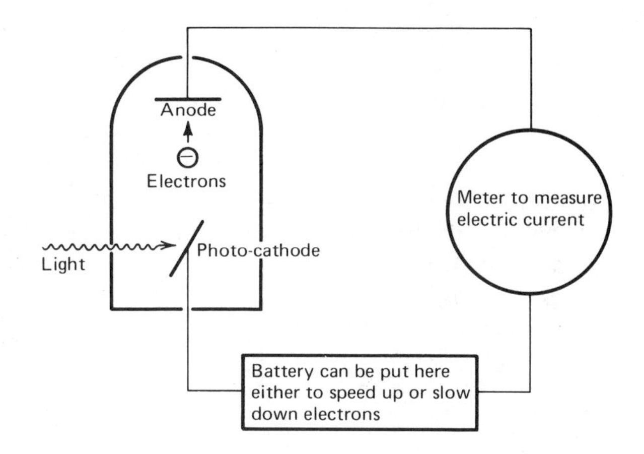

Figure 11-7 A photoelectric cell

The anode and the photocathode are connected, outside the bulb, by a wire. In the usual way, we can detect current flowing in this circuit. Such a current will flow when the right kind of light shines on the photocathode.

The electrons that come off the photocathode are quite energetic. They have enough kinetic energy to reach the anode even when a modest negative voltage is put there to repel them.

It is found that with any given photoelectric cell, light of too long a wavelength won't make current flow. If the wavelength is longer than a certain maximum, which is different for each cell, there will be no photoelectric current, no matter how bright the light is. Light whose wavelength is shorter than this maximum can make the photocurrent flow, even when the light is fairly dim.

The quantum idea

An explanation of the strange features of the photoelectric effect was proposed by Albert Einstein in 1905. He suggested that energy can be emitted or absorbed from electromagnetic radiation only in units of 1 *quantum*. The amount of energy that makes up 1 quantum is different, depending on the frequency of the radiation.

The higher the frequency, i.e, the shorter the wavelength, the more energy it takes to make a quantum of that frequency light. There is a simple formula that relates the energy of 1 quantum to the frequency of the radiation. We use a quantity known as *Planck's constant*, usually symbolized h, a very small number that keeps appearing in many places in atomic physics. The energy in each quantum is given by Planck's constant times the frequency. Planck's constant is so small that it takes 3 billion billion quanta of red light to make 1 joule!

At low frequencies, where the energy per quantum is small, even a very weak beam will have trillions of quanta in it. A radio receiver tuned to some radio station will absorb a few quanta (or even a few million!) from the beam sent out by that station. This does not keep the rest of the beam from continuing. A radio listener standing behind the first receiver can hear the same station on his own set. The loss of a few million quanta to the first receiver can hardly matter when there are so many to spare.

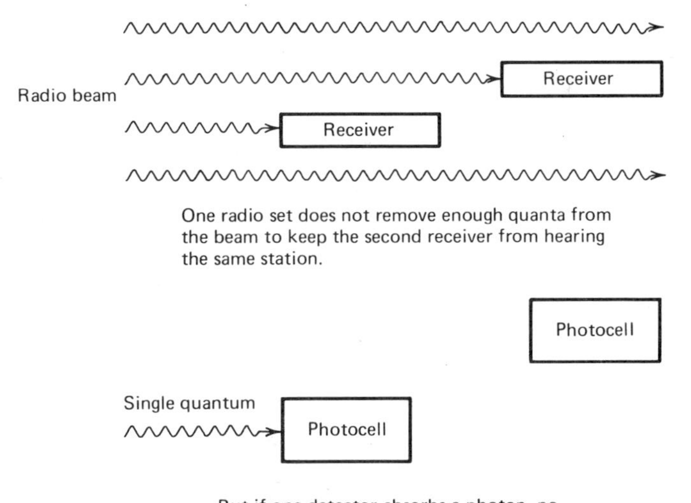

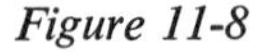
Figure 11-8

On the other hand, a low intensity beam of ultraviolet, or of some even shorter wavelength, radiation might consist of only a single quantum at a time. A beam of electromagnetic radiation whose total energy is only 1 quantum is known as a *photon*. There is an all-or-nothing aspect to such a beam. Either it is completely absorbed by some detector in the beam, or it passes through completely intact. No two eyes can see the same photon.

What does all this have to do with explaining the photoelectric effect? Consider a light beam striking a metal conductor. Its wavelength is a few thousand Ångstrom units. This means that the wave is spread over at least that distance in space. Within such an area on the surface of the metal plate, there are millions of electrons.

According to "classical" (before 1900) electromagnetic theory, all of the free electrons in the metal would respond to the changing voltages in the wave by dancing up and down a bit in rhythm with the wave. Each electron would thus absorb a small fraction of the energy in the wave.

We suppose that it takes a certain minimum amount of potential energy for the electron to escape from the metal. According to the classical picture, no single electron would ever accumulate enough energy to escape.

According to the quantum picture, however, if the energy of a quantum is less than what is required to remove an electron from the metal, it will still be true that no electrons will get out. This explains why long wavelengths (meaning low frequency, and hence low value of the quantum energy) don't work in making the photoelectric effect. But if the frequency is high enough, an electron can absorb 1 quantum, thus getting enough potential energy to escape from the metal, and even have some kinetic energy left over.

Of the millions of electrons in the path of the photon, one will be chosen to absorb all the energy. The other electrons are completely unaffected.

How can this happen? It is as if light was made up of little darts, aimed at random at the metal plate. Quite by accident, one of the electrons gets hit by a quantum of light, absorbs the whole quantum, and bounces out of the metal. This is the way light might behave if it were composed of a stream of particles.

But we have abundant evidence, which we have outlined already in this chapter, that light behaves like waves. How can we resolve the contradiction?

Waves versus particles

We see that there are some effects, like diffraction and interference, that are explained most easily if light behaves like a wave. On the other hand, there are experiments, most notably the photoelectric effect, that seem to prove that light has particle-like properties.

The wave versus particle controversy was not easy to resolve. The problem occupied physicists throughout the first quarter of this century. In the words of W. H Bragg, "On Mondays, Wednesdays, and Fridays we consider them waves. On Tuesdays, Thursdays and Saturdays they are particles."

The clue to this problem lay in noticing that the experiments that indicated wave-like properties for light mostly tested the way in which light moves from place to place. The experiments that showed particle-like properties mostly tested how light interacts with other kinds of matter. In short, light moves like a wave, but interacts like a particle.

This leads us back to the old question: Light waves are waves of what? After Einstein, we had rejected the idea of light as waves in the "ether." A picture had been substituted of electromagnetic waves as voltage differences in space, which oscillated back and forth, and could be detected with a suitable antenna. But such a detection system will work only if we are careful to extract only a small amount of energy from the beam. But if the beam consists of a single quantum, we cannot extract a "small" amount of energy from it. To detect it at all, we must absorb the whole quantum. So our electromagnetic model, while it may be useful in calculating some properties of a light beam, cannot really be directly verified when we get down to the quantum level.

Another problem: Consider the double-slit interference experiment. We shine light through a baffle with two slits in it. We observe a series of dark and bright parallel stripes on the screen. We can take a photograph of this pattern. We can even take a photograph with light so dim that we are sure that only one photon at a time goes through the baffle. This picture would have to be a long-time exposure, of course. The interference pattern persists, even in such dim light.

Very well, then, suppose a particular photon went through the slits and reached the film, and exposed a single grain in the emulsion. So we know the photon reached that particular grain and none of the others. What happens to the rest of the wave, when the quantum is absorbed?

Can we find out which slit the photon went through? Let us put a detector in front of one slit. Let it be a little mirror that reflects the light off to our eye every time a photon passes through that slit. If we do that, we absorb the whole photon whenever it goes through the slit we are watching. This is the same as if we closed the slit, so far as the interference pattern is concerned. Trying to see which slit the photon passed through leads to destruction of the interference pattern.

Max Born, in 1926, suggested that we interpret the waves as being waves of *probability*. The more intense the wave is at some point in space, the more likely we are to find a quantum absorbed at that position. But the photon is only at one place. When it interacts, it is then localized to one particular spot. Until that happens we do not have any knowledge of exactly where in the wave the photon can be found. We have only the probability of finding it at various places.

Philosophical problems

The idea of introducing probability into physics met with considerable intellectual resistance, from no less an intellect than Albert Einstein himself. Einstein said repeatedly that he could not believe that "God plays dice with the universe." The theory was strong enough to resist the attacks of the world's best-known scientist. Most physicists today would say that they accept the probabilistic interpretation of the wave function.

This is not probability in the sense of the Second Law. In the kinetic theory case, we could say, for example, that it was extremely improbable that all the molecules in a room would collect in one corner. But if such an event were ever to happen, we might in principle know it in advance if we made careful note of all the molecular motions. According to the quantum ideas, even the *information* about where a photon is about to materialize is not available.

The French mathematician Laplace, in about the year 1800, envisioned a "perfect mathematician" who could calculate, on the basis of the present position and speed of every molecule in the universe, using the principles of mechanics, exactly where every molecule would move next. Thus, the master computer could predict exactly the entire future. He could also work backward and reconstruct the entire past. Science in that era seemed to be indicating a world that was completely determined for all time, with no room for choice, no possibility of free will.

The advent of quantum physics brought science to the conclusion that the information Laplace required was not available. So the future cannot be predicted with absolute certainty. So there is room for free will after all.

For, suppose we desire to measure the position and speed of even a single electron. Let us use electromagnetic radiation as part of our detector. If we want to locate the electron precisely, we must use short wavelength light, since longer wavelengths will bend around small objects and lead to imprecision in locating the electron. But we can detect the electron only if it absorbs (or at least reflects) some of the light. But it cannot absorb or reflect a smaller amount than 1 quantum. Since we were using short wavelengths, the frequency of the radiation is high, so the quantum energy is high. Thus the electron gets a considerable "kick" from absorbing or reflecting 1 quantum. The price we pay for using short wavelengths to position the electron accurately, is that we give it such a kick that we are uncertain about its speed.

On the other hand, if we try to measure its speed more accurately, by using photons with low quantum energy (so there will be less of a kick when a quantum is absorbed), we pay the price of not knowing precisely where it is, since we are then using longer wavelength radiation.

The situation is summarized in the *Uncertainty Principle* of Werner Heisenberg. We may measure the position of a bit of matter as accurately as we wish, or else we may measure its *momentum* (the product of mass times velocity) as well as we desire, but we may not do both simultaneously. Quantitatively, the product of the uncertainty in position times the uncertainty in momentum, may be no smaller than Planck's constant.

Planck's constant is a very small quantity, so in most large-scale situations we can measure both position and speed closely enough for most purposes. But on an atomic scale the uncertainty principle becomes crucially important.

QUANTUM MECHANICS, THE 20TH CENTURY PHYSICS

Quantum mechanics is the set of principles that embody the wave-particle dual nature of matter and energy. In this century quantum mechanics has superseded Newton's principles of mechanics as the basic theory of physics. Of course, in any large-scale phenomenon—the motion of the planets or the design of airplanes, for example—quantum mechanics and Newton's principles give nearly identical results. An astronomer or an engineer would be likely to use Newton's laws, since they are simpler. But in atomic and subatomic calculations, quantum mechanics must be used.

According to the ideas of quantum mechanics all matter (electrons and atoms, as well as photons) exhibits both wave and particle properties in its motions and interactions. The advent of quantum mechanics has led to some important new features in the philosophical outlook science gives us. Among these are the ideas that:

1. Observation of the state of any system cannot be separated from the disturbance of that system.

2. Exact knowledge of the state of any system, in the sense of Laplace, is not obtainable.

3. We cannot predict the future, or reconstruct the past, with absolute certainty.

Quantum mechanics is not a negative theory, despite all the "cannots" in the last paragraph. It makes a great many important predictions about atoms, many of which have been verified. To illustrate the sort of new phenomena embraced by quantum mechanics, let us consider the idea of *metastability*.

In Newtonian mechanics a system in equilibrium may be stable, unstable, or neutrally stable. In a stable system (for example, a truck stuck in a ditch) any change in the system requires energy input. So the system tends to remain in its lowest energy state (in the example, at the bottom of the ditch). In an unstable system (for example, a stick balanced on its end) any small change releases energy, so the system is likely to move away from the balanced position. In a neutrally stable system (for example, a ball rolling on a flat table) changes can be made without any change in energy.

In a *metastable* system, any *small* change requires an input of energy. But there are states of the system with lower energy than the metastable state. So energy could be released if a sudden large change were to take place.

The example we will use is the water in a mountain lake. Energy from sunlight has caused water from sea level to evaporate and eventually reach the mountain lake as rainfall. Here the water is in a high state of gravitational

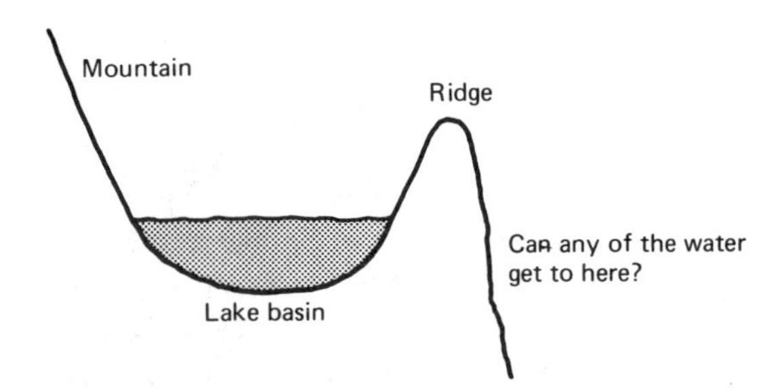

Figure 11-9 A metastable situation

potential energy. If the water were free to cascade down the mountain, much energy would be released in the fall. But there is a ridge blocking its fall. In order for the water to get over the ridge some energy must be supplied so that it can reach the top of the ridge first. Classically, the lake is in stable equilibrium. The winds that blow on the mountain may make the water slosh around a bit in its basin. But they do not supply enough energy for the water to reach the top of the ridge. So the water molecules, considered as particles, cannot escape from the basin of the lake.

Is it ever possible for a water molecule suddenly to appear on the far side of the ridge, at the same level as the lake surface? This position would require no energy input.

Surprisingly, quantum mechanics says that this may happen! The water molecules are governed by the uncertainty principle, which says that you cannot be completely certain that the water molecules are positioned inside the lake basin. There is at least some small probability that some molecule is on the far side of the ridge already. If that is so, the molecule may yet materialize there. If you insist that the wave function be so tightly constrained that there is no possibility that any molecules lie outside the lake basin, you do this only at the expense of leaving some probability that a molecule has enough energy to get over the top of the ridge anyway.

Glossary

angstrom unit (Å)—A unit of length commonly used in measuring wavelengths and sizes on the atomic scale. The diameter of a typical atom is about 1 angstrom unit. One hundred million such units make 1 centimeter.

anode—A metal conductor inside an electronic device (such as a photoelectric cell or a cathode-ray tube) that is kept positively charged, so that it attracts electrons.

Born, Max (1882–1970)—German theoretical physicist who did much work in the development of quantum mechanics. He made the suggestion that the wave that describes the motion of photons and of particles of matter, be interpreted as a wave of probability.

color—The sensation we get when our eyes detect visible light of various wavelengths. The color of an object is determined by which wavelengths of light it reflects most strongly.

diffraction—See Chapter 6.

Doppler effect—See Chapter 6.

electrode—A metal conductor inside an electronic device (such as a photoelectric cell or a cathode-ray tube). Depending on the charge kept on the electrode, it may attract or repel electrons.

Heisenberg, Werner (1901–1976)—German theoretical physicist who was one of the inventors of quantum mechanics. He is best known for the Uncertainty Principle, which sets limits on our ability to measure all the features of an atomic-sized system.

interference—See Chapter 6.

Laplace, Pierre Simon de (1749–1827)—French mathematician. He argued that, given a set of principles such as Newton's, the complete future of the world (and also the past) could be calculated from detailed knowledge of the present state. In such a scheme, there is no room for free will; everything that will ever happen is already determined.

metastability—The state of a system in which no *small* change can take place unless energy is supplied, but in which energy can be released if a *large* change (skipping all the positions in between) is allowed. Metastable states often occur in atomic systems under the laws of quantum mechanics.

ozone—A gas whose molecules have 3 atoms of oxygen, instead of the normal two. A layer of ozone is found at the top of our atmosphere, where it is effective in scattering light of shorter wavelengths. The ozone layer keeps ultraviolet radiation from reaching us. Its scattering is responsible for the blue color of the daytime sky.

photocathode—The metal conductor inside a photoelectric cell that releases electrons when struck by light of short enough wavelength.

photoelectric cell—A device that uses the photoelectric effect, putting out an electric current when it is exposed to light of the right wavelengths.

photoelectric effect—The release of electrons from atoms, metal surfaces, and other systems, when energy is absorbed from an electromagnetic wave.

photon—A beam of electromagnetic radiation so weak that its total energy is only 1 quantum, the minimum amount of energy that can be absorbed or emitted at the frequency and wavelength of that kind of radiation.

Planck's constant—A very small physical quantity that relates the energy of 1 quantum of any kind of radiation to the frequency of that radiation. Planck's constant is of such magnitude that the product of the wavelength of any kind of electromagnetic radiation (measured in angstrom units) times the quantum energy of that radiation (measured in electron-volts) is 12,400 ev-Å.

polarization—A property of transverse waves, that the disturbance can be either from side-to-side or up-and-down, with respect to the direction of motion of the wave. The fact that light waves can be polarized shows that they are transverse waves. This is what is expected of electromagnetic waves, according to Maxwell's theory.

quantum—The minimum amount of energy that can be absorbed from or emitted into a wave. Under the quantum idea, the energy in a wave can exist only in multiples of the quantum unit. When the last quantum is absorbed, the wave disappears completely.

quantum mechanics—The principles of the 20th century theory that have superseded Newton's principles of mechanics. Quantum mechanics takes account of the fact that both electromagnetic radiation (such as light beams) and material objects (such as electrons) move about like waves, but interact like particles.

refraction—The change in direction of a beam of light when it passes from one medium (such as air) to another (such as glass or water). Eyeglasses, telescopes, camera lenses, and microscopes all make use of refraction to focus and magnify optical images.

scattering—The random dispersal of part of a beam of light (or other kind of beam) by small objects in its path.

ultraviolet radiation—The range of electromagnetic radiation whose wavelength is shorter than 4000 angstrom units (the limit that our eyes can see) and longer than about 100 angstrom units. Ultraviolet radiation is readily absorbed by all kinds of materials, the more so the shorter its wavelength, because of the photoelectric effect.

Uncertainty Principle—One of the major principles of quantum mechanics: It is impossible to measure precisely both the position and the momentum of any object at the same time. The product of the uncertainty in one measurement times the uncertainty in the other is always at least equal to Planck's constant.

visible light—The range of electromagnetic radiation between 4000 and 7000 angstrom units in wavelength, which can be detected by our eyes.

Questions

1. Is an apple really red, or a leaf green? In the darkness, do they have any color at all? What do we mean when we say that an object has a certain color?
2. Why is it that when we mix red and yellow *light*, the mixture appears brown, but when we mix red and yellow *paint*, the mixture looks orange?
3. A painter trying to give the impression of many colors in a scene, will make separate spots of red, green, yellow, etc. Why doesn't he simply mix all these paint colors together?
4. How is *scattered* light different from *reflected* light?
5. The sun in the middle of the day has a yellowish cast. Why does it appear this color rather than white?
6. Pilots flying at extremely high altitudes report that the sky appears violet to them. Can you explain why?
7. How many cycles per second is the frequency associated with, say, orange light?
8. How can you use effects like interference and diffraction to measure the wavelength of a particular color of light?
9. When two light beams overlap in such a way as to cancel each other at one spot on the screen, what happens to the light energy that was in each beam?

10. If our eyes could see only ultraviolet light, what color would glass appear?

11. Do you find it strange that, in order to find the energy of a *quantum* of a particular kind of light, it is necessary to know the frequency of the *wave*?

12. Suppose a photon collides with a free electron at rest. The photon is not absorbed, but merely bounces off the electron, forming a scattered beam. (This is known as the Compton effect.) The electron recoils, thus taking up some kinetic energy. What will the relation be between the frequency of the scattered photon and the frequency of the original beam, taking into account the recoiled electron's energy? What would classical electromagnetic theory say would be the frequency of the scattered beam?

13. Give three pieces of evidence that light behaves like waves. What evidence is there that it sometimes behaves like a stream of particles?

14. Some physicists have tried to reintroduce certainty into quantum mechanics. They maintain that when a photon approaches a sheet of film, for example, there is some "hidden" pointer that determines exactly which grain in the film it will expose. Does such a theory have any advantages over conventional quantum mechanics?

15. Give two examples of metastable systems, one from the large-scale world, and one from the atomic realm.

12 x-rays, gamma rays, and $e = mc^2$

Introduction and summary

We continue our survey of the electromagnetic spectrum down to the shortest wavelengths. One hundred years ago, scientists did not suspect even the existence of such short wavelengths as we study in this chapter. The accidental discovery of *X rays* by Wilhelm C. Roentgen in 1895, and the almost immediate practical use of them, makes one of the great adventure stories in modern science.

X-rays occupy the region of the spectrum from roughly 100 angstrom units down to less than 0.1 Å. A dramatic change takes place when the wavelength is shorter than about 1 angstrom unit, because 1 Å is a typical size of an atom. The atoms make poor antennas for wavelengths smaller than the atom itself.

The photoelectric effect, by which quanta of light are absorbed in knocking electrons out of atoms, is very effective throughout the ultraviolet region. This is what makes ultraviolet so hard to use. It is absorbed by the first layer of atoms it meets.

But X-rays whose wavelength is shorter than atomic size are less effective at ejecting electrons. So they travel much farther through material. The penetrating nature of X-rays is their most obvious feature.

Gamma rays are not different from X-rays in their basic properties. They are however of even shorter wavelength. Any radiation with wavelength below about 0.1 Å is considered a gamma ray.

Whereas X-rays are usually produced by the motions, especially the sudden slowing down, of electrons, gamma rays are often created during the rearrangement of the charged particles within the nucleus of the atom. But gamma rays can also be made by decelerating electrons. Man-made gamma rays have been produced in this way with wavelengths shorter than 1 one-millionth of an angstrom unit. Even shorter wavelength may be found in outer space. There is no known lower limit to the wavelengths of the electromagnetic spectrum.

When the wavelength of a gamma ray is shorter than 0.01 Å, a new phenomenon can be seen. An electron and a positron *pair* can be *created* out of the energy of a gamma ray quantum. That is, a form of matter that starts with pure electromagnetic energy can be produced. This sort of experiment offers convincing proof that *matter itself is a form of energy*.

In his theory of relativity Einstein had made the statement that, as the kinetic energy of any object increased, its mass increased in proportion. This argued that mass and kinetic energy were just different aspects of the same thing, call it *mass-energy*.

But what of other forms of energy, and other forms of mass? Does a change in *potential* energy also reflect itself in a change in mass? Does the mass that an object has when its kinetic energy is zero, its *rest-mass*, also constitute a form of energy? Einstein speculated that the answer to these questions must also be *yes*. The observation of electron-positron *pair creation*, through the absorption of a quantum of gamma ray energy, gave dramatic proof that this answer was correct.

X-rays

The dividing line between very short wavelength ultraviolet radiation and what we commonly call *X-rays* is not sharply defined. We can place it roughly at a wavelength of 100 angstrom units. The lower limit of the X-ray spectrum is, again roughly, at a wavelength below one-tenth of an angstrom unit.

Like ultraviolet, the longer wavelength X-rays are quickly absorbed by any kind of matter. They can hardly penetrate even a few layers of molecules before they undergo the photoelectric effect with one of the atoms and lose a quantum of energy. Even to study radiation in this region of the spectrum requires that the whole apparatus be kept under vacuum.

A dramatic change sets in when the X-ray wavelength is less than a few angstrom units. The photoelectric effect still takes place. But the probability that any given atom will absorb a quantum drops by a factor of 1000 or more. So X-rays at shorter wavelengths can penetrate easily through thick layers of material. This penetrating feature is what we usually think of when we talk about X-rays.

Why is it that X-rays below 1 Å are so penetrating while X-rays above 10 Å are so quickly absorbed? The answer has to do with how the size of the atoms compares with the wavelength of the radiation. The size of most atoms is about 1 angstrom unit. We have discussed in an earlier chapter how an antenna becomes less effective when it is longer than one-quarter of the wavelength of the radiation it is detecting. So atoms, and the electrons in them, do not make good antennas when the wavelength is too short.

The reader may be surprised that we can estimate the probability that the atom will absorb a quantum, by treating it as if it were an antenna detecting waves. This is one of the ironies of the wave-particle dual nature of electromagnetic radiation.

For lighter atoms like hydrogen, which has only 1 electron, or carbon atoms with 6 electrons, nitrogen with 7 or oxygen with 8, the electrons are spread out loosely over the full diameter of the atoms. These are the kind of atoms that make up most of the material in living tissue. They make poor

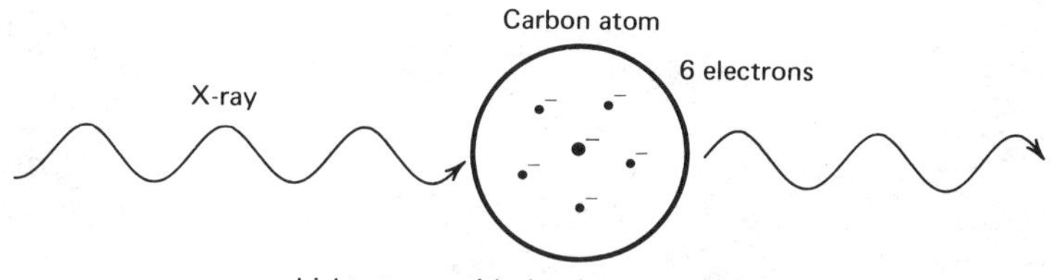

Light atoms, with the electrons widely spaced, make poor antennas for absorbing X-rays.

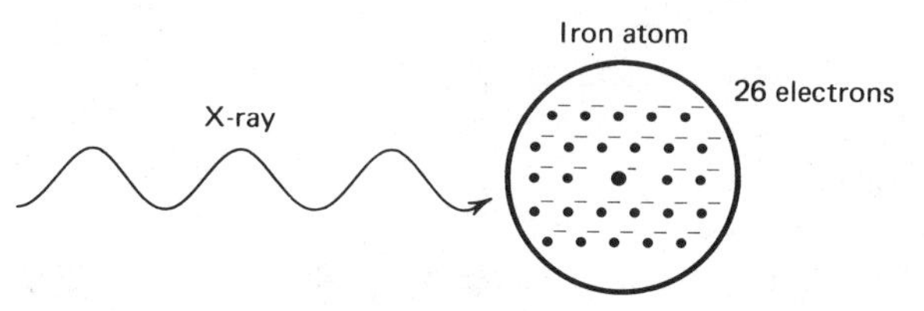

Heavier atoms, with the electrons closely packed, make better antennas. More X-rays are absorbed.

Figure 12-1

X-ray antennas below 1 Å, and so the X-rays pass through these materials largely undisturbed.

Heavier elements, like calcium, which has 20 electrons in 1 atom, iron with 26 electrons, and lead with 82, have concentrations of electrons close in to their center. These concentrations, being smaller than the typical X-ray wavelength, are better adapted to absorbing X-rays. So X-rays are more readily absorbed by bones (containing calcium), most metals, and other heavy elements. This explains why X-rays transmitted through the human body will expose the film where they pass through fleshy parts, but cast a shadow where they strike the bone.

The methods for detecting X-rays depend on the photoelectric effect. In a photographic film, a quantum of radiation is absorbed to knock one electron out of a silver atom. The silver atom that has lost one electron is more chemically active than when it was neutral. It can be used to form the center of a visible grain of silver when the film is developed. This effect works for X-rays as well as for ultraviolet and visible light.

In an electronic radiation detector, the X-rays knock electrons out of the atoms of a gas. With so many free electrons present, the gas can conduct an electric current that we can measure.

How to make X-rays The method used today to make X-rays is not very different from the way Roentgen first discovered them. We use an evacuated glass tube with some metal *electrodes* sealed inside. The electrodes are con-

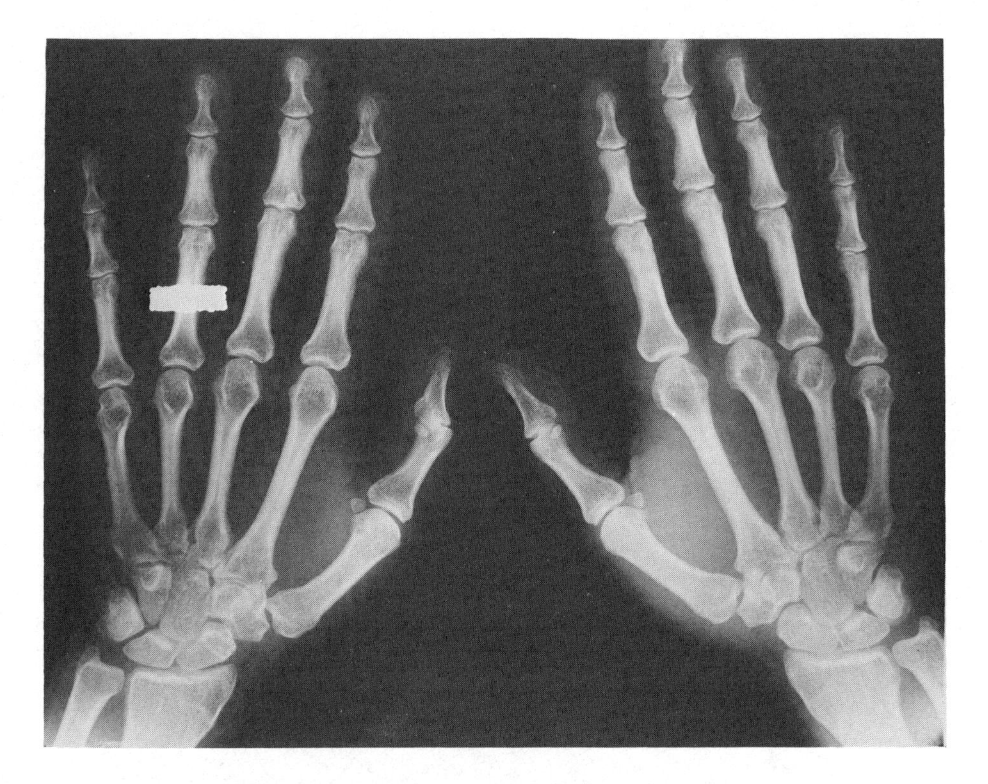

Figure 12-2 X-ray photograph of human hands. Note the ring. (Courtesy Lawrence Berkeley Laboratory)

nected, by wires that pass through the glass, to sources of high electric voltage.

One of the electrodes, the *anode*, is kept at 0 volts. That is, there is no voltage difference between the anode and the general level of metal conductors in the room. This level of zero voltage is known as the local *ground*, or common, voltage. For safety, the outside surface of all metal cabinets containing electrical equipment is kept at ground voltage.

THE DISCOVERY OF X-RAYS

If someone had asked Professor Wilhelm C. Roentgen to invent a way to examine bones or locate foreign objects in human bodies, it is very unlikely that he would have started by applying high voltages in evacuated glass bulbs.

Professor Roentgen was pursuing pure research. He performed his experiments for reasons of curiosity about nature, without any thought of what the applications might be. His discovery of X-rays is a classic example of *serendipity*, the art of finding what you were not looking for. Once the discovery was

Figure 12-3 Wilhelm C. Roentgen 1845–1923 (American Institute of Physics, Niels Bohr Library, W. G. Meggers Collection)

made, Roentgen was quick to recognize the uses that could be made of it.

The year was 1895. The place, Wurzburg, Germany. An important research topic at that time was *gaseous discharges*, the sort of thing that takes place in lightning strokes or electric sparks.

It was understood that atoms could occasionally be broken into electrically charged parts. We now know the negatively charged pieces as *electrons*, but in 1895 even the word "electron" had not been invented. The positively charged atom left behind when one or more electrons are split off is called an "*ion*."

Roentgen's equipment included a partially evacuated glass bulb. There was enough gas left in the bulb so that there would be a reasonable number of ions and electrons present. But the gas was not so dense that they would often collide with each other and recombine into neutral atoms.

Roentgen's purpose was to make the electric charges in the gas go very fast, by applying a high voltage across the bulb. He hoped to give them enough kinetic energy so that they could penetrate the thin window of the bulb. The

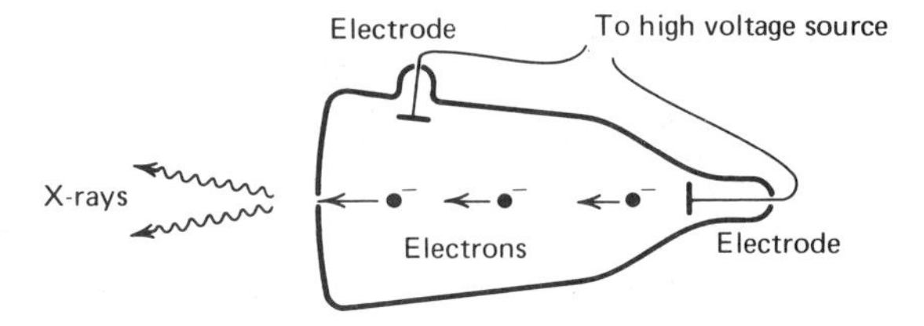

Figure 12-4 Diagram of the tube with which Roentgen discovered X-rays

fast particles would lose much of their energy by collisions with the molecules in the glass. But if they were going fast enough, he hoped, they would not lose all their energy in collisions. They would then come all the way through the window and into the air outside. There he could study the electrical discharges the fast charged particles caused in the air.

Roentgen had sealed two metal plates, or *electrodes*, inside the bulb, with wires from each coming out through the glass. Between these plates he applied a voltage difference of several thousand volts, produced by a high-voltage generator he had built. To see the glow of the discharge better, the room was kept in darkness. In order not to be confused by the glow of the gas inside the bulb, the whole bulb was covered with a shield of thin black cardboard.

In the room were some fluorescent ("glow-in-the-dark") crystals that had been used in other experiments. He noticed that whenever the high voltage was turned on, these crystals would glow. Roentgen began to look for the source of energy that was making the crystals light up. He put obstacles in the way, but the crystals continued to glow despite the obstacles.

Roentgen made a screen by depositing a layer of these crystals on a sheet of paper. The screen glowed equally well whether he turned the front or the back of it toward the glass bulb. The source of the mysterious radiation, which he called X-rays, seemed to be the spot where the electrons struck the glass.

He held his hand in front of the screen, and saw for the first time the outline of the bones now so familiar in X-ray photographs. In fact, it was he who discovered that photographic film was sensitive to X-rays. He took some exposures—his wife's hand with the wedding ring prominent, the body of a salamander—and published them. The announcement caused a world-wide sensation.

Within three months, X-rays were being used by a Vienna hospital in connection with surgical operations. Since Roentgen's time, X-rays have revolutionized certain phases of medical practice.

The value of pure research is illustrated by Roentgen's discovery. If he had set out deliberately to find a way to help surgeons set bones, does anyone suppose he would have gone about it as he did? What do high voltages and evacuated glass tubes have to do with broken bones? Yet this was the method of research that led to the discovery of X-rays.

The other electrode, the *cathode*, is kept at a high negative voltage, such as −10,000 volts. The cathode nowadays is usually a *filament* of tungsten wire. When such a filament is heated, electrons within the tungsten metal can get enough energy to overcome the potential energy barrier at the metal surface, and escape into the vacuum.

An electron that has just escaped from the filament, but is still close to the cathode, has electrostatic potential energy. It may also have a small amount of kinetic energy that it retains from being heated inside the filament. This is a very small amount of kinetic energy compared to what it will get later. We shall neglect the kinetic energy of the electron at this point. For all practical purposes, it is zero.

electron at cathode: energy = electrostatic potential energy
= (charge of electron) times voltage

The charge of an electron is −0.000 000 000 000 000 000 16 coulomb. If the voltage is −10,000 volts, then the energy is +0.000 000 000 000 0016 joule.

It is usually more convenient to express energy on an atomic scale in units of *electronvolts*. An electronvolt is equal to the electrostatic potential energy of an object with the same electric charge as a single electron, at a voltage of −1 volt.

So the energy of one electron at the cathode, when the voltage is −10,000 volts, is +10,000 electronvolts.

When the electron reaches the anode, which is at zero voltage, the electrostatic potential energy is zero (since the voltage is zero). The *total* energy,

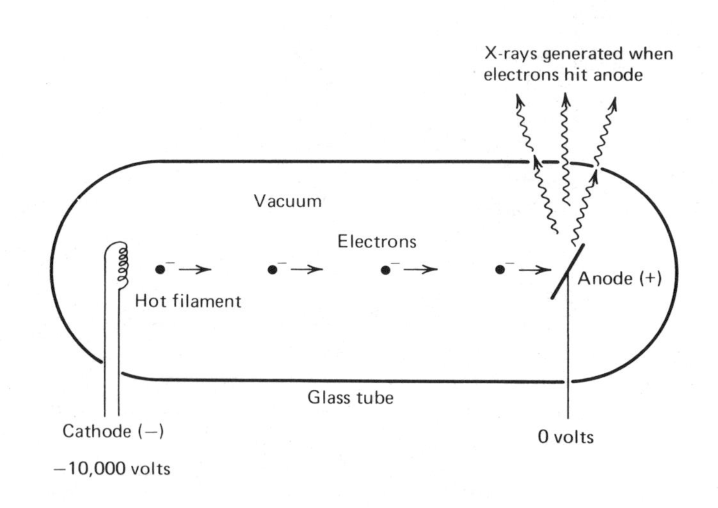

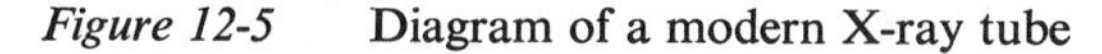
Figure 12-5 Diagram of a modern X-ray tube

by the law of conservation of energy, must be the same as before, namely, +10,000 electronvolts. It is now in the form of kinetic energy.

at anode: energy = kinetic energy
= +10,000 electronvolts

We could figure from this, and from the known mass of the electron, how fast it is moving. This calculation, however, is not necessary. The energy is soon transformed into another form.

Roentgen's apparatus was arranged so that the electrons could fly past the anode to the glass wall. In modern X-ray generators, the electrons usually collide with the anode itself.

When the electron reaches solid matter, it comes rapidly to rest by colliding with the atoms of the material. The 10,000 electronvolts is distributed among various forms of energy such as heat (mechanical energy of the atoms of the solid), electrostatic energy (electrons knocked loose from atoms), and chemical energy (molecules of the solid broken apart).

Some of the energy goes into electromagnetic radiation. Some such radiation must always be emitted, according to Maxwell's theory of electricity and magnetism, whenever a charged object is accelerated or, as in this case, is rapidly slowed down.

The ideas of quantum mechanics have not changed the prediction that electromagnetic radiation, including X-rays, is given off when electrons are brought rapidly to rest.

But the quantum idea does set a limit on the spectrum of wavelengths produced. The most energy that any single electron can put into X-rays is just the total energy the electron has. In this case, that is 10,000 electronvolts. It is of course not only possible but likely that the electron will put less than its total energy into electromagnetic energy, transferring the rest of it to other forms.

The shortest wavelength X-ray that an electron, with 10,000 electronvolts total energy, can produce is one whose quantum energy is just 10,000 electronvolts. This wavelength turns out to be 1.24 Å for this quantum energy. To produce shorter wavelength X-rays, we need higher energy electrons that will require higher voltage on the X-ray tube.

Examining small objects

We never expect to be able to see an atom directly. The visible light that our eyes can see has a wavelength thousands of times the size of any atom. Diffraction effects would completely blur out any image of atomic-sized objects we might try to form with visible light.

X-rays, on the other hand, have wavelengths that are comparable to, or shorter than, the sizes of atoms. So it should be possible to form X-ray images of atoms and molecules that can be photographed and examined. Much of what we know about the structure of crystals and large molecules comes from X-ray studies. X-rays were of prime importance in figuring the "double-helix" structure of the biological molecule DNA. This molecule, found in every cell of every living creature, holds the coded genetic information that enables the cells to reproduce themselves.

HOW MUCH IS A QUANTUM?

We wish to calculate what wavelength corresponds to a given quantum energy. We start with the relation:

quantum energy = Planck's constant times frequency

from which we can solve for the frequency:

$$\text{frequency} = \frac{\text{quantum energy}}{\text{Planck's constant}}$$

We also have the basic relation for waves that:

wavelength times frequency = wave speed

For electromagnetic radiation, the wave speed is, of course, the speed of light. Solving again for frequency:

$$\text{frequency} = \frac{\text{wave speed}}{\text{wavelength}}$$

We can equate the two expressions for frequency:

$$\frac{\text{quantum energy}}{\text{Planck's constant}} = \frac{\text{wave speed}}{\text{wavelength}}$$

which can be rewritten

quantum energy times wavelength =

Planck's constant times wave speed

Planck's constant, and the speed of light, are well-known constants. We can combine our knowledge on this topic to the simple relation:

The product of the quantum energy of a given type of electromagnetic radiation, expressed in electronvolts, times the wavelength of that radiation, expressed in angstrom units, is a universal constant whose numerical value is 12,400.

quantum energy times wavelength = 12,400 electronvolt Å

Electrons and other material objects are thought sometimes to show wave-like properties. For slow electrons, the formula relating quantum energy and wavelength is not the same as for photons. What is still true is that the product of frequency times wavelength equals the wave speed. It is also still true that the quantum energy equals Planck's constant times the frequency.

An electron with a kinetic energy of 1000 electronvolts has a speed of less than one-tenth the speed of light. Such an electron would have a wavelength less than one-tenth that of a photon with the same frequency, i.e., the same quantum energy. At 1000 electronvolts kinetic energy, the electron's wavelength is less than 1 angstrom unit. A photon with this energy has a wavelength, by the formula above, of 12.4 Å.

At very high energies, when the electron is moving at nearly the speed of light, the relation between wavelength and quantum energy is almost exactly the same for electrons as for photons.

It is hard to form X-ray images because it is very difficult to make a lens for X-rays. Except in very special circumstances, X-rays do not reflect from mirrors, nor change direction (refract) when passing from one material to another. Without lenses we cannot focus an X-ray beam nor magnify an image. So we usually do not make direct pictures of atomic systems with X-rays. Instead, scientists examine the diffraction patterns found when X-rays scatter from arrays of molecules. Then they use a computer to help reconstruct what the shape of the molecules must be.

Quantum mechanics teaches us that, just as light waves have some particle behavior, particles such as electrons can sometimes behave like waves. The importance of this fact is that electrons can be focussed by magnetic and electrostatic lenses. These are arrangements of currents and charges, designed to make a beam of electrons converge on a focal point, so that a beam of electrons can be used to take a real picture of atomic-sized objects. If the wavelength is short enough, diffraction effects do not blur the picture.

A device that uses an electron beam to make such pictures is known as an *electron microscope*. The present limit on examining small objects is set by how well the focussing system can be designed. It is claimed that electron microscope photographs have been made that show the outline of a single atom. Pictures that can separate objects 10 angstrom units apart are commonly made with this device.

The electron microscope has some drawbacks. Low-energy electrons have little ability to penetrate matter. So the samples to be photographed must be very thin and mounted in vacuum. Higher-energy electrons are more penetrating, but more difficult to focus. Some of these problems are avoided by

Figure 12-6 An electron microscope in operation (Courtesy Lawrence Berkeley Laboratory)

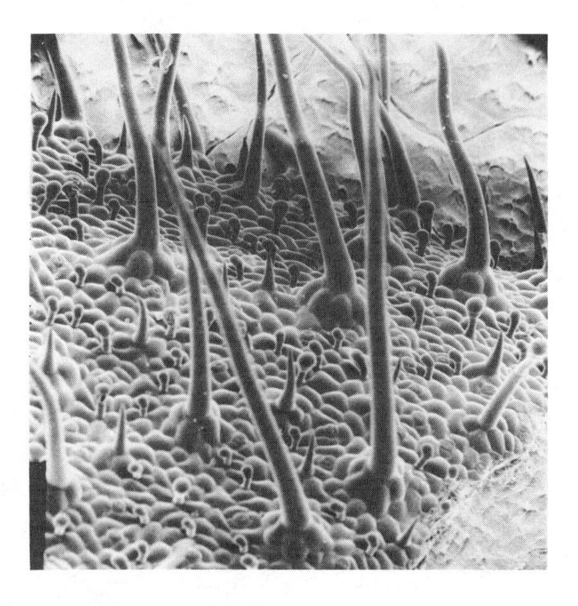

Figure 12-7 Photograph of plant cells taken with an electron microscope (Courtesy Lawrence Berkeley Laboratory)

the *scanning* electron microscope. This device sweeps a small beam spot across the sample. A direct photograph is not taken. Instead, the point-by-point detector response is used to make an image on a television screen.

High-energy accelerators

The nucleus of an atom, that incredibly tiny part in the center that contains nearly all the atom's mass, is 100,000 times smaller than the atom itself. If X-rays of 12,400 electronvolts quantum energy have a wavelength of 1 angstrom unit, it takes radiation with at least 100,000 times that energy to examine the atomic nucleus in detail. This means we are talking about quantum energies above 1 billion electronvolts.

At such high quantum energies, it is rare for any beam to contain more than 1 quantum at a time. So a beam of such electromagnetic radiation is just a succession of photons. Electrons with this much energy are travelling at 99.99999% of the speed of light. Except for the fact that electrons carry electric charge, electrons and photons are very much alike at such high energy. Either can be used to probe the nucleus of the atom.

It is not possible, with methods known today, to apply a direct voltage of 1 billion volts across the plates of something like an X-ray tube. The air, the glass, the insulating stands that separate the plates, would all fall victim to electrical breakdown. Giant sparks and streamers would jump across the gap, short-circuiting all the high-voltage equipment.

One solution is to give the electrons a moderate voltage "kick," say, several thousand volts, at a time. Suppose we have the electrons pass through a series of ring-shaped electrodes.

While the electrons are between the first and second rings, the second ring is kept several thousand volts higher than the first. The electrons are accelerated toward the second ring. They gain several thousand electronvolts kinetic energy in going from the first to the second ring.

As soon as the electrons pass the second ring, the voltages are switched. The third ring is now several thousand volts higher than the second. So the electrons are again accelerated while going from the second to the third ring.

These steps are repeated all down the line. As the electrons pass each ring, the voltage between rings is switched. The total effect is that the electrons get many millions of electronvolts kinetic energy. A machine that speeds up electrons in this way is called an *electron linear accelerator*, or "LINAC" for short.

The world's highest energy electron LINAC is in Stanford, California. The string of electrodes at Stanford is more than 3 kilometers long. The electrons coming out of the tube have kinetic energies of more than 20 billion electronvolts.

The Stanford Linear Accelerator began operation in 1967. Its beams have been used to study the structure of the particles that make up the atomic nucleus, such as protons and neutrons.

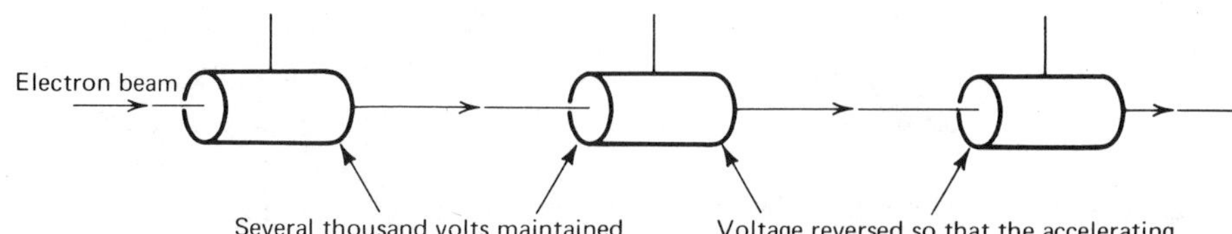

Figure 12-8 Part of a linear electron acceleration

Figure 12-9 The two-mile-long Stanford Linear Electron Accelerator (Photo courtesy SLAC)

Other machines to speed up sub-atomic particles to very high energies use magnets to keep them going in a circular path. This saves on real estate, and means that only a few electrodes are needed. The particles pass the same electrodes over and over, getting accelerated with each pass. Such circular accelerators are called *cyclotrons*, if the magnetic field is steady in time, or *synchrotrons*, if the magnets increase in strength as the beam energy goes up.

At the time of this writing, the world's most powerful synchrotron is the one at Fermi National Accelerator Laboratory in Batavia, Illinois. This machine, which began operation in 1972, has accelerated protons to energies as high as 500 billion electronvolts. In a typical experiment with these high-energy beams, the beam particles are made to collide with the nuclei of atoms in a piece of target material. The new forms of matter that can be produced

Figure 12-10 The proton synchrotron at the Fermi National Accelerator Laboratory, Batavia, Illinois (Photo courtesy FNAL)

in such collisions are of great interest to scientists working at the "frontier" of our understanding of the very small.

Synchrotron radiation Maxwell's theory has always said that charged particles being accelerated will send out electromagnetic waves. This effect poses a problem for high-energy particle accelerators. If too much energy is radiated away, its loss can keep the particles from reaching the desired energy. The problem is worse for circular than for straight-line acceleration. It is very much worse for electrons than for the more massive protons. This explains why the highest energy electron accelerator is a LINAC, and why the highest energy synchrotrons accelerate protons.

The "synchrotron radiation" sent out in this fashion comes in all wavelengths, down to the minimum set by the quantum effect. A high-energy electron beam moving in a circle makes a very strong source of ultraviolet and X-rays. This source can be used by scientists interested in studying those wavelengths.

Synchrotron radiation, at radio and microwave wavelengths, is often generated in outer space. Whenever fast electrons are moving around in circles, which they will do if magnetic fields are present, this radiation comes out. "Synchrotron" radio sources in space range from the planet Jupiter to distant exploding galaxies.

Gamma rays

Any form of electromagnetic radiation whose wavelength is shorter than about 0.1 Å is called a *gamma ray*. The quantum energy of gamma rays is thus 124,000 electronvolts or higher. Even some longer wavelength radiation can be called a gamma ray if it is emitted by an atomic nucleus.

There are no really sharp distinctions between very short wavelength ultraviolet rays, X-rays, and gamma rays. Perhaps the main differences among them is the size of the atomic and subatomic systems that typically emit and absorb each type of radiation.

The quantum of ultraviolet radiation has an energy sufficient to knock some of the more loosely bound "outer" electrons from most atoms. Hence, it is ultraviolet rays that usually cause the photoelectric effect.

The quantum of X-radiation has enough energy to knock loose the more tightly bound "inner" electrons from most atoms. Hence, X-rays are most readily absorbed by the heavier atoms that have more of the inner electrons.

The quantum of gamma radiation has enough energy to knock loose, or at least rearrange, the protons and neutrons that compose an atomic nucleus. Gamma rays are very often emitted by metastable nuclei as the latter drop from an energetic "excited" state to their lowest energy configuration.

Pair creation

An important new feature occurs for gamma rays whose energy is greater than 1,022,000 electronvolts. This corresponds to a wavelength less than one-eightieth of an angstrom unit.

When a photon of such gamma radiation passes close to an atomic nucleus it can be transformed into a pair of material particles, namely, an electron and a positron. A positron is a subatomic particle having the same mass as an electron, but a positive electric charge.

The process of *pair creation*, which was not discovered until the 1930s, would be enough to convince us that neither energy nor mass is separately conserved. Consider that, before the pair creation took place, there was energy in the form of the quantum energy of the photon of electromagnetic radiation, but there were no massive objects in the system.

After the pair creation takes place, the photon has disappeared. In its place there are two massive objects, the electron (e^-) and the positron (e^+). Furthermore, since the pair are usually created in motion, there is also kinetic energy.

In our previous discussions, whenever some form of energy was observed to disappear in some process, we always tried to deduce the existence of some new form of energy. If we were able to measure the new form of energy and determine the calibration of it, then we could conclude that energy, including the new form of energy, was still conserved. This was the procedure when we introduced various potential energies, heat energy, sound energy, etc.

So now, observing a known form of energy disappearing and matter appearing in its place, we might deduce that mass is a form of energy. We could then set out to measure the calibration constant: How many joules does it take to make a gram? It would turn out that it takes a very great many

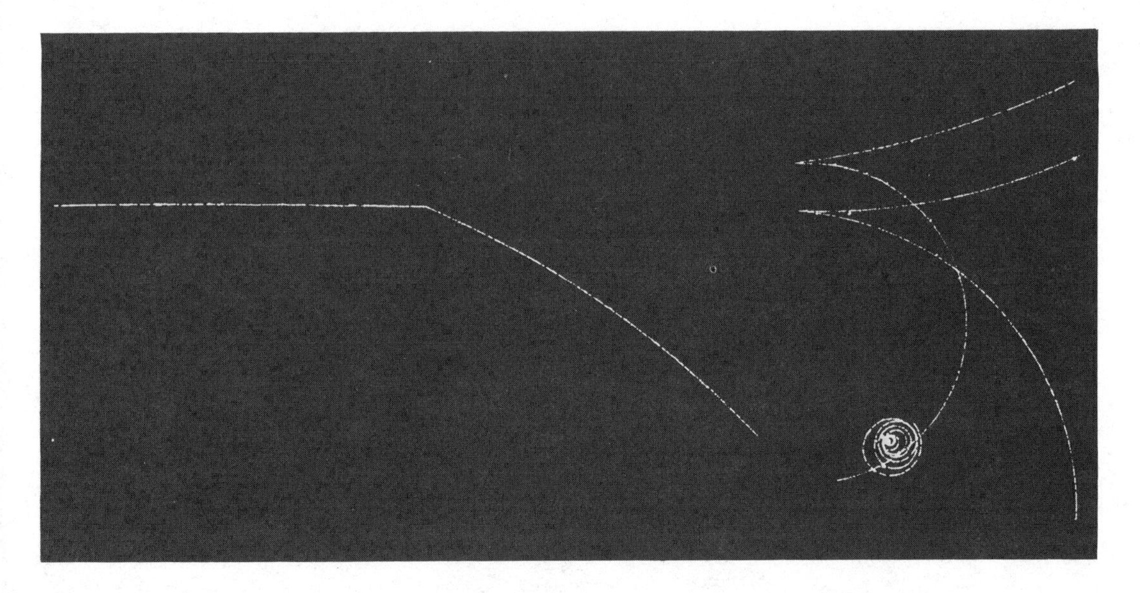

Figure 12-11 Pair creation. This is a photograph of the tracks made in a bubble chamber by, among other particles, two electron-positron pairs. The gamma rays that were converted into these pairs originated at the "kink" in the track entering from the left. The gamma rays themselves leave no tracks in the chamber. A strong electromagnet around the chamber causes the electron and the positron tracks to curve in opposite directions. (Courtesy Lawrence Berkeley Laboratory)

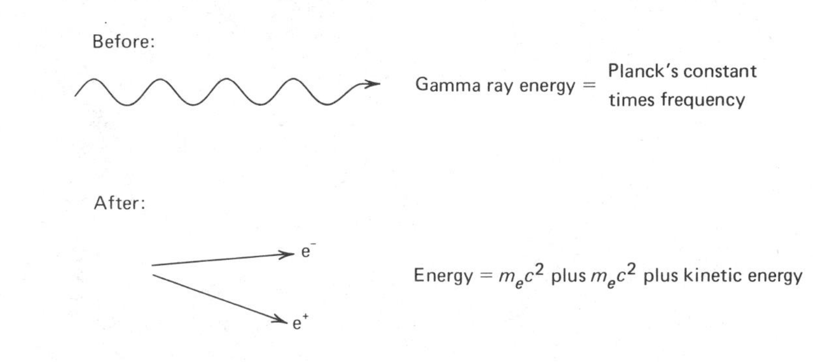

Figure 12-12 Pair creation

joules to make a gram; the calibration constant is a very large number. But this measurement can be made, and the results are consistent.

The inverse process to pair creation can also be observed. If a positron comes into contact with an electron, the two can *annihilate* each other. In this process, the electron and the positron disappear and electromagnetic radiation, in the form of two or more photons, is created in their place.

If the annihilation takes place *at rest*, with no kinetic energy for electron or positron before annihilation, and if only two photons are created, then each photon will have a quantum energy equal to the mass energy of one of the particles annihilated. For electrons and positrons, this turns out to be 511,000 electronvolts, which corresponds to a wavelength of one-fortieth of an angstrom unit. So when a positron annihilates with an electron, with both initially at rest, the two photons emitted will have this wavelength.

The annihilation need not take place at rest. Either the electron or the positron, or both, may be moving at high speed when the annihilating collision takes place. At the Stanford LINAC there is a *storage ring* into which high energy electrons and positrons can both be injected. They then circulate around the ring in opposite directions. Occasionally there is a head-on annihilating collision between one electron and one positron. In such high-energy annihilations, not only photons are created but any form of matter which there is sufficient energy to produce.

Figure 12-13 Photograph of the colliding electron-positron storage-ring facility at the Stanford Linear Electron Accelerator. Electrons and positrons from the main linear accelerator are injected into these rings, and revolve in opposite directions. Occasionally two of them collide and often annihilate each other. (Photo courtesy SLAC)

Mass and energy

Historically the understanding of the equivalence of mass and energy is older than the discovery of positrons. It is one of the most famous results of Einstein's theory of relativity. The argument goes as follows: There are two related effects that take place as we increase the speed of an object:

1. Its energy increases. The faster a particle goes the more kinetic energy it has. We can continue to "push" the object as long as desired. So there is no limit to amount of energy we can transfer to it. The formula for kinetic energy, $1/2\ mv^2$, turns out not to be valid at very high speeds. The formula was only an approximation, good at low speeds. As the speed gets close to c, a new, more exact formula is needed.
2. The object becomes increasingly difficult to accelerate. Since a material object can never attain speed c, it is very hard to speed it up when it is already going at nearly the speed of light. If we define mass in the sense of Newton's laws, in terms of the resistance of an object to acceleration, then we must say that the mass of any object increases greatly as the speed approaches c.

In Einstein's theory the increase in mass can be calculated, and it can be related to the kinetic energy. The relation is

kinetic energy = mass increase times calibration constant

Numerically the calibration constant turns out to be equal to the speed of light squared. Since the value of c is

300,000,000 meters per second,

the calibration constant is

90,000,000,000,000,000 $(\text{meter/sec})^2$.

In most everyday examples the increase in mass of a moving object is so small as to be unmeasurable. For example, suppose we impart to an object weighing 1 kilogram, a kinetic energy of 1 joule. The increase in its mass would be 1/(90,000,000,000,000,000) of a kilogram. So the total mass would then be 1.000 000 000 000 000 011 kilogram. We are quite justified in neglecting the small increase in mass for objects moving at ordinary speeds.

The formula $E = mc^2$ implies more than just that mass is a form of energy. It says that mass *is* energy and that energy *is* mass. Whenever we increase the energy of a system, we are increasing its mass as well. This applies not only to kinetic energy but also to other forms of energy.

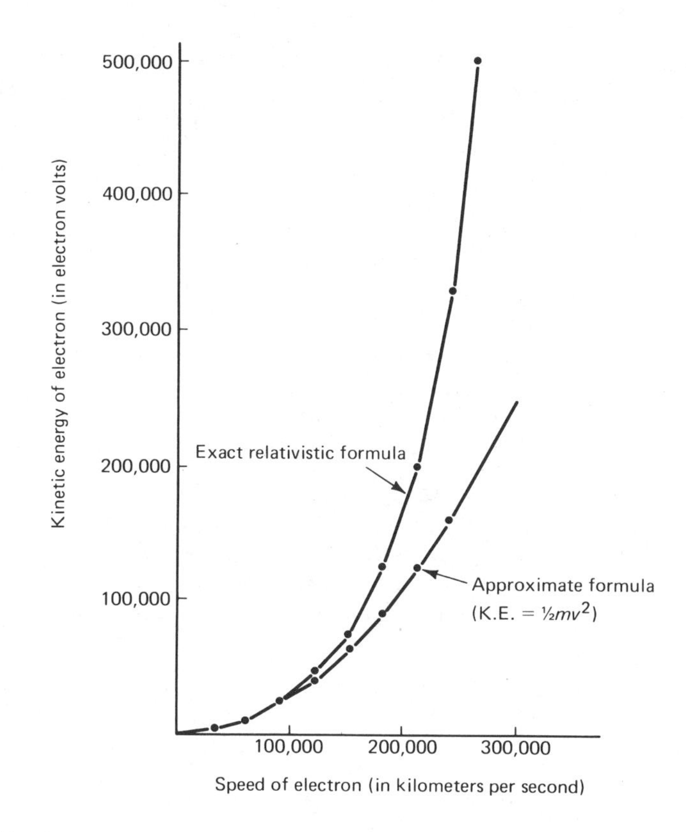

Figure 12-14

If we heat up an object, increasing its heat energy, then its mass must increase slightly. If we stretch a spring, increasing its elastic potential energy, we increase its mass, too. The gamma ray, because it carries energy, also possesses mass. The electron-positron pair, even at rest, because they possess mass, also possess energy. In this sense both energy and mass are conserved in the pair-creation or pair-annihilation processes, because both mass and energy are different aspects of the same thing.

Glossary

accelerator—A machine designed to make subatomic particles travel at very high speeds. Examples include cyclotrons, X-ray tubes, synchrotrons, and linear accelerators.

anode—See Chapter 11.

cathode—A metal conductor inside an electronic device, such as an X-ray tube, that is kept negatively charged to repel electrons. A cathode in the form of a hot, thin wire filament made of tungsten is often used as a source of electrons.

cyclotron—A machine designed to make subatomic particles travel very fast, keeping them moving in circular paths by means of a steady magnetic field. The accelerating voltage is reversed in step with the period of revolution, so that the particles get an increase of kinetic energy each time around.

electrode—See Chapter 11.

electron microscope—A microscope that uses fast electrons rather than light waves for illumination. The electrons can have much shorter wavelength than visible light. This reduces diffraction problems when we look at very small objects.

electronvolt—A unit of energy. The kinetic energy that a single electron (or any object with the same electric charge) gets when it is accelerated through a voltage of 1 volt. It takes 6 billion billion electronvolts to make 1 joule.

gamma rays—The range of electromagnetic radiation having a wavelength less than one-tenth of an angstrom unit.

ion—An atom that has acquired an electric charge through the gain or loss of electrons.

linear accelerator—A machine to make subatomic particles go very fast. The voltages between electrodes are reversed in step with the passage of the particles through them, so that they get an increase in kinetic energy at each electrode.

mass-energy—The single conserved quantity that shows itself as either mass or energy, or both. A system increases in mass whenever its energy increases. Energy is released whenever the mass of a system decreases.

pair annihilation—The disappearance of an electron and positron when they collide. The mass of the two particles is converted into energy, usually in the form of gamma rays.

pair creation—The appearance of a positron and electron pair, accompanied by the disappearance of some other form of energy, such as a gamma ray.

positron—A subatomic particle identical in mass and in many other ways to an electron, but having positive electric charge. Positrons do not last long on Earth because they soon collide with electrons and annihilate each other.

rest-mass—The mass an object has when it is at rest. In earlier chapters this was called the "amount of material," or simply its "mass." Under proper conditions the rest-mass, like mass-energy, can be converted into other forms of energy.

Roentgen, Wilhelm Conrad (1845–1923)—German physicist who discovered X-rays while investigating electrical discharges in gases.

storage ring—A machine to keep fast subatomic particles, such as electrons and positrons, moving in a circular beam where they can collide with each other.

synchrotron—A machine designed to make subatomic particles move very fast while keeping them in a circular path by means of a changing magnetic field. A synchrotron works much the same as a cyclotron. It has the advantage of occupying only a narrow ring, instead of a solid circular area. So synchrotrons can be built much larger than cyclotrons.

synchrotron radiation—The radiation emitted by charged particles when they are forced to travel in circular paths (i.e., accelerate) by magnetic fields. Synchrotron radiation is found not only at particle accelerators, but also in inter-stellar space.

X-rays—The range of electromagnetic radiation whose wavelength lies between 100 angstrom units and 1/10 of an angstrom unit.

Questions

1. What is the difference between pure and applied research? What is the value of each?
2. How many joules are in 1 electronvolt?
3. The mass of an electron is 0.000 000 000 000 000 000 000 000 000 000 9 kilogram. About how fast is it going when it has a kinetic energy of 12,500 electronvolts (0.000 000 000 000 002 joule)?
4. What is the minimum voltage difference needed across an X-ray tube to produce X-rays of wavelength 1/10 of an angstrom unit?
5. What is the quantum energy of orange light at a wavelength of 6200 Å? What is the quantum energy of X-rays at a wavelength of 1/2 an angstrom unit?
6. What is the minimum wavelength X-rays that can be produced by a typical dental X-ray machine that operates with a voltage of 62,000 volts?
7. Suggest a way to measure the wavelength of a beam of X-rays.
8. How might we measure the quantum energy of a photon of gamma radiation?
9. Choose three different parts of the electromagnetic spectrum. Tell one way we might produce each type of radiation, and one way we might detect it. Not all three methods should be the same.
10. Why are there so few positrons usually found on Earth? What is the fate of any positron that happens to be created here?
11. What would be the consequences if electrons could annihilate with *protons*, the positively charged massive particles found in the nucleus of all atoms? Would Earth and its people still be here if such annihilation could take place?

12. Would energy losses due to emission of synchrotron radiation cause a problem in keeping electrons in a storage ring? How might these losses be compensated for?
13. Show, from the old formula for kinetic energy, that any calibration factor that relates joules to kilograms must have units of a velocity squared.
14. What does the formula $E = mc^2$ mean?
15. If a photon has mass, shouldn't it be affected by gravitational forces? Suggest an experiment to detect this effect.

section four

THE FRONTIERS OF SCIENCE

13 nuclear energy

Introduction and summary

The study of *nuclear physics* has grown spectacularly in this century. One hundred years ago no one knew that every atom had a tiny, but heavy, *nucleus* at its center. No one even suspected then that there was such a form of energy as what we now call *nuclear energy*.

Today we know that the release of nuclear energy is what makes the sun and stars shine. We know the dangers of nuclear explosives. We know some of the benefits, and the hazards, of nuclear *reactors*. Some of us hope to make up for our dwindling supply of fossil fuels with more use of the controlled release of nuclear energy.

The first inkling that there were such things as a nucleus and nuclear energy came from experiments early in this century. These showed that most of the mass and all of the positive electric charge of an atom were concentrated into a tiny hard lump in its center, called the *nucleus*. The rest of the atom was mostly empty space, with some lightweight, negatively charged *electrons* circling around.

Some new force, the *nuclear force*, was needed to explain why all the positive charges in the nucleus did not fly apart from each other.

A second set of experiments, also in the early 1900s, showed that certain kinds of atoms could release rather large amounts of energy, suddenly and with no known source of energy. Such atoms were called *radioactive*.

Gradually it came to be understood that radioactive atoms were those with more than average amounts of nuclear energy stored in them. If a radioactive atom changes itself ("decays") into an atom with less nuclear energy (a more "stable" atom), the extra energy can be released.

This energy-releasing change does not always happen right away. A radioactive atom can live for quite some time in its unstable state. Then, suddenly and unpredictably, it releases its energy and changes its identity.

Each kind of radioactive atom has its own *half-life*, the time in which half the atoms of that kind will decay. The half-life for some kinds of atoms may be shorter than a second. For others the half-life may be billions of years.

By the middle of this century, a very simple picture of the nuclear atom had been painted. The nucleus is made up of a certain number of *protons*, fairly heavy objects with positive charge, and of *neutrons*, about the same mass as protons, but with no electric charge.

The number of protons in the nucleus is balanced by an equal number of electrons in orbit around the center. This number of protons and electrons is called the *atomic number* of the atom. The way the atom combines with other atoms to make molecules (the *chemical* behavior of the atom) is completely set by its atomic number.

The nucleus needs a certain number of neutrons in it to help hold the protons together. But extra neutrons can often be added to the nucleus of an atom. The extra neutrons don't change the chemical behavior of the atom. They only make it a bit heavier. Kinds of atoms that have the same atomic number, but differ only in the number of neutrons in the nucleus, are called *isotopes* of the chemical element they both represent.

Protons and neutrons are sometimes classed together under the name *nucleons*.

The amount of nuclear energy, per nucleon, stored in each atom is slightly but importantly different for each kind of atom. The atoms with the lowest level of nuclear energy per nucleon are those in the middle of the table of elements, atoms like iron and copper. These atoms are the most stable, with respect to nuclear energy. To change them into anything else, extra energy would have to be supplied.

Very light nuclei—free protons, free neutrons, or combinations of only a few of these—have a much higher level of nuclear energy. Energy can be released when these light nuclei come together to form heavier ones. This process is known as *fusion*.

Very heavy nuclei—such as uranium—also have a rather high level of nuclear energy. Energy can be released if the nucleus can be made to split apart into more stable nuclei in the medium-weight range. This process is known as *fission*.

Both fission and fusion have been used to release nuclear energy. We are all familiar with the *uncontrolled* release of energy in the explosion of nuclear bombs. We also know of the controlled release of energy in reactors based on nuclear fission. Perhaps soon we shall also find a way to release the energy of nuclear fusion in a controlled way.

In recent decades, the nuclear physicists have concentrated on studying *subnuclear particles*. They are interested in finding what the nucleons themselves are made of. In their quest they have predicted and searched for such strange objects as *neutrinos*, *antiparticles*, and *quarks*. The first two were found after 25 years of searching. Quarks have not yet been found.

The study of nuclear and subnuclear physics is still going on. The last chapter has still to be written.

The nuclear atom

A classic experiment in 1910 by Geiger and Marsden established the shape of the atom. They shot fast subatomic "bullets" at the atoms in a thin piece of foil. Most of the projectiles passed straight through, practically undeflected. A very few, perhaps one out of every 10,000, came bouncing back in the direction from which they had come. What did this all mean?

Ernest Rutherford (1871–1937) proposed a model to explain Geiger and Marsden's results. He suggested that most of an atom is empty space, with an occasional lightweight electron circulating about. These would offer little obstruction to the fast projectiles of Geiger and Marsden. At the very center, however, occupying a very small space, was a hard massive nucleus. If the fast projectiles happened to hit the nucleus, they bounced back.

> The atom can be compared to a bale of cotton, deep inside which are hidden some small steel objects. Bullets fired at random into the cotton will mostly pass through. If the bale is thick enough some bullets may lodge in the cotton. The occasional bullet that finds a hidden steel object ricochets back out of the bale at a wild angle.

The model of the atom that Rutherford suggested resembles a little solar system. The heavy nucleus at the center, like the sun in the heavens, keeps the light-weight electrons, like so many planets, in orbit about the center. The force that holds the atom together is electrostatic rather than gravitational. But since both forces follow an inverse-square law, the paths of the

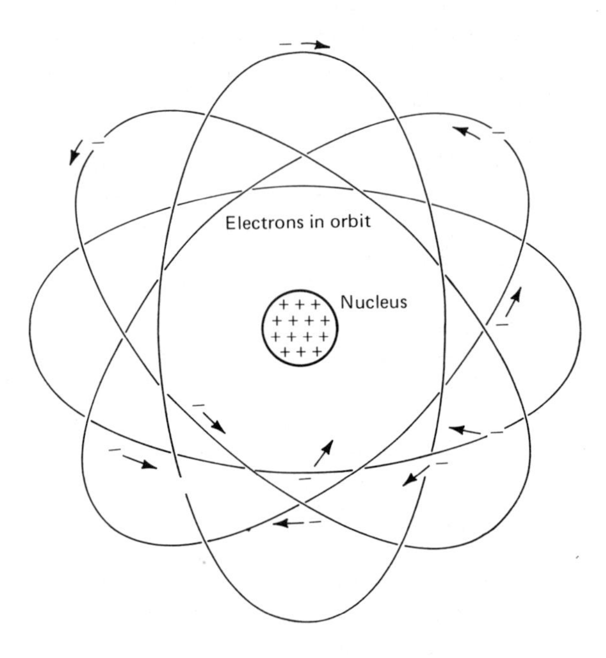

Figure 13-1 The nuclear model of the atom

electrons in an atom are expected to bear some resemblance to the orbits of planets and comets around the sun.

The atom as a whole is electrically neutral. So the total positive electric charge in the nucleus must just balance the negative charge of all the electrons revolving about it.

The number of charges in the nucleus (and the equal number of negative electrons circling it) determines what kind of atom it is. A *hydrogen* atom has a single electron in orbit around a nucleus with a single positive charge. An atom of uranium has 92 electrons circling a nucleus with 92 positive charges, just enough to balance the negative charges of the electrons.

There were two very strong objections to the nuclear model of the atom. The first question was why the electrons did not spiral down into the nucleus, emitting large amounts of energy as light and radiation. Calculations showed that every atom should collapse this way within a fraction of a second.

Quantum mechanics solved this problem. The uncertainty principle provided the answer. If an electron becomes localized closely in the center of the atom, the uncertainty in its speed must become very great. If that is so the electron has a good probability of escaping out to a wider orbit.

According to modern theory, an electron settles down into some minimum orbit within the atom. Here it emits no more radiation. It does not spiral any closer to the nucleus. So it took quantum mechanics to explain why Rutherford's nuclear atom does not collapse into its center.

A second objection to the nuclear atom was harder to meet. The many positive charges so close to each other within the nucleus should repel each other with enormous force. The laws of electrostatics say that like charges repel. The closer together they come, the stronger the repulsion.

How then can the 92 positive charges in the nucleus of an uranium atom be kept together? Why doesn't it fly apart immediately?

The nuclear force The only explanation we can offer is that there is another force, which we shall call the strong or the nuclear force, that holds the nucleus together.

Even after more than half a century of study, the details of the way the nuclear force acts are still obscure. But some of its important features can be summarized.

The strong force of attraction between two nuclear particles must be about 100 times greater than the electrostatic repulsion between them. This explains why a nucleus as charged as uranium can hold together. Nuclei with many more charges than that are unstable, and are not found in nature.

The nuclear force is very short-range. It operates only when the nuclear particles are so close together that they practically touch. When the charges are separated by as much as 0.001 angstrom unit (i.e. 10 to 100 times the size of an ordinary nucleus) they exert only electrostatic repulsion on each other, with no sign of nuclear attraction.

The electrons in an atom are not affected by the nuclear force. Just as an uncharged object is not subject to electrostatic force, so there are particles that are "turned off" to the strong nuclear force. We might say that an electron, while it possesses an *electric* charge, is *neutral* to the nuclear force. Particles, like electrons, that do not partake of the nuclear forces are classified as "*leptons*," from the Greek word for "weak." Objects that are subject to the strong forces are known as "*hadrons*," from the Greek word for "strong."

Neutrons and isotopes

In 1932 James Chadwick of Cambridge, England, discovered the neutron. The pieces of the puzzle were beginning to fall into place.

The neutron is massive. It is slightly heavier than a proton that, in turn, is more than 1800 times as heavy as an electron. The neutron has no electric charge. Most important, the neutron is a hadron; it responds to the strong forces that hold the nucleus together.

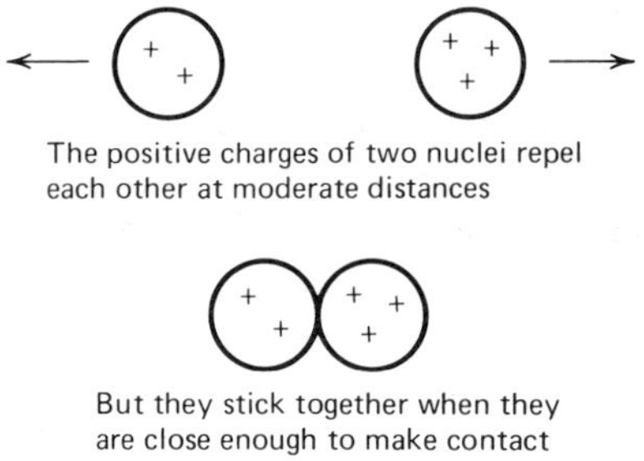

The positive charges of two nuclei repel each other at moderate distances

But they stick together when they are close enough to make contact

Figure 13-2

It is possible to make an extra neutron stick to the nucleus of almost any kind of atom. Since the neutron has no electric charge, it will not be repelled by the positive charge of the rest of the nucleus. Since the neutron is a hadron, it is likely to be attracted by the other hadrons—protons and neutrons already in the nucleus. Of course, the extra neutron must get close enough for the nuclear force to take effect.

An atom to which an extra neutron has been added is still very much like it was without the neutron. It has the same number of protons and electrons as before. It can make all the same reactions with other atoms as before. It can form all the same combinations. *Chemically*, the atom with the extra neutron is almost identical to an atom without it.

Two atoms of a given chemical element, which differ from each other only in the number of neutrons in the nucleus, are called *isotopes* of that element. The word comes from Greek words, meaning "the same place." Two different

isotopes of the same element will fit together into the *same place* in a table of chemical properties (such as Table 8-1 in Chapter 8). There can sometimes be several isotopes of a single element, each differing from the next by having one more neutron in its nucleus.

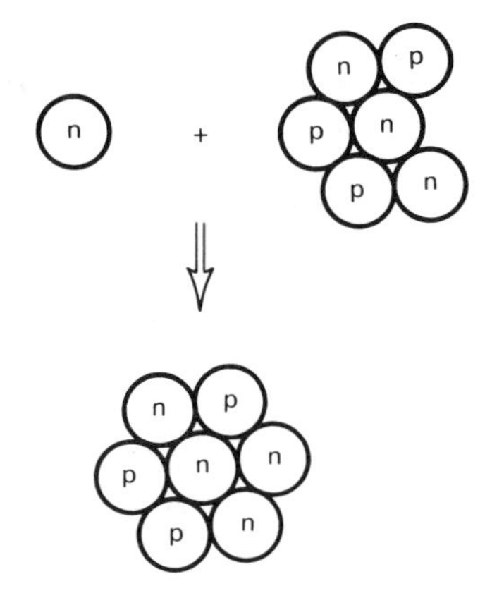

Figure 13-3 A nucleus absorbing an extra neutron to form a new isotope

The element tin has 10 stable isotopes. An atom of tin has 50 protons and 50 electrons in it. Depending on which isotope it is, it may have as few as 62 or as many as 74 neutrons. On the other hand there are some elements, such as aluminum or iodine, for which there is only one stable isotope.[1]

Of course there are *some* differences between the isotopes of an element. Extra neutrons make the atom heavier. A heavier atom moves more slowly than a lighter atom subject to the same forces. This effect is greatest when the atom has only a few neutrons and protons. When there are already 100 or more, the addition of one more neutron makes less than 1% difference in the mass.

There can be considerable difference in the *nuclear* properties of different isotopes. For example, the ability of the nucleus to absorb still one more neutron depends strongly on how many neutrons it already has.

A nucleus that has too many neutrons, or too few, for the number of protons in it, can be unstable or *radioactive*. A radioactive nucleus is one that can release energy by changing itself into a different nucleus. In some cases, when an isotope has too many neutrons, energy can be released if one

[1] There are some elements for which no combination of neutrons can form a stable nucleus. These elements are never found in nature, and can only be created artificially. Some examples are: technetium that would have 43 protons and electrons if it were stable, and promethium that would have 61 of each.

neutron can be changed into a proton. In other cases, when there are too few neutrons, energy can be released by changing one proton into a neutron. In still other cases, involving the very largest nuclei, energy is released when the nucleus splits into smaller pieces.

If a nucleus is radioactive, it may last for a time in its unstable form. Sooner or later, however, it must undergo the energy-releasing change that transforms it into a more stable nucleus. The time scale for this change to happen is different for each isotope. Some very unstable isotopes may disappear completely in less than a second. For an atom like uranium-238 (92 protons and electrons, 146 neutrons) the time scale is measured in billions of years, longer than the age of the Earth and the sun. So even though uranium-238 is radioactive, it takes so long for it to decay that most of the original uranium is still to be found on Earth in this unstable state.

One of the most important stable isotopes is the atom of *heavy hydrogen*, or *deuterium*. An ordinary hydrogen atom has a single proton and no neutrons in its nucleus. A deuterium atom has a nucleus containing a proton and a neutron. This nucleus is called a *deuteron* from the Greek word for "second." ("Proton" means "first.")

If an ordinary hydrogen atom absorbs a neutron, it becomes a deuterium atom. If we look carefully at the hydrogen we find on Earth—in sea-water, for example—1 hydrogen atom in 6000 is the deuterium isotope.

Deuterium atoms are twice as heavy as ordinary hydrogen atoms. This fact makes it fairly easy to separate samples of pure deuterium from ordinary hydrogen gas. Water in which all of the hydrogen atoms of the molecules are the deuterium isotope is called *heavy water*.

There is another known isotope of hydrogen whose nucleus has 1 proton and 2 neutrons. This isotope is called hydrogen-3, or tritium. Tritium is not a stable isotope. It is radioactive, with a half-life of 12 years. In the absence of nuclear reactions on Earth, there would be no tritium at all found in nature.

RADIOACTIVITY AND HALF-LIFE

An atom of strontium-90 contains 38 protons, 38 electrons, and 52 neutons. More than half a million electronvolts of energy can be released if one of the neutrons could somehow be changed into a proton (plus an electron, to keep electric charge balanced).

Sooner or later this will happen. The strontium-90 atom is not a stable isotope of strontium. It is *radioactive*.

The change does not happen right away. The odds are even that any given strontium-90 atom will last more than 29 years before giving up its energy. Put another way, if we started with a hundred atoms of strontium-90, after 29 years about half of them—50, give or take a few—would still be strontium-90. The rest would have released their energy and changed their form ("*decayed*" is the word often used).

Twenty-nine years is the *half-life* of strontium-90. The half-life of a radioactive isotope is the time over which half the atoms that we start with can be expected to decay. The half-life is different for each isotope. It can be shorter than 1 second or longer than a billion years.

Can we tell in advance exactly *when* a given atom will decay? Apparently not. We do not know how to identify an atom that is about to decay radioactively. Even if we knew what information to look for, we probably couldn't get it without breaking apart the nucleus. The uncertainty principle of quantum mechanics makes sure this is so.

The 50 or so atoms of strontium-90 that are left after 29 years look exactly the way they did 29 years before. In fact, they look no different from any brand-new strontium-90 atoms you could make. Given that they survived the first 29 years, the odds are now even that they will last another 29 years. After 58 years, there will be about 25 strontium-90 atoms left of the original 100. After another 29 years, about half of *them* will have decayed. The process continues until only one atom is left.

This one atom will have 50% chance of decaying within the next 29 years. If it survives one half-life, it will have an even chance of lasting 29 more years, and so on. When the last radioactive atom finally decays, there will be no strontium-90 left at all.

As long as the strontium-90 does not decay, it behaves just like any other atom of strontium. Strontium is a chemical element very much like calcium. If you eat it in your food, it is likely to get into your bones. There it will stay for the rest of your life— if it is a stable isotope—or for the rest of *its* life if it is radioactive.

That is the rub. When a radioactive isotope finally decays, its excess energy is released. This energy is delivered to the atoms and molecules nearby. These atoms and molecules are pulled apart, knocked out of place, or otherwise disturbed by the violent release of energy.

Usually your body is able to repair the damage. We are always being hit with many kinds of damaging energy releases. We would not have survived long as living creatures if our bodies could not recover from small injuries.

But sometimes someone gets such a large dose of radioactivity that his body cannot recover immediately. He then gets a *radiation burn*, or worse, *radiation sickness*. In extreme cases a victim of a very large radiation dose may die. Such overdoses can occur when a nuclear bomb explodes, or when there is a radiation accident at some nuclear plant. With the end of bomb-testing in the open air, and with careful control at all nuclear facilities, radiation burns and radiation sickness can be made extremely rare.

There are more subtle effects of radioactivity. Sometimes a key molecule in a living cell can be altered just enough that the cell doesn't die—most dead cells can be replaced—but it begins to grow out of control.

A cell that grows and multiplies out of control is said to be *malignant*. The growth and spread of malignant cells is the principal feature of the diseases called *cancers*.

The decay of radioactive atoms in the body is one possible cause of cancer. It is not the only cause; many other cancer sources are known. It is not even a

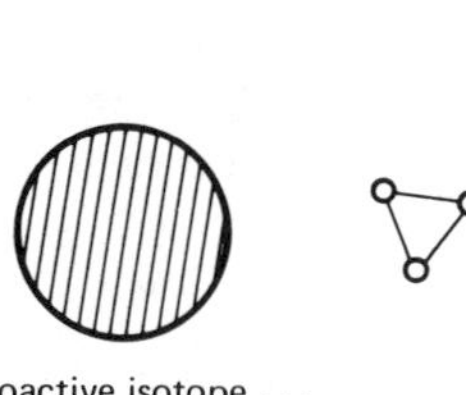

A radioactive isotope . . .

. . . among normal atoms and molecules

The isotope decays, leaving . . .

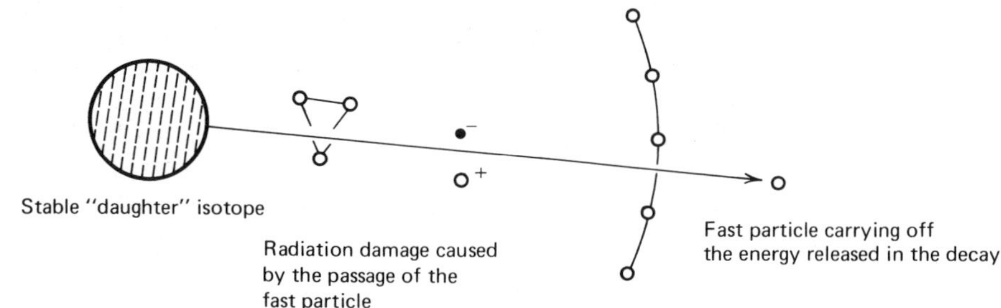

Figure 13-4

likely source. A random release of energy inside a living cell has a very small chance of causing the exact change that makes the cell malignant. Only when a large number of radioactive decays take place in a person's body do the chances become appreciable that one of those decays will cause a cancer.

The particular threat posed by strontium-90 is due to a combination of its chemical properties and its half-life. Its chemistry makes it likely to find its way to your bones and stay there. Radioactive isotopes of elements that don't get absorbed by your body are much less dangerous. An atom of radioactive argon, an inert gas that doesn't combine easily with other elements, would not lodge itself in your body.

If the half-life of strontium-90 were very short, only a few minutes or days, it would all decay before it could get to you. If the half-life were very long, thousands or millions of years, very little of it would decay even while it was lodged in your body. The half-life of 29 years matches well with a human life span. It gives the strontium-90 that might escape from a nuclear explosion or accident time to get into your bones, and then to decay gradually throughout your lifetime. Sooner or later one of the decays might just produce a malignancy.

We must not suppose that radioactive isotopes are a complete evil. We must respect their danger and handle them with care. But this is no different from many other kinds of dangerous, but useful materials.

A radioactive tracer is a small amount of radioactive material that is used to follow the course of some process. We use the fact that we can detect the decay of even a single radioactive atom, because of the large amount of energy released. So if a volunteer takes a drink containing only a few atoms of radioactive iodine, we can easily follow just where in the body the iodine goes, and how fast it gets there. Radioactive tracers can be used to help study the details of what is going on in many processes, such as in living bodies or in industry.

Radioactive sources can be used to treat cancer and other diseases. The energy release of the radioactive decay can also kill the malignant cells.

Radioactive atoms represent a way of storing energy. A quantity of radioactive material can be used as a power supply, slowly but reliably releasing its energy over a long period. This kind of energy source, while expensive, is useful in places that are hard to resupply. Stations on the moon and on satellites, and in remote Antarctica, have been powered by radioactive cells.

One serious problem with radioactive material is that it cannot be "turned off" when it is not wanted. Radioactive atoms continue to decay at their self-determined rate for as long as they exist. There is no known way to speed it up or slow it down.

An unwanted sample of radioactive material must be stored in a safe, shielded place until enough half-lives have passed for nearly all of it to disappear. When the half-life is many thousands of years, finding a storage place that will stay safe for all that time presents a serious problem.

Radioactive material can be both a hazard and a useful tool. There are benefits to be gained from the careful use of radioactive atoms. There are unique problems to be solved in handling them. It seems likely that we shall have to deal with many kinds of radioactivity from now on. We shall have to learn how best to cope with them.

At the other end of the table of elements there is an important isotope of uranium, called uranium-235. This atom has 92 protons in its nucleus, like other atoms of uranium, but only 143 neutrons. About one out of every 140 uranium atoms, as they are now found in nature, is uranium-235. (Most of them are uranium-238.) Uranium-235 is radioactive, with a half life of 7 hundred million years. It is disappearing faster than uranium-238, but the Earth is not so old that all the uranium-235 is gone. This is fortunate for us because uranium-235 plays a very important role as a source of energy, which we shall discuss later.

Stable isotopes The total number of neutrons and protons in the nucleus of an atom is called the *nucleon number*. We shall use the letter A to symbolize it. This is the number that we have been attaching to the names of isotopes (e.g., strontium-90, uranium-238). For any given value of A, there is one combination of neutrons and protons that gives the most stable nucleus.

For most values of A there is only *one* combination of neutrons and protons that is stable. This is always true when A is an odd number. It is true about half the time when A is even. In most of the rest of the cases there are two stable isotopes with the same value of A. Very rarely are there three. In a few cases, notably $A = 5$ and $A = 8$, there are no stable isotopes.

If there are too many neutrons or too many protons for that value of A, the nucleus is unstable and the isotope is radioactive.

For A up to about 40, the most stable isotope has an equal number of protons and neutrons. Thus, carbon-12 (6 protons, 6 neutrons), oxygen-16 (8 of each), silicon-28 (14 and 14) are examples of particularly stable nuclei.

An unstable nucleus like nitrogen-16 (7 protons, 9 neutrons) will decay very rapidly into oxygen-16, with one neutron turning itself into a proton. Similarly nitrogen-12 (7 protons, 5 neutrons) will quickly release its excess energy by changing one proton into a neutron, to become a stable carbon-12 atom.

In heavier atoms, the most stable nuclei tend to have more neutrons than protons. This is because the protons in the nucleus are repelling each other, while the neutrons are not. So there is more room for extra neutrons than for protons.

As we go up in nucleon number the neutrons tend to outnumber protons more and more. The most stable isotope of iron has 26 protons and 30 neutrons. Tin-118 has 68 neutrons to only 50 protons. By the time we reach uranium, the neutrons outnumber the protons by 50%. It is the force of those extra neutrons, pulling on all the protons, that keeps the nucleus from flying apart.

The curve of binding energy

A carbon-12 nucleus does not weigh quite as much as the 6 protons and 6 neutrons that make it up. The whole is not equal to the sum of all its parts.

If we were to mix 6 kilograms of protons with 6 kilograms of neutrons, and somehow get them all to stick together as carbon-12 nuclei, the aggregate would weigh only 11.9 kilograms, rather than the 12 kilograms of matter we started with.

There is one-tenth of a kilogram less mass in the system than we started with. We have learned that mass is a form of energy. So there is also less energy than we started with. The energy lost in the formation of all the carbon-12 must have been converted into other forms. In this case it may have been emitted as gamma rays.

The energy given off in this supposed reaction is really quite enormous. Using the formula $E = mc^2$, the loss of one-tenth of a kilogram of mass results in the release of

$$0.1 \text{ kilogram} \times (300{,}000{,}000 \text{ meters/sec})^2 = 9{,}000{,}000{,}000{,}000{,}000 \text{ joules}$$

This is enough energy to power a large city completely for several days. This is also a lot more energy than we could get by, say, burning the 11.9 kilograms of carbon. If the energy were released all at once, the explosion would destroy the whole city where it happened.

The fact that carbon-12 weighs less than the protons and neutrons in it explains why it is such a stable nucleus. In order to break up this nucleus we would have to resupply all the energy that the gamma rays carried off when it was formed. This would be even harder to do than making the nucleons stick together in the first place. A carbon-12 nucleus does not fall apart by itself, because so much energy would have to be supplied from outside to allow it to do so.

The argument holds equally well for any other nucleus. If a nucleus had even slightly more mass than the sum of all its protons and neutrons, energy could be released by letting the nucleus fly apart. Even if its mass were exactly equal to the sum of its nucleons, there would be no reason why it couldn't fall apart by itself. But if its mass is only slightly less than that, the nucleus cannot be broken up without putting in energy from outside. For this reason we should not be surprised to learn that the mass of *every* known nucleus is less than the sum of the masses of the protons and neutrons that make it up.

Figure 13-5 plots the masses of many of the stable nuclei, and a few radioactive ones, versus A, the number of neutrons and protons. In order to get them all on the same graph, we divide the mass by A in each case.

The mass per nucleon is nearly the same for all nuclei. In energy units, it is always between 930 and 940 million electronvolts per nucleon. In order to show the differences between nuclei more clearly, we display only the "tip of the iceberg," the top 1% of the graph. Zero on this scale is far off the bottom of the page.

The nuclei that lie lowest on this graph are the most tightly bound. It would require the most energy per nucleon to tear them apart. Atoms that lie higher on the graph are energy-rich. If their nucleons could somehow be transformed into the lower-lying atoms, energy could be released.

Free neutrons and protons are by far the most energy-rich particles on the graph. As we have said in a different way before, all of the known nuclei have less mass per nucleon than the free neutrons and protons. To a lesser extent, there is much energy content in such light nuclei as deuterons, tritons, helium-3, and lithium atoms.

Notice how tightly bound the helium-4 nucleus is, compared to other atoms of such low mass. This unusual stability of helium-4 is recognized by its having a special "nickname," the *alpha particle*.

If two deuterons, with masses of 938 MeV per nucleon, come together to form an alpha particle, at 932 MeV per nucleon, we figure to release 6 MeV per nucleon by the reaction. The energy so released might be carried off by a

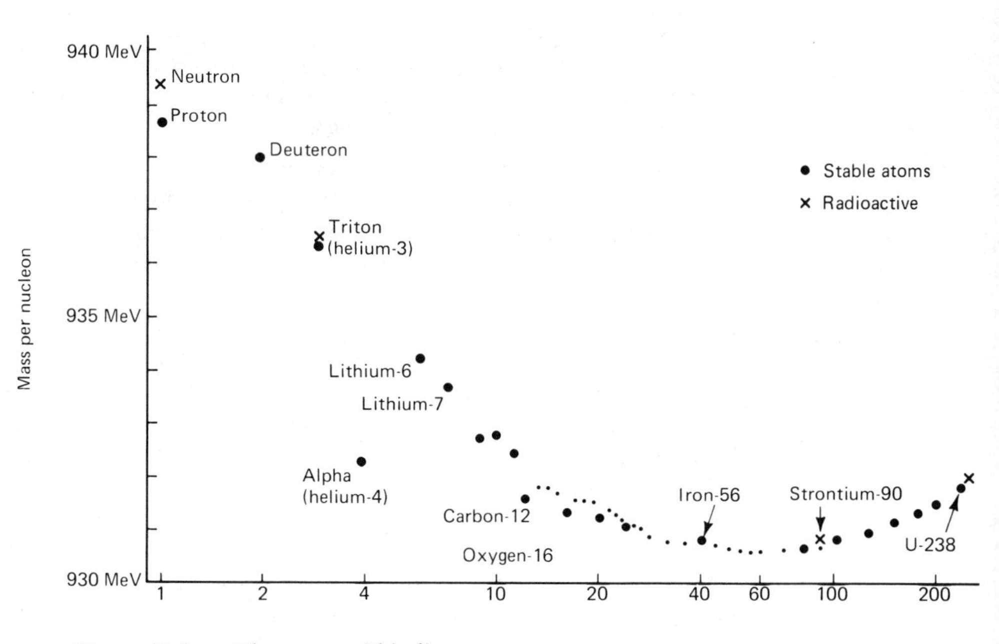

Figure 13-5 The curve of binding energy

gamma ray. This kind of reaction, in which energy is released by the combination of light nuclei, is known as *fusion*.

The fusion reactions in which alpha particles are formed from more energy-rich light nuclei (like protons and deuterons) are called *thermonuclear* reactions. They are very common in nature. The sun and most other stars derive their energy from such reactions. In recent years it has been possible to set off thermonuclear explosions on Earth, also known as "hydrogen bombs."

It would be very useful to release this kind of energy in a controlled way. The difficulty is that two deuterons, both with positive electric charges, repel each other. Only when they are going very fast can they get close enough together for fusion to happen. The usual solution is to try to make the deuterium very hot, so the atoms have high kinetic energy. The temperature needed is so high that any container would melt.

A large research effort is now underway to find how to release the energy of thermonuclear fusion in a controlled way. It may take many years for this effort to succeed, but the rewards will be very great. There is enough deuterium in the oceans of the Earth to supply energy that, even at our present rate of energy use, would last us billions of years!

You will also notice from Fig. 13-5 that carbon-12 lies even lower in mass per nucleon than helium-4. This means that even further energy can be released if 3 alpha particles came together to form a carbon-12 nucleus. Some very old stars are known to get their energy from this reaction.

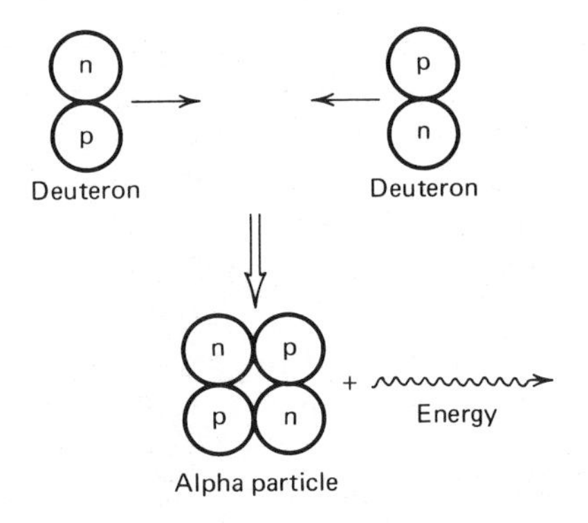

Figure 13-6 Two deuterons fusing to release energy by forming an alpha particle

We can squeeze even more energy out of the nuclei by forming still heavier nuclei, such as oxygen-16, but the returns are diminishing. At a nucleon number of about $A = 60$, the mass per nucleon is at its minimum. To form heavier atoms we have to put in energy.

The fact that the nuclei at the right of the graph are not as tightly bound as those near $A = 60$ has to do with the electrical repulsions between the protons. It is not so easy to bind a nucleus together, with 80 or 90 or more protons pushing each other apart, as when there are only 20 or 30 protons.

Fission

If energy cannot be released beyond $A = 60$ by sticking nuclei together, perhaps we can do it by breaking heavy atoms apart.

If uranium nuclei, at more than 931.5 MeV per nucleon, can be broken into atoms near the minimum of 930.5 MeV per nucleon, we can release about 1 million electron volts per nucleon. This is not as much energy *per nucleon* as in thermonuclear fusion, but since a uranium atom has over 200 nucleons in it, it is much more energy *per atom*. In any case, it is a great deal of energy to be released. It is much more, for a given amount of fuel, than in any chemical reaction, such as burning.

The problem was that most heavy atoms do not simply split in half on their own. If they did, there would be few of them left on Earth by now.

Several of the heaviest nuclei are so energy-rich that they can release energy by splitting off 2 protons and 2 neutrons as an alpha particle, leaving behind a slightly smaller nucleus. This process is known as *alpha decay*. It doesn't happen very often. It is not very likely that the 4 nucleons inside the big nucleus will come together in just the right configuration to make an alpha particle. Even if they do, it is even less likely that the alpha will hold together and escape in the face of all the other nucleons pulling it back. Once in a while the odds are overcome, and an alpha particle does escape. How seldom this happens is shown by the case of uranium-238, whose half-life is 5 billion years. The odds are even that not once in 5 billion years will an alpha particle manage to escape from a uranium-238 atom.

If it is so hard to break off an alpha-particle, how much less likely is it to form a carbon-12 nucleus, much less a fragment with $A = 60$ or more? Moreover, if there were a special nucleus that could split apart easily, it wouldn't last very long. You would have to make it right there in the laboratory.

The surprising discovery of the years around 1940 was that it was possible to make a nucleus that could split in half fairly easily. You made such a nucleus by attaching a slow neutron, one with very little kinetic energy, to a fairly stable but special "fuel" nucleus. There are three such fuel isotopes known. The first one discovered was uranium-235. The others are thorium-232 (90 protons, 142 neutrons) and plutonium-239 (94 neutrons, 145 neutrons). When a slow neutron is absorbed by any of these nuclei, it turns into a new nucleus that is likely to split in half by itself, or *fission*.

Suppose a uranium-235 atom absorbs a neutron. It might split in many different ways. A typical split is shown in Fig. 13-7.

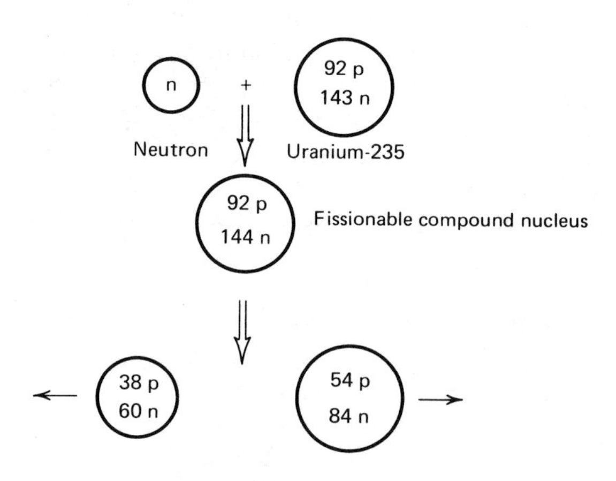

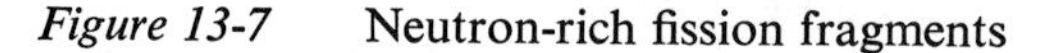

Figure 13-7 Neutron-rich fission fragments

The fission fragments come off with high kinetic energy. Most of the energy of the fission is released in this form. A glance at the table of stable isotopes shows us that both fragments have too many neutrons. This will usually be the case. Uranium atoms have a higher proportion of neutrons than the stable isotopes with A around 100 to 150, characteristic of the fragments.

The balance can be adjusted if the fragments release a few neutrons. There is usually enough energy left over to allow this to happen. But this means that there are neutrons available to be absorbed by still other uranium atoms. And so we have the makings of a *chain reaction*. (See the box.)

The energy released by the fission of uranium has been demonstrated by the successful operation of many nuclear reactors. There is enough uranium-235 on Earth to last at least as long as the fossil fuels as an energy source. The possibility of creating plutonium-239 from the more plentiful uranium-238 would increase the amount of fuel from this source by a hundred-fold. But, of course, the use of these fuels creates problems of radioactive waste disposal that still have to be solved.

CHAIN REACTIONS AND NUCLEAR REACTORS

Demonstration: Mousetrap Chain Reaction. Inside a wire frame cage 30 or 40 mousetraps are set. Each mousetrap is mounted with two corks in such a way that when the trap is sprung the corks are thrown into the air. A cork falling on any trap can spring it.

A single cork is dropped into the cage. Within seconds, all of the traps are sprung. As each trap is sprung it releases flying corks that can spring still other traps. This process is a very graphic illustration of *chain reaction.*

A nuclear chain reaction can come about because: (1) a neutron can make an isotope like uranium-235 split apart, and (2) each time such a fission takes place, more neutrons are released. These neutrons can then make other atoms undergo fission, and so forth.

If the fuel material is pure uranium-235, or pure plutonium-239, the reaction is likely to go on uncontrollably until all the fuel is used up. When this happens, we have a nuclear explosion, a so-called "atom bomb."

Even a single stray neutron, which is almost always present, can set off the whole chain reaction. The explosion can be avoided, however, if the material is spread out so that most of the neutrons get away. This will happen if the sample of fissionable atoms is too small. A sample that is so small that too many neutrons escape is said to be subcritical.

When enough fissionable material is put together in one place, the neutrons do not get away. The chain reaction then goes to completion. The size of the sample needed for this to happen is called the *critical mass.* The exact size of a critical mass depends on the design and shape of the device. It is estimated to be a few kilograms of U-235 or plutonium. Whenever a critical mass of fissionable material is forced together, the chain reaction will go off almost immediately.

The first nuclear bomb was made of pure uranium-235. The problem of separating U-235 from the much more common U-238, with which it is always found mixed, was a tough one. All isotopes of uranium react the same way with other atoms. So no chemical reaction can separate them. The best that can be done is to make use of the fact that U-235 atoms, being about 1% lighter than U-238 atoms, might move a little faster in getting through a filter. So if such a step is carried out many times, the fraction of U-235 can be increased.

Controlled nuclear reactors do not require *pure* U-235, but they do need an enriched sample. It is enough if the fraction of U-235 is 1 in 20 or 30 atoms,

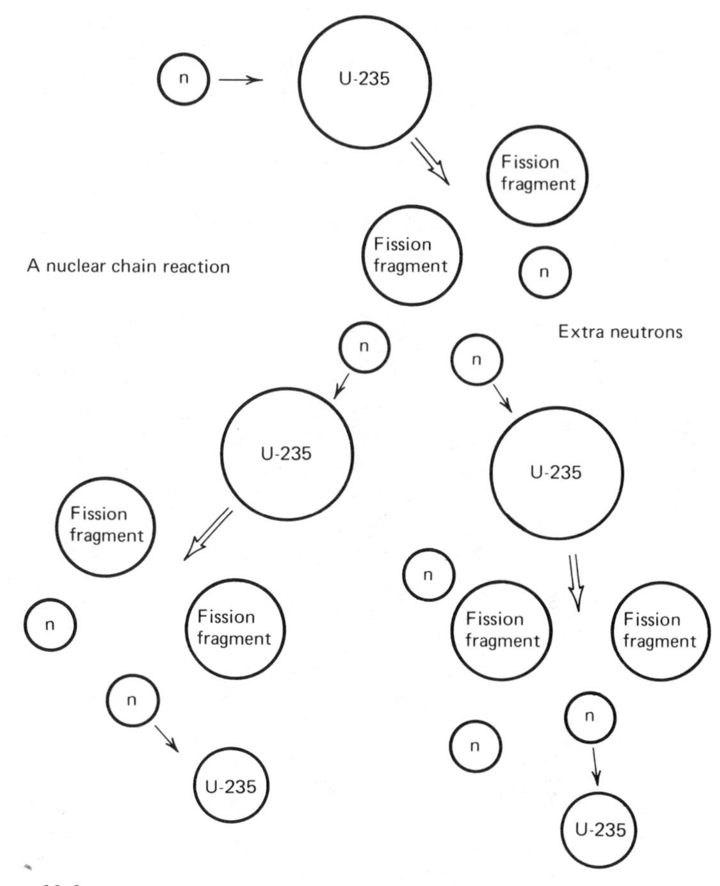

Figure 13-8

rather than the one in 140 mixture found in nature. It is, of course, much easier to make this *enriched* uranium than to make *pure* U-235.

Plutonium, being a different chemical element, is much easier to separate from uranium. You can dissolve plutonium in certain liquids that will not affect uranium. Plutonium is not found in nature, however. It must be made.

When uranium-238 absorbs a neutron, it becomes uranium-239. This atom is radioactive, with a half-life of 23 minutes, decaying into neptunium-239 (93 protons, 146 neutrons). This isotope is also radioactive, and in a few days decays into plutonium-239. Plutonium-239 has a half-life of 24,000 years, quite long enough to be separated from the other atoms, and worked on mechanically. Plutonium is the material now used for most nuclear explosives.

In a controlled nuclear reactor, we do not use pure uranium-235. Several other kinds of atoms are present to absorb some of the neutrons. This includes the uranium-238 atoms that come mixed with the U-235. There are also graphite or water, or sometimes liquid sodium and lithium. These atoms serve to slow down the neutrons, and to carry away the heat energy released by the reactions.

There are also control elements, like cobalt, that are good at absorbing neutrons. If the reaction is going too fast, the control rods can be inserted deeper into the reactor to slow it down. If the reaction seems to be dying down, the control rods can be pulled out some more. A careful balance is maintained between the neutrons that escape, those that are absorbed, and those that continue the chain reaction.

Nuclear engineers are quite certain that a nuclear reactor cannot explode like a bomb. If it overheats, the control rods fall in, absorb the neutrons, and stop the reaction. What is more of a danger is that some of the radioactive materials formed in the reactor can escape into the environment. When a reactor is working properly, essentially none of the radioactivity gets out. Whether an *accident* can allow such an escape, and how likely such an accident is, are matters of concern and debate.

The neutrino

The story of the invention of the neutrino, and its eventual discovery, is one of the more bizarre episodes in the history of science. So convinced were physicists of the law of conservation of energy, that they could not admit that energy appeared to be disappearing in some radioactive decays. Instead, they insisted that the missing energy was being carried off by particles so elusive that no apparatus then existing could detect them. These ghostly objects were given the name "neutrinos" ("little neutral ones," in Italian).

Twenty-five years were to pass from the time that neutrinos were thought of until apparatus sensitive enough to catch a few of them could confirm their existence. During all those years physicists continued to talk about neutrinos as if their existence was already certain. The law of conservation of energy was not easily given up.

A free neutron is radioactive. With a half-life of 12 minutes it will, if undisturbed, turn into a proton, emitting an electron as it does so, to balance electric charge. This decay is not the usual fate of free neutrons, however. Most of them are absorbed by the nuclei of atoms they pass close to, long before the decay takes place.

Neutrons inside radioactive nuclei, those that have too many neutrons to be stable, can decay the same way. The fast electron that gets emitted when this happens is sometimes called a "beta particle." This form of radioactive process is called "beta decay."

The energy released when a beta decay occurs is just equivalent to the difference in mass between the radioactive atom, and the more stable atom into which it decays. It was expected that this released energy would show up as the kinetic energy of the emitted fast electron.

Such was not the case. The beta particle would sometimes come off with nearly all the released energy. Just as often it would come off with hardly any kinetic energy at all. Most of the time its kinetic energy was somewhere in between. What was happening to the rest of the energy?

In one experiment, a radioactive source was placed inside a water bath. The experimenters knew exactly how many atoms were decaying. From the masses of the atoms, before and after decay, they knew how much energy was released with each decay. If all of that energy was being transformed into heat, they could measure it by the rise in temperature of the water.

Only half the expected energy showed up as heat. Whatever form the rest of the energy was taking, it was escaping completely from the bath. The bath and its walls were thick enough to absorb X-rays, neutrons, and any other known carrier of energy. Here were the elements of a mystery.

Wolfgang Pauli, a Swiss theoretical physicist, proposed in 1931 a solution almost as mysterious as the problem it set out to solve. When a neutron (bound or free) decayed into a proton, it emitted not only a fast electron but also another particle. Enrico Fermi, the Italian physicist, soon suggested the name neutrino for this object.

The neutrino was to be electrically neutral, so that electric charge stayed balanced. It was also to be neutral to nuclear forces. It was, like the electron, a *lepton*. Since the neutrino responded to neither nuclear nor electromagnetic forces, it had practically no interactions with any other form of matter. Such an object could pass easily through the water bath. It could pass right through the center of Earth!

The neutrino was supposed to carry off the missing energy from every beta decay. No detector then existing had a chance of detecting a neutrino. It would pass right through everything as though it wasn't there.

Mystery piled upon mystery. The neutrino was believed to have very little, if any, mass. Certainly, it was at least a thousand times lighter than even the electron. Possibly, it had no rest mass at all. It was an object of pure kinetic energy. It travelled, like photons, at the speed of light.

The only forces to which the neutrino was subject were gravitation—which is so weak on the tiny neutrinos we can forget about it—and the so-called "weak interactions." These latter are the forces that acted when the neutrino was created in the first place.

We can work backward from the rates for beta decay to calculate the probability of reverse reaction: a neutrino interacting with a bound neutron to produce, say, a proton and an electron. The rates are very small. If a neutrino passed straight through the center of Earth, the odds are 1 chance in 10 billion that it would react with even a single neutron on the way.

Nevertheless, the odds are not zero. If we put a massive detector near the world's largest nuclear reactor, there is a possibility that once in a while you might detect a neutrino. Our apparatus would have to be buried behind many meters of shielding to make sure that no other particle but a neutrino could get through and fool you. The experiment was finally done, in 1956, and the detection of neutrinos was announced, by F. Reines and G. Cowan.

The law of conservation of energy was verified once again. Even if half the energy released in every beta decay was escaping undetected, physicists could rest confident that the energy had not really disappeared. It was simply being carried off by neutrinos.

Antiparticles

The nuclear atom has always seemed a little unsymmetric. Why should positive charges always be carried by the heavy protons in the nucleus, and negative charges always by the light, non-nuclear electrons? Why can't it be the other way around?

The perhaps surprising answer is that it can be. The fact that all the atoms we know are the way they are may only be an accident of the way matter is distributed in this corner of the universe.

We have already encountered the positron, a particle with positive electric charge, but otherwise very much like the electron. The positron is the *antiparticle*, or mirror image, of the electron.

Antiparticles have the same mass as each other. If they are unstable, they have the same half-life. If they are electrically charged, they have equal and opposite charges.

Particles and antiparticles can be created in pairs, if enough energy is available. If they come in contact, they can annihilate each other.

A theory proposed by the British theorist, P.A.M. Dirac, in 1928 (later revised) suggested that *every* particle in nature has an antiparticle.

The same year that the neutrino was first detected, the antiproton and the antineutron were also discovered, by a group at the University of California, in Berkeley.

An antiproton has a negative electric charge. It has the same mass as a proton. An antineutron is electrically neutral. But it is not identical to the ordinary neutron. When an antineutron undergoes beta decay, it turns into an antiproton, never into a positive proton. A neutron and an antineutron will annihilate each other, whereas two ordinary neutrons will not.

Many other kinds of subnuclear particles have been discovered. In every case the corresponding antiparticle has always been found.

We can speculate about a planet somewhere whose atoms all have positrons circling nuclei made up of antiprotons and antineutrons. So complete is the symmetry that people living on such a planet, if there are any, would not suspect that their world was any different from ours. Only on the dramatic day when one of their spacecraft came to pay us a visit would we discover the difference.

Subnuclear particles

The description of matter given so far in this chapter, which was the way most physicists accepted it about the year 1945, was the simplest we have ever had. All materials were made up of molecules. All molecules were composed of atoms, of which there were only about 100 different kinds. Atoms were made of electrons circling nuclei composed of protons and neutrons.

Protons, neutrons, electrons and their respective antiparticles. Photons of electromagnetic radiation. The mysterious neutrinos. And that was it.

In the 1940s and following years, nuclear physicists began a program of examining the nucleus in even greater detail. The requirements of the uncertainty principle were clear. If you wanted to see the details of a nucleus down to very small distances, you needed to use ever higher-energy projectiles.

So scientists began to build high-energy *accelerators*, of the kind described in Chapter 12. The idea was to shoot very fast protons or electrons at various kinds of atomic nuclei to try to learn how they were put together. It is a very crude method. It is like shooting bullets at a fine watch, and trying to figure out how the watch worked from the shape of the pieces that get knocked out. But this method is the only way we have of studying the nucleus.

The results of these experiments were surprising. The "pieces" that got knocked out of a nucleus were not just protons and neutrons. New objects, whose existence had not been suspected before, were being created in these high-energy collisions. The world of *subnuclear particles* was being discovered.

The subnuclear particles are created out of the energy of the colliding nuclei. This is very much like the way electron-positron pairs are created out of pure gamma-ray energy. The higher energy the projectile has, the more numerous the subnuclear particles can be. Also, the more different kinds of particles can be made.

Subnuclear particles come in many sizes. Such a particle might have barely one-tenth the mass of a proton. It might be ten times the proton's mass. Or it might have many values in between.

Some of them are electrically charged, and some are neutral. When a subnuclear particle has a charge, it is exactly equal in strength to the charge of an electron or of a proton. Why this should always be so has not yet been explained.

All of the new subnuclear particles are radioactive. The half-lives are extremely short. None of them has a half-life as long as a millionth of a second. That is why they are so very rare in nature.

Usually the subnuclear particles are created travelling at very high speeds, almost at the speed of light. They have enough energy to knock many electrons loose from the atoms they pass close to. This trail of *radiation damage* can be used to trace their paths. The particles themselves are never seen—only their "footprints."

If the half-life of a particle is as long as 1 ten-billionth of a second, it can have a track long enough to measure. Travelling at nearly the speed of light, it can move a centimeter or more in that time. If the half-life is much shorter than that, it can be detected only through the characteristic form of the

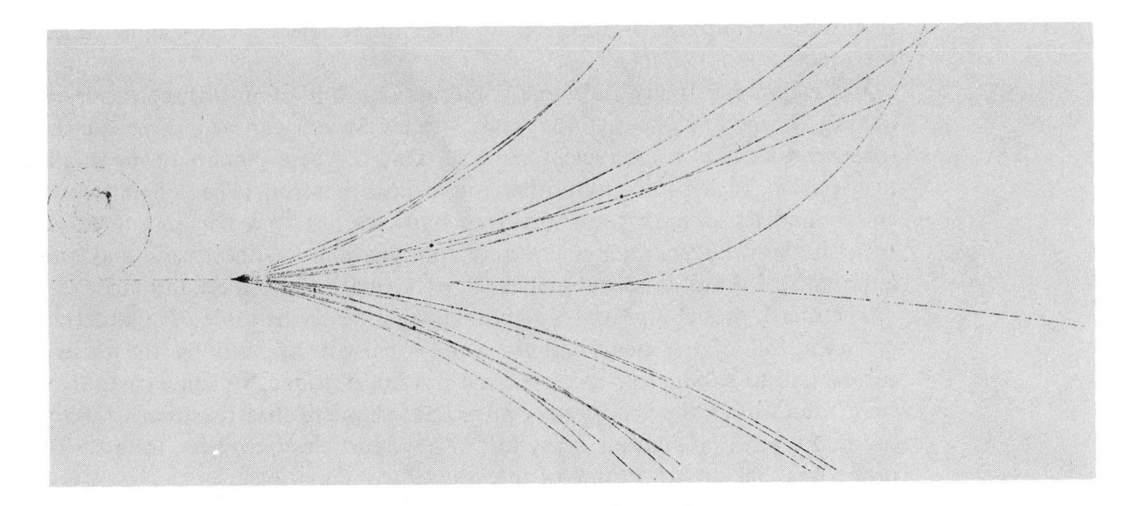

Figure 13-9 Subnuclear particles being created in a high-energy collision in a bubble chamber. (Courtesy Lawrence Berkeley Laboratory)

lighter particles it decays into. There are some subnuclear particles whose lifetime is believed to be so short that their track is no longer than the size of the nucleus itself.

The list of subnuclear particles grows longer year by year. Each has its own properties: its rest mass, its electric charge, its half-life, the particles it decays into. They have names ("mesons," "hyperons," "strange particles") or Greek letters ("pi," "mu," "omega," "psi," etc.) to identify them. The Greek alphabet has been nearly exhausted in naming them all. Sometimes a number, giving its mass-energy in millions of electronvolts, is also used (e.g., the "Delta-1920"). There are now more subatomic particles known than there are chemical elements.

Quarks The idea of the nuclear atom had helped to explain the existence of so many different chemical elements. In a similar way, a new idea has been proposed to simplify the table of subnuclear particles.

In 1961, Murray Gell-Mann and George Zweig, American physicists, suggested that all the subnuclear particles could be put together out of three basic building blocks. Every atom can be built out of protons, neutrons, and electrons. The protons and neutrons and other subnuclear particles might be themselves composed of these new elementary objects. Gell-Mann called these new things "*quarks.*"

We can make many different combinations out of just three kinds of quarks. There are 27 ways (3 times 3 times 3) you can put three quarks together to make a subnuclear particle. One of these combinations might turn out to be a proton. Another might be a neutron. The others would correspond to some of the short-lived particles discovered in recent years. In a similar way you can make nine combinations of one quark and one antiquark. (Like all other subnuclear objects, quarks have their antiparticles.)

The quark model gives a very good accounting of the table of subnuclear particles. All of the quark combinations that are allowed by the theory correspond to subnuclear particles that have been found. In some cases they were found *after* the theory was proposed, showing that the quark theory can make valid predictions. Very few, if any, particles have been found with properties that do not fit one of the allowed combinations.

There are some troubles with the quark theory, nevertheless.

One of them is that none of the quarks themselves have ever been detected. The search has gone on for more than 15 years now. As each new high-energy accelerator is completed, one of the first experiments is to try to create free quarks. Searches have been made in places from the bottom of the ocean to the moon, from the Antarctic to outer space, with no success. We cannot easily explain why quarks can fit so easily into combinations, can jump from one atom to another in a collision, but never seem to break completely free of each other.

The other main problem with the quark theory is that it is getting more complicated. Some very recent experiments (1974–1976) seem to require a *fourth* (and perhaps a fifth and a sixth) kind of quark. Meanwhile some other versions of the theory require *three* copies of *each* kind of quark. The pendulum that swings between complication and simplification in subatomic physics may be swinging toward complication again.

We have not reached the end of the trail in research into the very small. Physics is still alive. There is more work to be done. We are working here on one of the frontiers of our state of knowledge. Many fascinating discoveries have been made on this frontier—all within the lifetime of people still alive! We cannot help but believe that many more such discoveries lie ahead of us.

Glossary

alpha particle—The nucleus of the helium atom, made up of 2 protons and 2 neutrons. Alpha particles are sometimes released in the decay of the heavier radioactive nuclei.

antiparticle—A "mirror image" of a subatomic particle, having the same mass and other properties, but opposite electric charge. The positron is the antiparticle of an electron. When a particle and an antiparticle come together they can annihilate each other, turning into pure energy. It is thought that every kind of particle in nature has an antiparticle.

atomic number—The number of electrons in a neutral atom of a given kind, equal to the number of protons in its nucleus. The atomic number determines how the atom can combine with other atoms, that is, its chemistry.

chain reaction—A reaction that goes explosively, when each collision releases energy or particles that help induce more such reactions to take place.

deuteron—A nucleus made up of 1 neutron and 1 proton. It is important in fusion reactions. An atom with a deuteron as its nucleus is called deuterium, or heavy hydrogen. It is an isotope of hydrogen, found (at a rate of 1 in 6000) everywhere that hydrogen is found, particularly in the ocean.

fission—The release of nuclear energy by the splitting apart of heavier nuclei, such as that of uranium.

fusion—The release of nuclear energy by the combining of light nuclei, such as those of hydrogen into helium and larger nuclei.

hadron—A subnuclear particle that is affected by the strong nuclear forces. Protons and neutrons are the most common hadrons.

half-life—The period of time, for a given kind of radioactive nucleus, such that the odds are even that it will release its energy and undergo decay. If the nucleus survives one half-life, the odds are still even that it will last another half-life, and so on.

isotope—A species of nucleus with a certain number of protons *and* a certain number of neutrons in it. Nuclei with the same number of protons, but differing numbers of neutrons, are said to be isotopes of the same chemical element.

lepton—A subnuclear particle that is not affected by the strong nuclear forces. Electrons and neutrinos are examples of leptons.

neutrino—A subnuclear particle with no electric charge; it is not affected by the strong nuclear forces. Neutrinos are produced in certain kinds of radioactive decay (notably those in which electrons are also produced). Neutrinos are very difficult to detect, since they pass easily through everything. They were invented in 1931 to explain how energy was being carried away in certain reactions. They were finally detected in 1956.

neutron—A particle having about the same mass as a hydrogen atom, no electric charge, but subject to the strong nuclear forces. A free neutron is radioactive, with a half-life of about 12 minutes. However, neutrons bound into nuclei are quite stable themselves, and are needed to help hold the nucleus together.

nuclear energy—A form of energy, not yet completely understood, whose existence is needed to explain why nuclei hold together.

nuclear force—The strong force that holds protons and neutrons together in nuclei. The nuclear force also governs the reactions between many kinds of subnuclear particles.

nuclear physics—The branch of physics that studies the structure of nuclei, the particles of which they are composed, the interactions between them, and the forces that hold the nucleus together.

nucleon—A proton or a neutron.

nucleon number—The number of protons plus the number of neutrons in a given nucleus. The nucleon number is the nearest whole number to what is called in chemistry the atomic weight.

nucleus—The small, massive, positively charged bit of matter found at the center of every atom.

proton—The nucleus of a normal hydrogen atom; also, the positively charged particles that combine with other protons and neutrons to make up every kind of atomic nucleus. The charge of a proton is equal but opposite to that of an electron. Its mass is such that 6 trillion trillion protons would weigh 10 grams.

quark—A hypothetical kind of particle that is supposed to go into making up protons, neutrons, and other subnuclear particles. No single quarks have been detected, as of 1978.

radiation damage—The disruption caused in matter when fast subatomic particles pass through it and deposit their energy. When this happens, large numbers of atoms can be broken loose from the molecules and crystals in which they were bound. The effects often show up as the weakening of metals, the darkening of glass, and sickness and mutations in living cells.

radioactivity—The condition of a nucleus when it can release energy by rearranging its neutrons and protons, but it does not do it immediately. Radioactive nuclei release their energy after an almost random interval, whose time scale is set by the half-life of the particular nucleus.

reactor—A nuclear reactor is an installation in which nuclear energy is released in a slow, controlled manner. Fission reactors work by the splitting of uranium nuclei within them. A nuclear reactor can be used as the source of heat energy for the warm reservoir of a heat engine.

subnuclear particle—One of a whole class of short-lived objects that can be created in the collision of two nuclei, thrown at each other at very high energy. The study of subnuclear particles is at the frontier of our knowledge today.

thermonuclear—Said of energy that is released in the fusion of light nuclei, such as isotopes of hydrogen, into helium and heavier nuclei. Thermonuclear reactions power the sun and the stars, and dominate the energy released in hydrogen bombs. The design of a *controlled* thermonuclear reactor is one of the aims of modern research.

Questions

1. What do we now think an atom looks like? What experimental evidence makes us think this?
2. What objections were raised to the nuclear model of the atom? How were these objections overcome?
3. In what ways can different isotopes of the same element behave differently from each other? In what ways do they behave similarly?
4. What chance is there that a single atom of strontium-90 will last 58 years? 116 years? Forever?
5. Trace how strontium-90, released into the air and settling on the grass, might find its way into your bones.
6. Name three benefits that arise from the use of radioactive materials. Name some hazards.
7. Why can't a carbon-12 nucleus left by itself fly apart?
8. What is the difference between fusion and fission?
9. Compare the amount of energy available on Earth from (a) burning fossil fuels, (b) fission of heavy elements like uranium, (c) fusion of light isotopes like deuterium.
10. The interior of the Earth and similar planets is thought to be made of iron. Explain why iron is so abundant in the world.
11. In the mouse-trap demonstration, where did all the energy come from that was released when the traps were sprung? What is the source of the energy released in a nuclear chain reaction?
12. What is the difference between a controlled nuclear reactor and a nuclear bomb? How is the reactor kept under control?
13. If you had been a scientist during the years before the neutrino was found, would you have believed in its existence? Why, or why not?
14. What would happen if a space-craft from a planet made of antimatter arrived on Earth? What steps could we take to prevent this catastrophe?
15. What do we mean by an "elementary" particle? Give three different stages of history when scientists thought they had defined the basic elements out of which all nature is made.

14

beyond the solar system

Introduction and summary

This book started by talking about scientific explanations of nature. People have always been curious about the world we live in. Any explanation of nature, scientific or not, has to answer certain questions about the world.

How big is the world? How did it get the way it is now? What *is* "the way it is now"? How long has it been here? When and how will it end?

There is no lack of answers to these questions. They are the subject matter of myth and legend, of sacred books and the writings of wise men. Modern science, too, has its model of the universe.

This model may very well be wrong. It is certain to be revised many times. It has already gone through many revisions. But people want to know the answers to their questions about the world. And science tries to provide the best answers it can.

What makes it hard to answer questions about the universe is that we are going so far beyond our normal experience. We must deal with objects so far away that we can never hope to travel to them. We talk about events that happened so long ago that no living creature could have been there to see them. We observe temperatures far higher than have ever been felt on Earth. We have to discuss conditions so different from what we know on Earth that we can hardly imagine them.

It is no surprise that science often makes mistakes in this realm. What scientists try to do is take the laws and principles that we already know, and try to apply them to the extreme conditions of the universe. But theories are on shaky ground when we try to apply them beyond the area where they have been tested. We have seen how scientific theories have foundered, and had to be replaced, when they had to deal with the very small or the very fast. We cannot be sure that our present ideas are good enough to handle the very large or the very distant.

We cannot do experiments with the universe. We can't control what happens in the stars. Most processes in the distant heavens take place so slowly that we cannot even wait for changes to happen. All we can do is observe, and try to explain.

In this chapter we give the best explanation we can about the universe.

We expect our picture of the world to change as new discoveries are made and new ideas are presented. Nevertheless, the picture we have fits together remarkably well. Science in the late 20th century has provided us with a view of the world that is fascinating and exciting. You must decide for yourself whether the answers it gives to our questions about the world are satisfying.

The age of the sun

As the sun shines it pours enormous amounts of energy out into space. It has been doing this at a steady rate for all recorded history, and for long ages before there were any men on Earth.

The only source we know of that can provide so much energy over so long a time is nuclear energy. We think that the sun, and most of the stars, get their energy from nuclear fusion. In the hot interior of the sun free protons are able to combine to form nuclei of atoms like helium. The energy released in this fusion comes out of the sun mainly in the form of visible light and infrared radiation.

We can figure about how long the sun has already been shining. We can measure how much fuel has already been burned, how much helium has already been formed. Suppose the sun has been been shining all those years about as brightly as it is now. How long must it have taken to consume all that fuel?

The best estimate we can make now is that the sun has been shining for at least 5 billion years. There is enough fuel left to keep it burning at the same rate at least another 5 billion years. There is no danger of the sun burning out very soon.

There is another way of finding out the age of the solar system. Certain rocks that are thought to be very old contain a few radioactive atoms. These particular atoms have half-lives of several billion years. They were in the rock when it was formed and have not all decayed yet. We can measure how many of the radioactive atoms have already decayed and how many are still left. From these measurements we can get the age of the rock.

The oldest rocks found on Earth were formed about $4\frac{1}{2}$ billion years ago. Rocks brought back from the moon are also that old.

Meteors are chunks of rock and iron that fly about in the space between the planets. They may be pieces of a planet that once broke up. Or the meteors may be bits of material that never got together to form a planet. Occasionally a meteor crashes into Earth, leaving a fiery trail as it roars through the air. Some meteors that didn't burn up completely have been found on the ground.

The latter, called *meteorites*, can be dated by how many radioactive atoms they have in them. The oldest meteors are also found to be about $4\frac{1}{2}$ billion years old.

The pieces of the picture fit together. Sometime, about 5 billion years ago, the sun began to shine. At about the same time, or not long after, Earth, the moon, and the meteors, shaped up in close to their present form.

We can speculate that the other planets are also the same age. Some day space vehicles may measure the age of rocks on the distant planets and their moons and test this idea.

The birth of stars

What was there before the solar system was formed? We have no direct information, but we can invent models and theories. We can try to find clues by looking at places in the sky where perhaps we can see new suns being formed even now.

There seems to be quite a bit of gas and dust in some regions between the stars. It is not very thick by our normal standards, maybe 1 atom per cubic

Figure 14-1 The Great Nebula in Orion, where new stars are forming out of interstellar dust. (Lick Observatory Photograph)

centimeter. But there is enough of it that the pull of gravity, with which each atom attracts all the others, can eventually clump a large mass together into a thick cloud.

According to this model, a star is formed out of a swirling cloud of gas and dust. Most of the material gets pulled closer and closer together by its own gravity. When the atoms get close enough together, and move fast enough, nuclear fusion can begin. And the star begins to shine.

Not all of the material winds up in the center. If the cloud has any spin to it, some of the material will be left in orbit around the new star. Very often there will be enough of the cloud left out to make a second, and even a third, star. Perhaps half of the stars we can see turn out on close inspection to be double or multiple star systems, in orbit around each other.

It does not seem unlikely that planets can form, much of the time, when a star condenses out of a gas cloud. We do not know of any other star that has planets. They would be too dim to see with even our best telescopes. But if this model is correct, every star that is not a double or multiple star system is a good candidate for having planets like our own sun.

The death of stars

What will be the end of our sun?

To answer this question we need to use all the predictive powers of our theories of nature. We don't have any experience at all with the subject. We don't have time enough to wait and see what happens. But the question has interested humans since man first began to think.

We can look again to the stars for clues. Most of the stars seem to be shining as steadily as our own sun. Some are a bit brighter, some are dimmer. In no way does our sun seem out of the ordinary.

This is reassuring. If the sun is just like most of the other stars, if they all have long careers both behind and ahead of them, then we can look forward to no disturbing changes for ages to come.

It would be different if the sun were unusual. An unusual kind of star is probably one that is changing. Or it is going through a rather brief stage in its life. If the stars we see are of all different ages, then the way we see most stars represents the way they spent most of their lives. If every star spends 5% of its life in a certain unusual stage, then 5% of the stars we see at any moment are likely to be going through that stage when we see them.

We can guess at the future of our sun by listing some of the unusual kinds of stars that can be seen in the sky. We can use some of our models of how stars behave to see how these various kinds of stars might relate to the possible future of the sun.

Red giants There are quite a few stars, some of them very prominent in the night sky, that fall into the *red giant* category. Their dull color shows that their surfaces are not very hot (for a star). Yet their brightness means that they must have a very large surface area to make up for the dullness.

The star Betelgeuse, in the shoulder of Orion and easily seen in the winter sky, is an example of a red giant. It is estimated to be so large that, if our sun were that size, Earth and Mars would be *inside* the star.

The model says that red giants get much of their energy from the fusion of helium nuclei into carbon-12. It takes 3 helium-4 nuclei ("alpha particles") to make a carbon-12 nucleus. This reaction does not happen in ordinary stars because it is so rare to find 3 helium nuclei together.

An ordinary star becomes a red giant when it has been shining so long that a large concentration of helium has built up. Once the fusion of helium begins, it goes very fast. So much energy is released in the center of the star that the outer mantle is forced to expand. This causes the characteristic large size of a red giant star.

The red giant stage does not last very long, as star lifetimes go. After a few tens or hundreds of millions of years, the helium is used up. The stages that follow are even shorter.

Supernovas Every few centuries a star will flare up so brightly that it outshines the combined light of all the other stars in the sky. This happens in a spot where no star had been noticed before. For a few days or weeks this *supernova* shines so brightly that it can be seen in broad daylight. Then the star dims and disappears, leaving behind a large expanding cloud of material that can still be seen, with a telescope, hundreds of years later.

It has been nearly 400 years since a supernova appeared that was visible to the naked eye. Before that a very famous supernova was recorded by Chinese historians in 1054 A.D. Its remnant, known as the Crab Nebula in the constellation Taurus, can still be observed. Many other supernovae have been spotted by telescope in distant galaxies; the supernova is as bright as the whole galaxy combined.

A supernova is a star that is literally blowing itself apart. It uses up nuclear fuel at a fantastic rate. Possibly the remaining helium and carbon-12 is being fused into oxygen and silicon and iron, and all the other elements, to release the final scraps of nuclear energy in one great blow-up.

We can only guess at what processes are taking place inside an exploding supernova. It is possible that some of the heavier elements, from tin to lead to uranium, are being made inside the supernova, even though this process would actually *consume* nuclear energy.

The fact that much of the material in a supernova is blown off into space is important. Eventually this debris becomes mingled with the gas and dust of interstellar space, forming the material out of which new stars are born.

We notice that in the solar system, the sun seems to be mostly hydrogen and helium, but on Earth there are plenty of heavier elements. When did all the iron and oxygen and silicon and uranium get made? Is the whole solar system made of "recycled" material, the remnants of past supernova explosions? There are some scientists who think so.

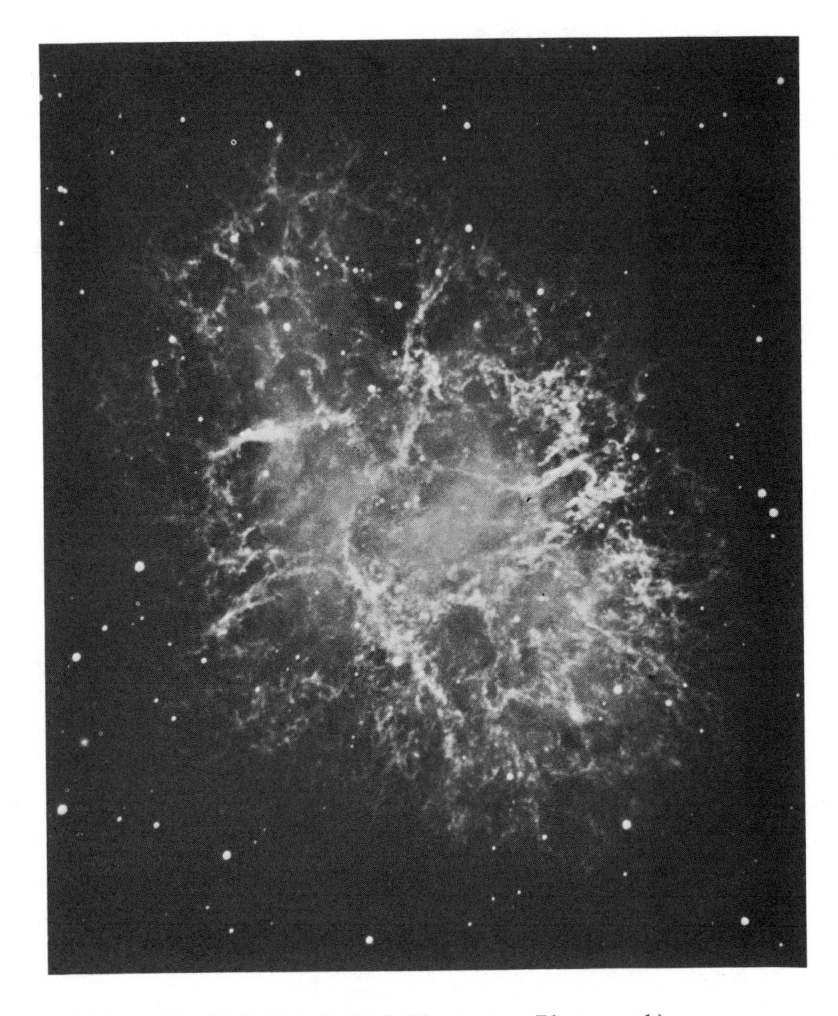

Figure 14-2 The Crab Nebula (Lick Observatory Photograph)

If this model is correct, the sun is a second- or third-generation star. The original stars and their planets, if any, were a sterile lot. They were mostly made of hydrogen and helium, out of which few kinds of solid structures can be built. But the original stars have long since burned themselves out. Their materials, particularly the heavy elements made in their dying spasms, were hurled out again into space, where they coalesced again into new stars and new planets. Among these is our sun and the planets of our solar system.

Dead stars Most stars do not go through a supernova stage. Even if a star does explode and blow off most of its outer shells, a small core at the center may still hold together.

There are several fates that can befall a star that is no longer releasing energy in large amounts. Which of these fates comes about depends mostly on how much mass is left in the star.

Gravitational forces are always working to make a star collapse into a smaller and smaller bundle. As long as the star is burning nuclear fuel, this collapse will be resisted. The energy being released, and trying to escape, will generate enough internal pressure to keep the star from collapsing. Once the fuel is gone, the star will collapse until the atoms are pressed very closed together. Atoms resist being pushed too close to each other. This resistance may be strong enough to balance the gravitational forces. If it is, we have a stable dead star. If not, there may be runaway gravitational collapse.

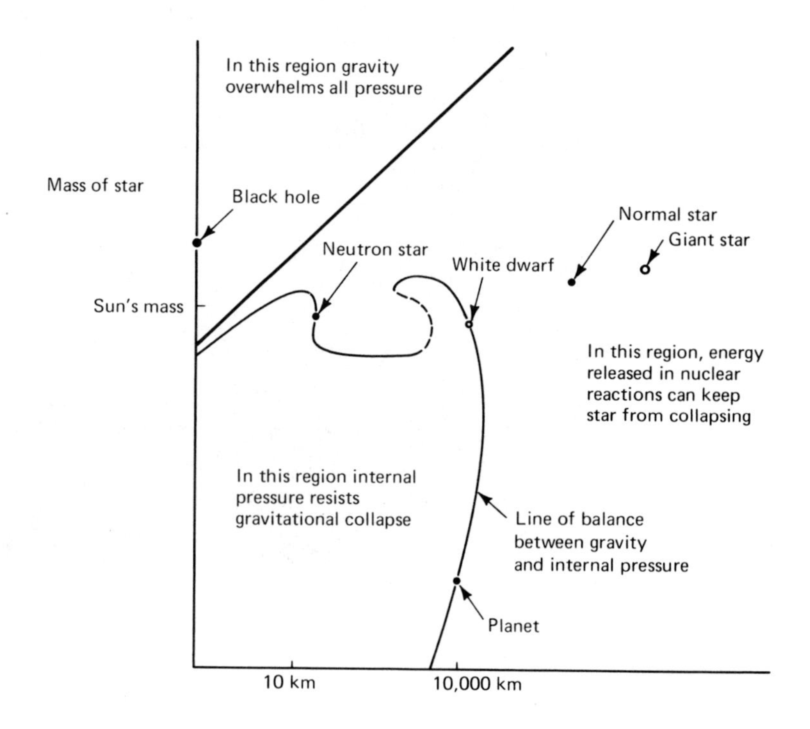

Figure 14-3 Mass versus radius of stellar objects, showing stable and unstable regions

At one extreme, there are low-mass objects like planets. There is no doubt that the weight (that is, the gravitational force) of Earth's mantle presses very hard on its core. But the strength of the rocks and metals of which Earth is made is enough to hold up the weight. So Earth does not collapse.

Even as large a planet as Jupiter does not collapse under gravity. It does not even get compact enough to ignite the nuclear fuel, hydrogen, which Jupiter has in abundance.

> The planets are not, and never were, stars. There must be a minimum mass, bigger than Jupiter, less than the sun, that is needed to turn on a star. Every star must be at least that minimum mass.

When an object that is big enough to have once been a star collapses, new things begin to happen.

White dwarfs White dwarf stars are dim objects. They glow because they still have some heat energy left. Their temperature is fairly high, but they are not releasing any more nuclear energy.

White dwarfs are not much bigger than Earth in size. But their mass is closer to that of the sun. And the sun is 300,000 times as massive as Earth.

The material in a white dwarf must be extremely tightly packed. A teaspoonful of it would weigh several tons. It is unlike anything we have ever experienced.

In such material the atoms are so close together that it cannot be decided which electron belongs to which atom. Rather, all the electrons are stripped from their parent atoms and go running around freely, and at high speed, through the whole material.

It is the electrons themselves that resist being compressed even more. The more closely an electron is confined, the more energy it must have, on the average. This is to satisfy the uncertainty principle of quantum mechanics. The white dwarf reaches a balance when the gravitational energy that would be released by contracting a bit more would not equal the energy needed to confine the electrons that much closer together.

Neutron stars, pulsars An even more collapsed kind of star than a white dwarf was discovered in 1967. Because we receive short bursts of radio waves from this kind of star at regular intervals, they are called *pulsars*.

The bursts repeat themselves at such exact periods that you can keep time by them. Some pulsars repeat at a rate of about once per second. There is at least one that pulses 30 times a second. The pulses themselves may last only a few thousandths of a second.

The best explanation we have is that these stars are rotating at the rate of once or several times per second, and that they have a "radio beacon" fixed in the star which points toward Earth once during each turn.

These stars must be very small to be able to spin so fast. Even an object as large as a white dwarf would surely fly to pieces if it were spinning 30 times per second.

The pulsar with the fastest rate of rotation is located at the center of the Crab Nebula. That is exactly where the Chinese saw a supernova explode 900 years ago. Perhaps pulsars are the remnants of the core of a supernova, the part that was not blown off. Perhaps the other pulsars are much older and have slowed down over the years. The Crab Nebula pulsar is observed to be slowing down also.

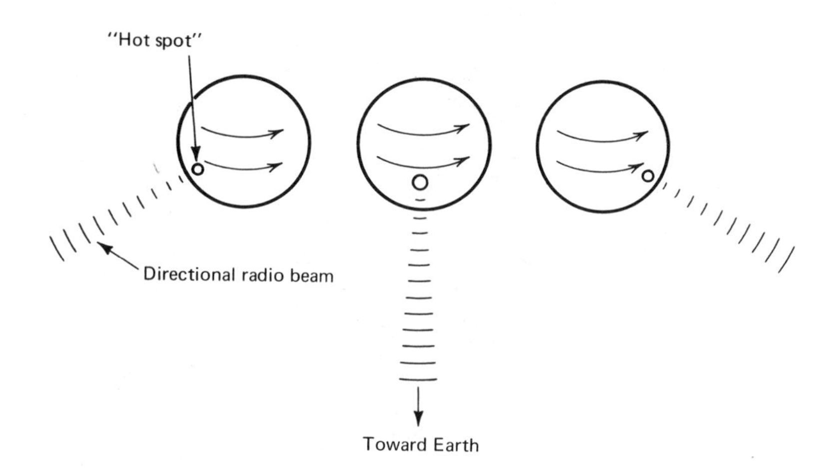

Figure 14-4 Three positions in the rapid rotation of a pulsar. Only for a brief period during each cycle does the beacon point toward Earth. We observe a short burst of radio waves once during each cycle.

The theoretical idea that most nearly fits what we observe about pulsars is that of a *neutron star*, proposed by J.R. Oppenheimer in 1939. A neutron star has the mass of about that of the sun. But it may be only 10 kilometers across!

In a neutron star there are no electrons. All of them have been forced to combine with protons in the nuclei to form neutrons. A neutron star is like a giant nucleus of a single atom. It is so unbelievably dense that we have difficulty even imagining what this kind of material must be like.

THE GENERAL THEORY OF RELATIVITY

When we left Albert Einstein, back in Chapter 3, we had seen how his "Special" Theory of Relativity had helped explain some difficulties in Maxwell's theory of electromagnetism.

In Chapter 12, we pointed out that some changes were needed in Newton's Principles of Mechanics. Objects moving at close to the speed of light are hard to accelerate. We said that they gained mass as their energy increased.

There are also some problems reconciling the laws of gravitation with the ideas of relativity.

For one thing, the effects of gravitation cannot spread out faster than the speed of light. If the Earth changes its position, the moon will continue to be attracted toward the old position, at least for the $1\frac{1}{2}$ seconds it takes for the information to travel from the Earth to the moon.

This delay in the onset of gravitation leads to the existence of *waves* of gravitation. The reasoning is similar to the explanation given in Chapter 10, for electromagnetic waves. Einstein predicted that gravitational waves can be found. However, they are weak and hard to detect. Searches are being made for such waves, but none have been reliably reported yet.

A second problem with the law of gravity has to do with the mass of the attracting body. The mass of a star or planet depends on its speed, as we have seen. The speed is different in different frames of reference. Will its gravitational attraction for other bodies also be different in each such frame?

Einstein's General Theory was an effort to reconcile gravitation with the principle of relativity.

He started with the observation, made by Galileo, that all objects, no matter how massive, fall at the same rate under gravity.

He noted a similarity with a situation on a fast-turning merry-go-round. If you put a marble on the turntable floor, it quickly flies off on a tangent. This is not surprising to someone on the ground. The marble is just following the law of inertia. But someone on the merry-go-round, not realizing he is turning, might claim that there is a "gravitational force" acting, pulling the marble outward. He would note that, just as in the case of other gravitational forces, all marbles of whatever masses get accelerated at the same rate.

We, of course, know that if the observer would just get off the merry-go-round, and look at things from the more natural frame of the solid ground, there would be no need to invoke special "forces."

By the same token, Einstein argued, the reason we claim an apple falls from a tree is because we insist that the surface of Earth is standing still. A more "natural" frame might be inside a freely falling elevator. In such a frame everything, falling at the same rate, would appear to be not falling at all.

Such a falling elevator can be used to "transform away" the effects of gravity at a single place and time. But we cannot stay in such an elevator too long, for it will crash. At other places on or near the Earth, we would have to use elevators falling in different directions, and at different rates.

Einstein was led into a complicated problem of geometry that requires advanced mathematics to solve. He achieved a beautiful solution by applying his principle of relativity in a more general form:

> "*The laws of physics must be expressed in such a way that they have the same mathematical form, no matter which coordinate system is used.*"

This idea led him directly to a unique mathematical equation.

The solutions of this equation, in most ordinary cases, are almost the same as the results of Newton's laws of gravitation. This is what we expect, from our first criterion for acceptance of a new scientific theory. Newton's laws had been tested in so many ways, over so long a time, that any major differences in results would have led us to reject the new theory.

The differences that do show up between Einstein's and Newton's predictions are either very tiny and hard to measure, or show up in such inaccessible places as the inside of collapsing stars.

So far, the verified predictions of general relativity include:

- Light, and other electromagnetic waves, are deflected by gravity.
- A photon that "falls down" under gravity will gain in energy, and thus increase in frequency. A photon that rises against gravity will lose energy, and thus decrease in frequency.
- A radar signal sent from Earth to Mercury and back will take longer to make the trip, over the same appearent distance, when it has to pass close to the sun than when it does not.
- A planet in an elliptical orbit about the sun will have its motion affected in such a way that the second focus of the ellipse (the one not at the sun) will slowly rotate around the sun.

The more dramatic predictions of general relativity, about gravitational waves and black holes, about the "big bang" or the expansion of the universe, are harder to confirm. We are only now, more than 60 years after the theory was proposed, getting to the point where we can test some of the predictions. We cannot be sure the theory is correct in all details, nor do we expect that it will be the last word. At this stage of our knowledge, the general theory of relativity is the best basis we have in our ideas about the universe.

Runaway gravitational collapse, black holes If a star is too massive, it cannot find stability as either a white dwarf or a neutron star. When such a large star runs out of fuel, there is nothing to stop it from collapsing to practically nothing.

How much mass is "too massive"? Calculations show it is not out of range. A star only one-third more massive than the sun could not be stable as a white dwarf. We are not exactly sure what the most massive neutron star could be. It is almost certainly less than twice the mass of the sun. Any star heavier than this limit must, unless it loses some mass, eventually collapse completely.

The speculation grows more and more outside our realm of experience. For the complete gravitational collapse of a star we have only theories to describe the possible situation. A very nice theory, but one that has not been tested in this realm, is Einstein's general theory of relativity (see box).

In a situation of gravitational collapse, the general theory describes the star as "digging itself a hole in space-time and pulling the hole in after it." At some point the space itself around the star is collapsing so rapidly that nothing, not even a light signal, is fast enough to escape.

This kind of star has been called a "black hole." We cannot see a black hole, since light cannot escape from it. Hence, the name.

We can detect the presence of a black hole in a couple of ways. For one thing, its gravitational pull on objects outside itself is the same as ever. If one of the stars in a double star system has become a black hole, the two will continue to revolve around each other. We can watch the normal companion star, and use its motion to learn about the black hole.

We can also see some of the energy released when some new material falls into the black hole. This energy comes out in the form of X-rays that can be detected with instruments above our atmosphere.

Scientists have been looking for signs of black holes for many years now. In some cases they may even have found such an object. If they have indeed done so, it would be a striking confirmation of a prediction of the general theory of relativity.

The galaxy

It isn't easy to measure the distance to a star.

For some of the closer stars we can reckon the way a surveyor would. We can look for the slight shift in angle when we look at the star, first from one side of Earth's orbit, and then 6 months later when Earth is on the opposite side of the sun. From this shift we can figure how far away the star is.

For more distant stars it is more of a guessing game. We estimate how brightly the star is actually shining. Then we observe how much light reaches us. This gives us an idea how far away the star must be. We may not guess right on the true brightness of any given star. We assume that *on the average* all stars of a given type will be about the same brightness.

Kilometers are not very useful units for giving the distance between stars. Even the astronomical unit, the distance from Earth to the sun, is not big enough. So we use as our unit of length the *light-year*.

One light year is the distance a signal, moving at the speed of light, can travel in 1 year. It takes light 8 minutes to get from the sun to Earth. It takes radio signals a few hours to get from Earth to the most distant planets. A light-year is thus more than a thousand times the size of our solar system.

The nearest star to our sun, Alpha Centauri, is more than 4 light-years away. This star is visible from south of the equator on Earth. It is actually a triple star system, not likely to have any stable planets around it.

The nearest star that can be seen in the northern hemisphere is Sirius ("the dog star"), 8 light-years away. Most of the stars that we can see easily are a few tens or hundreds of light-years away.

In one region of the sky ("the Milky Way") there are so many stars that we do not see them separately, but as a white patch in the sky. The sun and the nearby stars are actually part of this great system, which we call the *Galaxy* (written with capital G), from the Greek word for "milk."

It is not easy to map out the Galaxy from our position inside it. More than half of it is blocked from our vision by clouds of dust between the stars. We can imagine what its shape must be by comparing our Galaxy with other star systems outside of it that we can see more clearly.

The stars in the Galaxy mostly lie within a region shaped like a flat disk. This disk is about 100,000 light-years across, and about 10,000 light-years

Figure 14-5 The great spiral galaxy in Andromeda (Lick Observatory Photograph)

thick. There must be over 100 billion stars in the Galaxy. They are arranged into a densely packed center with long spiral arms stretching out, much like the spiral galaxy in Fig. 14-5.

The sun and our solar system lie along one of the spiral arms, about two-thirds of the way out from the center of the Galaxy. The whole system is spinning about its center. The sun revolves around the center of the Galaxy once in about 200 million years. To make this journey the solar system has to be moving at a speed of 300 kilometers per second through the Galaxy.

The *halo* of the Galaxy is the spherical region above and below the main disk. Some stars in the halo are gathered together into what are called *globular clusters*. From the amount of fuel they have used up we can estimate that the stars in the globular clusters are older than average for the Galaxy. From this we infer that the Galaxy was once more sphere-shaped than it is now, filling up the halo as well as the main disk.

The expanding universe

With telescopes astronomers can look beyond the Galaxy and find other star systems very similar to ours. These are known as *galaxies* (with a small g).

Some galaxies have spiral arms like our Galaxy. Others are shaped like footballs or like bars. Some have no definite shape at all.

The nearest galaxies outside our own are roughly 1 million light-years away. Several galaxies besides our own form a group known as the Local Cluster.

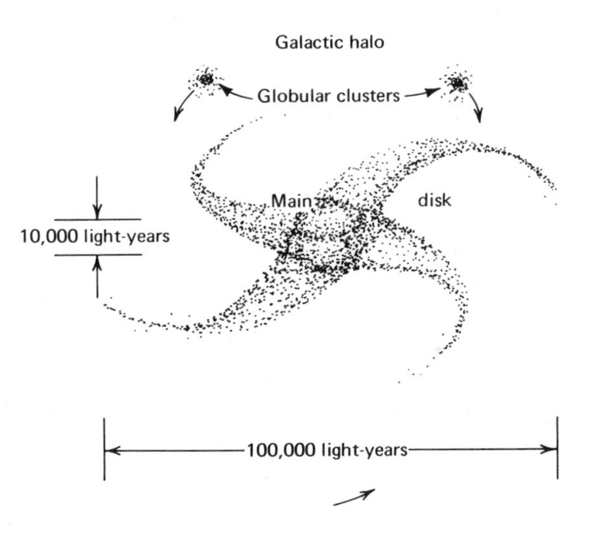

Figure 14-6 Our Galaxy

Beyond the Local Cluster galaxies are visible all over the sky. Thousands of them have been charted and listed in star catalogs. Many, many more galaxies are no doubt visible through strong telescopes. Only the lack of time to chart them all limits our knowledge of all these galaxies.

The distance to other galaxies ranges from the million light-years of the Local Cluster up to more than a billion light-years away. There are probably galaxies even farther away, but at a distance of a billion light-years even a huge galaxy looks like a tiny point of light in our best telescopes. It would be hard to tell the difference between such a galaxy and some dim star close by in our own Galaxy. So the limit is set by our telescopes, and by the air above them which limits how well they can make out tiny features. As far away as any telescope can make them out, there are plenty of galaxies visible.

The light from stars—and from galaxies, that are made of many stars—is often extra bright at certain special wavelengths. They may be somewhat dimmer than average at other special wavelengths. These special colors are characteristic of the atoms that are usually found in the outer layers of the stars.

If the star or galaxy is moving away from us, these special wavelengths will be a bit longer than if the star is standing still. This is the *Doppler effect*, explained in Chapter 6. The faster the source is moving away from us, the more the special colors will be shifted toward the red (longer wavelength) end of the spectrum.

Edwin Hubble, the American astronomer, found in the 1920s that all the galaxies beyond the Local Cluster had such a red shift. All appeared to be moving away from us. The farther away a galaxy is, the greater its red shift, i.e., the faster it appears to be speeding away.

The model that explains Hubble's findings most easily is that *the universe is expanding*. All the galaxies are flying apart from each other.

> Take a sheet of thin rubber. Paint dots on it to represent the galaxies. Stretch the sheet. The dots will move apart from each other. The farther apart the dots are, the faster the distance between them is increasing. From the point of view of any single dot-galaxy, the picture is the same. All the other dots are moving away from it. The farther away to the other dot, the faster it is receding.

What does this mean for the future of the universe? Will the galaxies continue to fly apart forever? Or will they someday turn around and come back toward each other? We can only speculate on the answer. And we will not be around to find out whether we were right.

The big bang We can take Hubble's results and work backward in time. If all the galaxies are flying apart, then in the past they must have been closer together. If we assume that the rate of expansion has always been the same, we can even figure a time when they were all close together. There are some uncertainties in the exact measurements, however, but we can make a reasonable estimate that the expansion began about 20 billion years ago.

If we had difficulty imagining the conditions inside a white dwarf or a neutron star or a black hole, we are really up against it now. If we take Hubble's findings backward to the extreme, we find that 20 billion or so years ago there was a great cataclysm. All the matter in the universe, all the stars, all the galaxies, all of what are now planets and comets and interstellar dust, were all close together in one small spot. This state of affairs is known as "the big bang."

How all the matter in the world was put into the big bang, we have no way of knowing. What made it blow apart and send what became the future galaxies flying apart, we can only guess. Whether it all really happened this way, we cannot be sure.

Can we see the big bang today? In a way, we can, but we must know what to look for.

At the moment before the big bang blew itself apart, the sky must have been blazing with light. This was coming equally from all directions, because we were surrounded by matter on all sides. Any light that we now see from that time will also then be coming equally from all directions.

The light that we see now from the big bang will be coming from the farthest stretches of the universe. It will come from beyond the farthest galaxy. The light we see now will be that part of the big bang that started out so far away that it has been travelling for 20 billion years to reach us. Like a man running up a "down escalator," the light has been struggling against the expansion of the very space it had to cross.

By the time it reaches us now the light from the big bang has been red-shifted so much that is not visible light any more. Its wavelength is now a few centimeters, in the middle of the microwave region of the spectrum.

> Another way to explain why the light from the big bang now has such a long wavelength is to say that space itself has expanded. The distance between crests of the light waves, once a few thousand angstroms, has expanded with the universe, and is now 10,000 times as long as when it was first emitted.

Interestingly enough, this microwave radiation, supposedly from the big bang, has been found. Many different experiments have detected this *cosmic background radiation*. As expected, it comes in equally brightly from all directions. The cosmic background has been found not only close to the earth, but in some cases it has been shown to be warming the gases between the stars. If our model is right, the cosmic background is present everywhere, between galaxies as well as among the stars.

Quasars

The discussion of the universe would not be complete without mentioning one kind of object for which we have no explanation at all.

They are called "quasi-stellar objects", or *quasars*, because they look as point-like as any star through even the best telescopes. Sometimes they are associated with large clouds alongside of them that emit radio waves.

The quasars themselves are certainly quite small. Some quasars get brighter or dimmer over a period of weeks or months. This proves that they cannot be as large as 1 light-year across. (Why? See Question 9 at the end of this chapter.)

All quasars have rather large red shifts. If we interpret the red shift according to Hubble's law, we would guess that they must be billions of light years away, farther than any galaxy we can identify.

If they are, how can we see them at all? The quasars must be shining a thousand times brighter than any galaxy. We have no idea what their energy source must be.

On the other hand, if the quasars are not really that far away, why do they have such large red shifts? And why don't we see anything that looks like a quasar closer in?

We have no answers to these questions yet. The world is full of mysteries, and quasars are some of the biggest mysteries we have met.

Conclusion

The universe is full of very strange things. Red giants, white dwarfs, and black holes form a spectrum of mysteries. The big bang, pulsars, and quasars stagger the imagination. What will the astronomers find next?

In an earlier chapter we mentioned C.P. Snow's comment that an educated person should be as familiar with the second law of thermodynamics as he is with the plays of Shakespeare. We have dealt with the second law already. Perhaps here is the place to invoke a pertinent quotation from Shakespeare:

> "There are more things in heaven and earth, Horatio,
> than are dreamt of in your philosophy."

Glossary

big bang—A model of the universe that supposes that 15 or 20 billion years ago all the matter which is now in all the stars and galaxies was concentrated very close together. For some unknown reason, it all began to fly apart (according to the model), resulting in the universe which we see today.

black hole—A star so massive and so compact that it is undergoing runaway gravitational collapse. The space around a black hole is contracting so rapidly that nothing, not even a light beam, can escape from it.

cosmic background radiation—A low but significant level of microwave radiation that seems to be always present around us. It comes at us with equal intensity from all directions, and perhaps fills all of interstellar and intergalactic space. According to the big bang model, the cosmic background is the light left over from the days when the universe started.

Galaxy—The group of about 100 billion stars to which our solar system belongs. Most of the stars in our Galaxy lie in the white patch of sky known as the Milky Way (for which "galaxy" is the Greek word). The Galaxy lies mostly within a disk-shaped volume, of diameter 100,000 light-years and thickness about 10,000 light-years. We are in one of its spiral arms, about 30,000 light-years from the center.

galaxy—A grouping of billions of stars, similar to our own Galaxy, bound together by gravity and separated from other galaxies by millions of light-years. Thousands of galaxies have been identified by telescope, and no doubt many billions of them are out there.

globular cluster—A grouping of millions of stars, found in the halo of our Galaxy. The stars in these globular clusters are believed to be 10 to 15 billion years old, among the oldest in the Galaxy.

halo—The spherical region, about 100,000 light years in diameter, centered around our Galaxy. Perhaps in its early days the Galaxy was ball-shaped. Now most of its material has collapsed into a disk, but some stars and other material (particularly the globular clusters) remain in the halo.

Hubble, Edwin (1889–1953)—American astronomer who discovered that: 1) the galaxies are at least millions of light-years away, and not part of our own star system; and 2) the distant galaxies are all flying away from us at high speed.

Hubble's law—The discovery by Edwin Hubble that the other galaxies are all flying away from us (and from each other) at high speeds, and that the more distant the galaxy is, the faster it is receding. This law suggests that the whole universe is expanding at a fairly uniform rate.

light-year—The distance travelled by a light beam, moving at 300,000 kilometers per second, in one year. A light-year is about a thousand times the size of our solar system. It is comparable to the distance between stars in our neighborhood.

local cluster—The group of galaxies nearest to our own. It includes the Andromeda nebula, the Magellanic Clouds, and a few smaller galaxies.

meteor—A chunk of solid material that has been in orbit in the solar system. One that lands on Earth is called a meteorite. Meteors may be fragments of a planet that broke up long ago. Or they may be pieces of rock that formed in the early days of our solar system and hadn't yet been collected into a large planet.

neutron star—A collapsed star that has about the same mass as our sun, but only a few kilometers in size. The matter is so closely packed together in a neutron star that we may think of it as one giant atomic nucleus.

pulsar—An astronomical object that emits short bursts of energy at regular intervals of 1 second or less. Pulsars are believed to be rapidly rotating neutron stars.

quasars ("quasi-stellar objects")—Compact astronomical objects that have large Doppler shifts. If this shift is interpreted according to Hubble's law, the quasars are billions of light-years away. The fact that we can see them at all means they must be very bright, perhaps a thousand times as bright as a whole normal galaxy.

red giant—A star in a late stage of its life when it is using helium fusion as its fuel. Red giant stars can be as large as our whole solar system. The time spent as a red giant is fairly short as star lifetimes go, perhaps a few hundred million years.

supernova—A star in a final stage of collapes. For a few days or weeks, a supernova can be as bright as a whole galaxy. No supernovas have been seen in our Galaxy for a few centuries (visible from Earth, that is), but they can be seen fairly often from scanning a large number of other galaxies.

white dwarf—A star that has used up most of its nuclear energy and has collapsed into a size not much larger than Earth. At this density, most of the electrons have been stripped from the atoms within the star. The pressure of these electrons keeps the star from collapsing further.

Questions

1. In an old rock containing radioactive atoms, how can we estimate how many atoms have already decayed?
2. If an average star lives 10 billion years as a normal star, and then goes through a red giant stage that lasts only 100 million years, what fraction of the stars that we see will be red giants?
3. Why does it *consume* energy to make atoms like tin and uranium?
4. How many kilometers are there in a light-year?
5. How does the speed of the sun through the Galaxy compare with the speed of Earth around the sun?
6. If the average distance between neighboring galaxies is a million light-years, how many galaxies do we expect there to be within a *billion* light-years of our own?
7. If we take Hubble's findings to their extreme, we might expect some galaxies to be so far away that they are receding from us at faster than the speed of light. Is this possible? Could we ever see such a galaxy?
8. If the sun is really moving at 300 kilometers per second around the Galaxy, shouldn't the cosmic background look a little brighter in the direction we are heading than behind us? Does this suggest a new version of the Michelson-Morley experiment?
9. If a quasar were more than a light-year across, could the information that one part of it had started to flare up affect the rest of the quasar within a few weeks? Is it possible that the whole quasar can change coherently in less than a year? Why or why not?
10. If we know how brightly a star is really shining and we see how bright it looks to us, how can we figure how far away it is?
11. The special theory of relativity says that no *information* can travel at faster than the speed of light. In this chapter we have met cases (such as the inside of black holes) where the *distance between* certain objects is increasing faster than the speed of light. Can you reconcile these two statements?

12. Using the criteria for acceptance of a new scientific theory, from Chapter 1, compare Einstein's general theory of relativity with Newton's theory of gravitation.
13. What might you conclude from the fact that the sky is dark at night? How would it look if we had an unblocked view of an infinite number of stars spread out evenly over space?
14. Trace the history of the atoms that now make up your body from the time of the "big bang" until now.
15. If things continue to happen at their present rate, what do you expect the world will be like when it is twice as old as it is now?

list of suggested readings

List of suggested readings

Asterisks indicate books that are especially good.

ALFVEN, H., *Atoms, Man and the Universe* (Freeman, 1969), *Living on the Third Planet* (Freeman, 1972), *Worlds-Antiworlds* (Freeman, 1966)

ANDRADE, E.N., *Rutherford and the Nature of the Atom* (Doubleday), *Sir Isaac Newton* (Anchor), a short biography

ASIMOV, I., Any title (non-fiction)

BAKER, A., *Physics and Anti-Physics* (Addison-Wesley)

BARNETT, L., **The Universe and Dr. Einstein* (Signet). A well-written account for non-scientists.

BERENDZEN, R., *Man Discovers the Galaxies* (Science History Publ., 1977)

BONDI, H., **Relativity and Common Sense* (Anchor)

BORN, M., *The Restless Universe* (Dover)

BRACEWELL, R.M., *Intelligent Life in Outer Space* (Freeman, 1975)

BRONOWSKI, J., *The Ascent of Man* (Little, Brown), *The Common Sense of Science* (Torchbooks), *Science and Human Values* (Vintage), A scientist gives a wordy, classical account of the science-society relationship

BROWN, S.C., **Count Rumford* (Anchor)

BUTTERFIELD, H., *The Origins of Modern Science* (Anchor) The book by a famous historian in which the concept of "Scientific Revolution" was first introduced

CALDER, N., *Unless Peace Comes* (Viking Press) Essays on the role of science in war

CHARON, J., *Cosmology: Theories of the Universe* (McGraw-Hill, 1970) paperback

COHEN, I., **The Birth of a New Physics* (Anchor)

COMMONER, B., **Science and Survival* (Ballantine), **The Closing Circle* (Bantam), **The Poverty of Power* (Knopf) On the role of science in society; contemporary; excellent

D'ABRO, A., *The Rise of the New Physics* (Dover)

DAVIS, N., *Lawrence and Oppenheimer* (Simon and Shuster)

DE SANTILLANA, *Origins of Scientific Thought* (Dover) Authoritative tome on the rise of modern science

EINSTEIN, A., **Evolution of Physics* (Clarion) Non-mathematical. On the conceptual development of physics

ELLUL, J., *Technological Society* (Knopf) A pessimistic view

FEATHER, N., *Lord Rutherford* (Crane, Russak)

FEINBERG, G., *What is the World Made of?* (Doubleday, 1977) About atoms, nuclei, particles, quarks

FERRIS, T., *The Red Limit: The Search for the Edge of the Universe* (Morrow, 1977) History of 20th Century astronomy

FEYNMAN, R., **The Character of Physical Law* (MIT Press) A brilliant, non-mathematical account by one of the greatest contemporary physicists

FLEMING, S.J., *Authenticity in Art* (Crane, Russak) How science is used in the art world

FRIEDRICH, L., *The Nature of Physical Knowledge* (Indiana Univ. Press)

GAMOW, G., *Thirty Years that Shook Physics* (Anchor) A delightful look at the development of quantum physics

GHISELIN, B., *The Creative Process* (Mentor)

GREENBERG, D., **The Politics of Pure Science* A lively contemporary account by a science journalist

GARDNER, M., **The Relativity Explosion* (Pocket Books) An entertaining romp through relativity

HEISENBERG, W., *Physics and Philosophy* (Torchbooks) A well-known physicist tries his luck at philosophy

HILL, R.D., *Tracking Down Particles* (Benjamin) On elementary particles

HOOK, S., *Determinism and Freedom in the Age of Modern Science* (Macmillan)

HODGE, P.W., *Concepts of the Universe* (McGraw-Hill, 1969) paperback.

HOYLE, FRED, Any title (non-fiction)

JAFFE, B., *Michelson and the Speed of Light* (Anchor) Good biography, with a description of a famous experiment

JUNGK, R., *The Big Machine* (Scribner) Story of the big European particle accelerator. **Brighter than a Thousand Suns* Beautifully written account of scientists and the atom bomb

KEARNEY, H., *Science and Change* (McGraw-Hill, 1971) paperback. A historian takes a critical look at the birth of modern science

KLAW, S., *The New Brahmins* (Apollo) Authoritative account of the science community

KOESTLER, A., *The Act of Creation* (Macmillan), **The Watershed: Biography of Johannes Kepler* (Anchor), **The Sleepwalkers* brilliantly written

KUHN, T., **Structure of Scientific Revolutions* (Univ. of Chicago Press) A famous classic on the history and philosophy of science

LINDSAY, R., *The Nature of Physics* (Brown Univ. Press)

MCPHEE, J., *The Curve of Binding Energy* (Ballantine 1975)

NEYMAN, J., (ed), *The Heritage of Copernicus* (MIT Press)

PAGE, (ed), *Stars and Galaxies* (Spectrum)

SAGAN, C., (ed), *Communication with Extraterrestrial Intelligence* (MIT Press) Some famous scientists in a hot debate

SAGAN, C., **The Cosmic Connection* (Doubleday, 1973; Dell paperback, 1975) Speculations about life on other planets

SAGAN, C., AND SHKLOVSKY, IS., *Intelligent Life in the Universe* (Dell, 1968)

SCIAMA, D., *The Physical Foundations of General Relativity* (Doubleday)

Scientific American, *Energy and Power* (Freeman, 1971) The starting point for the new field of Energy/Environment research

SHANNON, C., AND WEAVER, W., *Mathematical Theory of Communications* (1st part)

SULLIVAN, W., *We Are Not Alone* (McGraw-Hill)

ULAM, S., *Adventures of a Mathematician* (Scribners, 1976). Autobiography of a talented mathematician, who was also the co-inventor of the H-bomb

WATSON, J., *The Double Helix* (Atheneum) One of the discoverers of DNA gives a human account of the discovery process.

WATSON, W., *On Understanding Physics* (Cambridge Univ. Press)

WEINBERG, S., **The First Three Minutes* (Basic Books, 1977) An exciting account of the beginning of the universe

WEISSKOPF, V., *Knowledge and Wonder* (Anchor)

WEIZENBAUM, J., *Computer Power and Human Reason* (Freeman, 1976)

YORK, H., **The Advisors: Oppenheimer, Teller and the Superbomb* (Freeman, 1976) An interesting account by someone who was there

ZIMAN, J., **Public Knowledge: The Social Dimensions of Science*. By a well-known scientist

index

Page numbers in *italics* represent Glossary entries